Water Quality, Water Security and Risk Assessment

Water Quality, Water Security and Risk Assessment

Editor

Imokhai Theophilus Tenebe

MDPI

Basel • Beijing • Wuhan • Barcelona • Belgrade • Novi Sad • Cluj • Manchester

Editor
Imokhai Theophilus Tenebe
Mineta Institute of
Technology
San Jose State University
California
United States

Editorial Office
MDPI
St. Alban-Anlage 66
4052 Basel, Switzerland

This is a reprint of articles from the Special Issue published online in the open access journal *Water* (ISSN 2073-4441) (available at: www.mdpi.com/journal/water/special_issues/U8QHLQN3LZ).

For citation purposes, cite each article independently as indicated on the article page online and as indicated below:

Lastname, A.A.; Lastname, B.B. Article Title. *Journal Name* **Year**, *Volume Number*, Page Range.

ISBN 978-3-7258-0448-1 (Hbk)
ISBN 978-3-7258-0447-4 (PDF)
doi.org/10.3390/books978-3-7258-0447-4

Contents

water

MDPI

Article

Bacterial Contamination Levels and Brand Perception of Sachet Water: A Case Study in Some Nigerian Urban Neighborhoods

Imokhai T. Tenebe [1,*], Eunice O. Babatunde [2], Chinonso C. Eddy-Ugorji [3], Egbe-Etu E. Etu [4], Nkpa M. Ogarekpe [5], Chikodinaka V. Ekeanyanwu [6], Oladapo A. Diekola [7], Oluwarotimi S. Oladele [8] and Obiora B. Ezeudu [9]

1 Mineta Transportation Institute, San Jose State University, San Jose, CA 95112, USA
2 Civil Engineering Program, Ingram School of Engineering, Texas State University, San Marcos, TX 78666, USA; eunicebabs@gmail.com
3 Faculty of Engineering and Applied Science, Memorial University of Newfoundland, St. John's, NL A1B 0G5, Canada; ceddyugorji@gmail.com
4 Department of Marketing and Business Analytics, San Jose State University, San Jose, CA 95192, USA
5 Department of Civil Engineering, University of Cross River State, Calabar 540281, Nigeria; nkpaogarekpe@unicross.edu.ng
6 Department of Geography, Texas State University, San Marcos, TX 78666, USA; cve9@txstate.edu
7 Department of Computer Information Systems, University of Houston—Victoria, Victoria, TX 77901, USA; diekolao@uhv.edu
8 Department of Demography and Social Statistics, Federal University Oye-Ekiti, Ekiti 370112, Nigeria; sammyladele@gmail.com
9 Centre for Environmental Management and Control, University of Nigeria, Enugu Campus, Enugu 410001, Nigeria; obiezeudu@yahoo.com
* Correspondence: yoshearer@gmail.com

Abstract: Sachet water (SW) is a major source of drinking in most Nigerian homes, thus making it a possible conveyance medium for health risks due to contamination if persist rather than for replenishment of the body. This study collected SW from three busy neighborhoods in South-West Nigeria and investigated for the presence of indicator bacteria (*Escherichia coli* (*E. coli*), Total Coliform (TC), Total Heterophilic Bacteria (THB), Staphylococcus (Staph)) and some physio-chemical parameters (total dissolved solids (TDS), pH, electrical conductivity (EC), and salinity). Multi-variable and exploratory statistical methods were applied to the results to determine correlations between bacterial contamination levels and perceived brand reputation. Bacteriological tests with raw SW samples appeared too numerous to count (TNC) and thus required serial dilutions. After seven-fold serial dilutions, results obtained revealed that SW brands with good reputations had no TC and *E. coli* and was statistically significant with groupings of other SW brands ($\chi^2 = 12.28$; $p < 0.05$ and $\chi^2 = 37.96$; $p < 0.05$). Additionally, SW brands with poor reputations had mean values of TC (19.7×10^8 cfu/mL; 14×10^8 cfu/mL 1.15×10^8 cfu/mL) and *E. coli* (18.2×10^8 cfu/mL; 38.7×10^8 cfu/mL, 32.4×10^8 cfu/mL) exceeding the threshold value of zero set by the World Health Organization (WHO). Only one sample from a poor reputation brand tested positive for Staphylococcus and was not statistically significant ($\chi^2 = 5.2191$; $p = 0.074$). Principal Component Analysis (PCA)/Factor Analysis (FA) revealed that most of the SW had fecal contamination was the major source. Therefore, this study suggests that periodic cleaning of distribution lines, location-specific treatment, and other quality control (QC) measures should be enforced to reduce water security risk for SW consumption in the region.

Keywords: sachet water; water quality; bacteriological quality; public health; drinking water

Citation: Tenebe, I.T.; Babatunde, E.O.; Eddy-Ugorji, C.C.; Etu, E.-E.E.; Ogarekpe, N.M.; Ekeanyanwu, C.V.; Diekola, O.A.; Oladele, O.S.; Ezeudu, O.B. Bacterial Contamination Levels and Brand Perception of Sachet Water: A Case Study in Some Nigerian Urban Neighborhoods. *Water* **2023**, *15*, 1762. https://doi.org/10.3390/w15091762

Academic Editor: Abasiofiok Mark Ibekwe

Received: 21 March 2023
Revised: 25 April 2023
Accepted: 29 April 2023
Published: 3 May 2023

1. Introduction

The quality of drinking water plays a significant role in the quality of health of a populace. Therefore, water that is meant for consumption should undergo various treatment stages (from source to consumer) to qualify as safe for humans and animals.

With an increase in anthropogenic activities around the globe, many source-water bodies have been exposed to one level of contaminant or another [1]. These contaminants could be naturally occurring compounds discovered during exploratory activities for diverse applications in our environment and introduced to source-water bodies via runoff or groundwater infiltration [2]. In other cases, groundwater contamination may occur from underground fabricated structures or naturally occurring elements and gases in the earth's core [3]. Regardless of the primary water source, it is likely that there is one form of contamination or another that makes treatment a necessity to make water fit for human consumption [4–6].

In West African countries like Nigeria and Ghana, water is packaged in sachets as a low-cost option compared to bottled water [7,8]. Sachet water (SW) also serves as a source of drinking water for on-the-go commuters and other consumers. It is widely used in households with no access to clean drinking water sources. Producers of packaged water obtain water from surface water bodies and groundwater wells [9,10], hence it is important to investigate the quality of water before packaging and after sales to retailers or direct consumers to mitigate health risks posed by the consumption of untreated or poorly treated water. Packaged water is generally perceived to be treated water that is safe for drinking, however, studies have shown that public perception and water quality are not always directly related [7,11–13]. In a 2017 study conducted to determine and compare the quality of packaged water in different regions, it was discovered that 65 percent of the representative samples tested were contaminated. This result applied to both forms of packaged drinking water available—sachets and bottles [7]. Other studies have attempted to determine contamination levels of packaged drinking water in different parts of Nigeria and Ghana and have found the presence of heavy metals, and microbial contamination prevalent [10,14–16].

To reduce the exposure of humans to water-borne illnesses, it is essential to monitor the source quality of drinking water. In the case of civilizations with decentralized municipal water supply, regular monitoring of water bodies would be beneficial to preventing outbreaks and health hazards [17–19]. The detection of pathogens and contaminants in source water is not sufficient to prevent outbreaks, hence proper treatment must be ensured. In many regions where SW is consumed, there is a lack of regulatory enforcement of drinking water quality. Hence, private companies producing SW are in control of deciding the source of water used in production processes and the extent of water treatment before packaging [9]. In many cases, pretreatment tests are not conducted to determine the nature of pollutants in the water. Often, only chlorination is applied before water is packaged and as such, many contaminants go untreated, and very low levels of residual chlorine remain in the packaged water [20]. Another challenge with the quality of SW is storage and method of sale. Once the SW has been packaged in the factory, it is often transported in open trucks that allow exposure to sunlight, some vendors store the bags of water in outdoor cages that allow sunlight exposure while others hawk the SW in bowls containing ice and allowing sunlight exposure [21]. It is a known fact that residual chlorine in water protects public health by limiting the regrowth of microbes, however when water is exposed to sunlight, chlorine levels are likely to deplete resulting in poor water quality. Sunlight exposure to SW has also been linked to increased contaminant concentrations and turbidity at post-packaging stage [21,22].

These studies presented above show that the presence of bacterial contaminants in SW can exist due to poor quality source water, inadequate treatment processes, or improper handling after packaging. Considering the above, this study aims to investigate the physiochemical and microbial quality of SW sold in three busy neighborhoods in Ota, Ogun State, and to determine if the public perception of brand quality matches the results obtained from laboratory tests. The study also includes statistical and communality tests to predict the possible sources of bacteriological contamination. This information will add to the body of knowledge in Nigeria regarding sachet water production and consumption and will also inform and/or reiterate to the various agencies the seeming plights surrounding SW consumption in Nigeria.

2. Materials and Methods

2.1. Study Area and Sampling Procedure

The sampling of SW took place in the Ado-Odo community (6.6117° N, 3.0576° E) of Ota in Ogun State, Nigeria. As of the 2006 population census, the community had about 526,565 residents living in the region. While we expect this number to have increased beyond this due to heavy rural-urban migration, this value represents an estimate as of the time of sampling. It consists of both rural and urban communities and has a large concentration of industries. The presence of industries in the area poses a challenge to water quality due to illicit effluent dumping in surface water bodies and near groundwater recharge zones [23].

The study was centered on the bacteriological quality of SW sold and consumed in the community. SW samples were collected between November and December 2017. The sampling took place in Ado-Odo, Ota in Ogun State because the region receives about three million people yearly concentrating around Canaan-land for numerous Christian activities. During this period, it is envisaged that there will be an increased need for SW. Therefore, SW producers may decide to compromise on water treatment quality to meet up the demand. Canaan-land and its environs constitute low to medium-income households, however, irrespective of the household income category, almost everyone depends on SW to meet drinking water demands. SW-producing companies source water from boreholes due to the belief that groundwater is generally cleaner than surface water, yet studies show that water from boreholes may be of questionable quality [5]. Previous studies investigating groundwater quality in the region showed that the water posed no major health risk to consumers based on physio-chemical characteristics. However, bacteriological tests were not performed thus resulting in a partial view of the quality of groundwater as source water for SW producers in the region [24]. Major rivers in the region have been sampled and have shown elevated levels of impairment by pathogens and heavy metals [23,25–27].

In this study, eighty (80) SW of different brands were purchased around Canaan land, the meeting place for most Christians during the November–December period. The areas from which these brands were purchased were Iyana Iyesi, Elegushi, and Igbogbo area (Figure 1). During the sampling periods, these SW samples were purchased from either a store or a kiosk. According to [15], eliminating samples from hawkers i.e., those who move from one point to another, was required to eliminate weather variability such as sunlight, which can impact the water quality. In the study area, we observed two distinct sets of SW producers. The first conforms to producers who have large orders and thus produce more SW while the second set of producers have fewer orders or are less preferred by SW consumers. This a priori knowledge aided our sampling design. With this information, we ensured that we sampled a mixture of SW not exceeding 1 week old from both categories of SW producers. We also ensured that during the sampling period, SW produced on the day of sampling were also purchased. Similarly, we categorized SW based on the preference of consumers, storage duration, and packing quality to gain more insight into how significantly these categories can affect consumers. This further informed the study of two similar SW groupings (good or poor SW brands). Furthermore, in order to gather information about consumers' brand perceptions, we created an informal questionnaire that was used during a focus group discussion, one-on-one discussions, and retailer interviews as a guide while sampling was taking place in the study area. This allowed us to gain insight into how consumers perceive different brands and what influences their opinions when they purchase SW. In addition, we consulted with retailers to gather data on consumer behavior, including which brands are most frequently purchased and any patterns of brand selectivity.

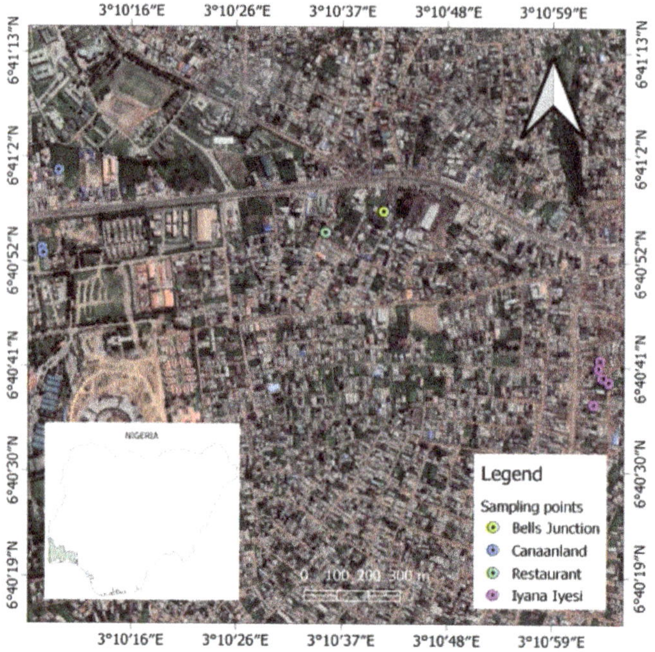

Figure 1. Clustered location of stores for Sachet Water purchase.

Furthermore, SW collected were immediately kept in an iced cooler and transported to the laboratory for some physio-chemical and bacteriological testing. We ensured that all tests were conducted within the first 24 h of purchase and samples were kept in ice at a constant temperature of 4 °C before bacteriological testing.

2.2. Physico-Chemical Analysis

The physio-chemical parameters measured include pH, Total Dissolved Solids (TDS), Salinity, and Electrical Conductivity (EC). It is worth mentioning that all physico-chemical tests were carried out using Hanna Edge Multi-parameter Dissolved Oxygen Meter (HI2040) and was frequently calibrated based on the manufacturers' description using appropriate calibration solutions. The results obtained was useful in determining the suitability of the samples collected for drinking and other forms of consumption per WHO guidelines for drinking.

2.3. Bacteriological Analysis

Bacteriological testing was conducted for the detection of *E. coli*, Total Heterotrophic bacteria, Total Coliform, and Staphylococcus Aureus. The Multiple Tube Fermentation Test was used to evaluate the number of active bacteria specie present in all samples collected. The total coliform count in SW samples was determined using the Most Probable Number (MPN) assay. This test is also known as the presence or absence test since it focuses on detecting the presence or absence of coliform bacteria in a sample. This test is based on the premise that for every 100 mL of drinking water, there should be no coliform bacteria present. However, if one viable cell of coliform bacteria is present in the sample being tested, it will reproduce to give a population of lactose fermenting cells that release acid and gas [28,29].

Most Probable Number Assay

The MPN Assay is widely used for the estimation of viable microbes in a known quantity of samples (food or water) by creating replicates of liquid broth growth in ten-fold dilutions [30,31]. Based on statistical probabilities, the number of viable microbes in the original sample is approximated using an MPN table [32,33]. The MacConkey broth used for these tests was prepared in single and double-strength concentrations following standard procedures and samples were added according to standard methods to determine the presence of bacteria communities (Figure 2) [31]. The resulting solution was incubated at 37 °C for 24 h. The fermentation tubes were inspected for microbial growth after 24 h as well. Fermentation tubes with no indication of microbial growth were incubated for an additional 48 h [34]. Once the reaction time was completed, the number of fermentation tubes indicating microbial growth per sample was compared to the standard MPN charts and the number of microbes present was recorded.

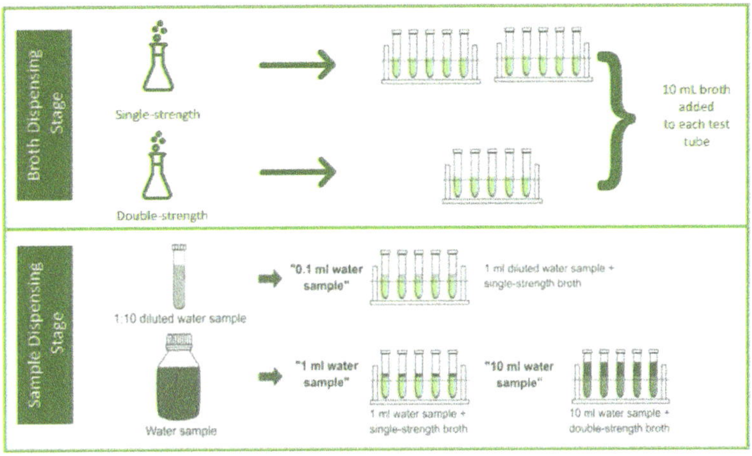

Figure 2. Most Probable Number Test Procedure.

Following the presumptive tests, the fermentation tubes that produced gas were inoculated on a variety of agar mediums. The streak-plate method was used to isolate pure cultures on eosin methylene blue agar plates, nutritional agar plates, and MacConkey agar, which were incubated at 37 °C for 24 h. For the detection of thermotolerant *E. coli*, high-temperature incubation at 44.5 ± 0.2 °C was used. The colonies that grew on each agar were then counted and recorded. Most of the results for the first set of samples yielded colonies that were too numerous to count (TNC). This prompted the use of serial dilution in the second and third tests of the samples before plating and incubation in order to obtain results for statistical analyses. After incubation was complete, the plates were physically and microscopically examined for the formation of colonies [35].

3. Results

The 80 samples tested in this study were collected from three main sampling location clusters—Iyana Iyesi, Igbogbo and Elegushi. In the process of sample collection, we observed that some consumers across different income levels purchased SW that was cheaper and packed in larger sachets despite all brands claiming to sell a uniform 50 mL of water. Thus, we can infer that to the consumers, all SW is the same by constantly reciting the parlance that "water is water" as "water has no enemy". Despite this inference, responses from focus groups, one-on-one conversations with consumers and retailer perspective informed us that other chunk of consumers have a perception of which SW brands are likely to be of higher quality compared to others.

The results presented in this section are based on three levels of comparison. The first level highlights the samples categorized as good or poor brand based on data collected from focus groups and other consumer-retailer assessments from each location cluster (Table 1). The second level highlights the samples originally categorized by consumers as being good brand whose bacteriological results match public perception. All other samples that do not meet bacteriological test standards are categorized as poor brands in level 2 as well (Table 2). The third level of comparison highlights SW samples that are of good quality based on bacteriological tests and regardless of consumer brand perception (Table 3).

Table 1. Sachet Water Characteristics based on the neighborhood of purchase.

Variables	Iyana Iyesi n = 23	Elegushi n = 44	Igbogbo n = 13	χ^2	*p* Value
Nominal measures, count (%) *					
Good brand reputation	23 (100)	41 (93.18)	7 (53.85)	19.6435	<0.001
Collected before 1 day old	6 (26.09)	21 (47.73)	6 (46.15)	3.0728	0.215
Poor sachet wrapper	11 (47.83)	14 (31.82)	10 (76.92)	8.5139	0.0014
E. coli > 0	9 (39.13)	30 (68.18)	13 (100)	16.1175	<0.001
THB > 500	21 (91.3)	42 (95.45)	13 (100)	1.3647	0.505
TC > 0	13 (56.52)	15 (34.09)	1 (7.69)	8.7665	0.012
Staph > 0	0 (0)	0 (0)	1 (7.69)	5.2191	0.074
Continuous measure, mean (S.E) *					
Salinity	0.1 (0.006)	0.13 (0.006)	0.12 (0.010)		0.2600
EC	48 (5.35)	51 (3.92)	51 (2.97)		0.8978
TDS	25 (1.38)	33 (8.73)	24 (2.66)		0.7030
pH	5.57 (0.16)	6.05 (0.09)	5.79 (0.08)		0.0156
Staph (CFU/mL) × 10^8	0 (0)	0 (0)	2.46 (2.46)		0.0745
THB (CFU/mL)) × 10^8	46.7 (5.14)	30.4 (5.07)	21.0 (3.78)		0.0349
TC (CFU/mL) × 10^8	19.7 (5.56)	14 (51.8)	1.15 (1.15)		0.0716
E. coli (CFU/mL) × 10^8	18.2 (5.79)	38.7 (12.9)	32.4 (7.12)		0.4872

Note: * Values in brackets represent percentages for nominal and mean for continuous measures.

Table 2. Sachet water characteristics based on classification of brand reputation matching SW quality and collection period.

Characteristics	Samples Matching Brand Reputation		*p*-Value	Collection Period		*p*-Value
	Good n = 11	Poor n = 69		>1 day n = 47	≤1 day n = 33	
Nominal measures, count (%) *						
Collected after 1 day old	6 (54.54)	41 (59.42)	0.609			
THB > 500	9 (81.81)	67 (97.10)	0.465	45 (95.74)	31 (93.94)	0.715
E. coli > 0	0 (0)	51 (73.91)	0.016	34 (72.34)	17 (51.51)	0.056
TC > 0	0 (0)	29 (42.03)	0.096	12 (25.53)	17 (51.51)	0.0017
Staph > 0	0 (0)	1 (1.45)	0.720	0 (0)	1 (3.03)	0.230

Note: * Values in brackets represent percentages.

Table 3. Comparison between good and poor-quality sachet water based on bacteriological test results.

Characteristics	Hebron and Medicx	Other Brands	χ^2	*p*-Value
	n = 17	n = 63		
Nominal measures, count (%)				
Collected after 1 day old	8 (47.06)	39 (61.90)	1.218	0.270
THB > 500	13 (76.47)	63 (100)	15.604	0.000
TC > 0	0 (0)	29 (46.03)	12.275	0.000

Table 3. *Cont.*

Characteristics	Hebron and Medicx	Other Brands	χ^2	*p*-Value
	n = 17	n = 63		
E. coli > 0	0 (0)	51 (80.95)	37.964	0.000
Staph > 0	0 (0)	1 (1.59)	0.2733	0/601

3.1. Contamination Dominance and Brand Reputation

From the three (3) sampling cluster locations, 23 (100%) SW samples from Iyana Iyesi, 41 SW samples (53.85%) from Elegushi, and 7 SW (53.85%) from Igbogbo were classified as the brand of perceived good reputation based on direct responses from field studies. Of these SW of good reputation, 6 SW were purchased from Iyana Iyesi (26.09%), 21 SW (47.73%) from Elegushi, and 6 (46.15%) from Igbogbo were purchased on the day of production with a statistically significant difference between sets of SW purchased ($p < 0.05$). Cross tabulation was used to determine the relationship between variables and their corresponding chi-square and *p*-values were used to establish statistical significance between sets. The multi-variable comparison suggests that perceived brands of SW of poor reputation were sold in Iyana Iyesi and Elegushi combined. From which, fifty-two SW samples had *E. coli*. Specifically, 9 SW from Iyana Iyesi (39.13%), 30 SW (68.18%) from Elegushi, and all 13 (100%) samples from Igbogbo with a strong statistical difference ($p = 0.001$). These SW samples had mean count values of 18.2×10^8 cfu/mL, 38.7×10^8 cfu/mL and 32.4×10^8 cfu/mL respectively for these locations with no statistically significant difference ($p > 0.05$). These values recorded exceeded the world health organization's (WHO) limits of *E. coli* of zero tolerance. For THB, the values measured from SW samples from these three locations were significantly higher than 500 cfu/mL. Specifically, 21 SW (91.38) from Iyana Iyesi, 42 SW (95.45%) from Elegushi, and all 13 (100%) samples from Igbogbo with no statistically significant difference ($\chi^2 = 1.365$, $p = 0.505$). In addition, more information on mean *E. coli*, TC, THB, and physicochemical values were estimated and recorded in Table 1.

The actual mean values recorded revealed that THB values from these locations exceeded the threshold for THB with values at 46.7×10^8 cfu/mL, 30.4×10^8 cfu/mL and 21.0×10^8 cfu/mL for each location respectively and were statistically different ($p = 0.034$). TC on the other hand was not different in terms of count from the three locations as SW samples were prevalent with TC contamination. To mention, 13 SW (56.52%) from Iyana Iyesi, 15 SW (34.09%) from Elegushi, and 1 SW (7.69%) from Igbogbo ($\chi^2 = 8.7665$, $p = 0.012$) were recorded to have TC in SW in the three locations. Staphylococcus was present in only one sample and was purchased from Igbogbo which may have been due to chance ($\chi^2 = 5.219$, $p > 0.05$). In addition, physio-chemical properties measured for all SW samples from the three locations revealed no statistical difference between them except for pH. Electrical conductivity (EC) which can be a measure of the ionic content in water showed no statistical difference between SW obtained at different locations ($p = 0.8978$) with mean values of 48 µs/cm, 51 µs/cm, and 51 µs/cm for Iyana Iyesi, Elegushi, and Igbogbo respectively. The pH values measured from SW suggest that they are acidic with mean values of 5.57, 6.05, and 5.79 for Iyana Iyesi, Elegushi, and Igbogbo respectively while TDS values were also at 25 mg/L, 33 mg/L, and 24 mg/L with no significant difference ($p = 0.7030$) (Table 1).

For the second level of comparison, the 80 SW samples tested in this study were classified into brands of good and poor reputation after being tested in the laboratory. Of the 80 samples, sixty-nine (69) were placed in the category of SW brand with poor reputation while eleven (11) samples were placed under good reputation SW brand. The quantity of SW classified as good was based on samples whose quality matched public perception. All other samples with poor bacteriological results, were categorized as poor brands (Table 2). Additionally, the SW samples collected were classified based on the time of purchase after production and was recorded in Table 2. 9 samples (81.81%) of SW containing THB were of good reputation and 67 samples (97.10%) were brands of the poor reputation and the values of THB were above the threshold ($\chi^2 = 0.5337$; $p = 0.465$). None of the perceived good brands of SW contained *E. coli* while 51(73.91%) of the perceived poor brands had *E. coli* present and were statistically significant. ($\chi^2 = 5.766$; $p = 0.016$). In addition, no SW from the good reputation brand category had TC present while 29 SW samples of poor reputation brands recorded TC values with statistical significance ($\chi^2 = 2.773$; $p = 0.096$). Table 1 highlights the values of THB, *E. coli*, TC, and Staph based on the above-mentioned classifications.

Only one brand from a poor reputation SW brand reported staphylococcus, which overall was not statistically significant ($\chi^2 = 0.1284$; $p = 0.720$). Surprisingly, some samples collected on the day of production reported the presence of THB, TC, and *E. coli*. As seen in Table 2, THB values were present from 31 (93.94%) SW produced on the same day while 45 (95.74%) were seen from SW samples stored over time with no significant statistical difference between them ($\chi^2 = 0.1330$; $p = 0.713$). This simply suggests that storing them over time was not a reason for the presence of THB, rather these bacteria were inherent in the distribution system of water supply. Similarly, 17 (51.51%) SW had both TC and *E. coli* and were produced on the same day while more samples 34 (72.34%) SW stored over time had a significantly high value of ($\chi^2 = 5.664$; $p = 0.77$) for TC.

During the bacteriological investigation of the SW samples, it was observed that two SW brands had no *E. coli* and Total coliform (TC) present (Table 3). These results informed the classification of samples in the third level of comparison. To mention, one of the brands was already classified as a brand with a good reputation while the other was a brand perceived to have a poor reputation by the consumers living within the vicinity where the SW sample was sold. Therefore, for this study, we stratified these SW into two groups which comprised of both SW brands (Hebron and Medicx) in one group and other brands as a second group to determine whether a significant difference exist in their contamination levels. The findings of his process have been captured in Table 3. The results showed the SW samples yield on TC ($\chi^2 = 12.28$; $p < 0.05$), *E. coli* ($\chi^2 = 37.96$; $p < 0.05$), and Staph ($\chi^2 = 0.273$; $p > 0.05$) over time with these values showing statistical significance.

3.2. *Contamination Dominance and Explanatory Analysis of Bacteriological and Physio-Chemical Stressors*

In this study, a correlation was carried out on 80 SW samples to determine the relationship between bacteriological and physio-chemical variables called stressors. Spearman's correlation coefficient was used in place of Pearson's correlation coefficient based on the normality test conducted using Shapiro-wilk's test for normality. The test revealed a non-normal distribution of stressors measured in SW samples ($p < 0.05$). Table 4 shows the relationship and the statistical significance of the 2-tailed test conducted. Statistical significance but negative correlations were observed between conductivity with [TC (r = −0.247; $p < 0.01$); *E. coli* (r = −0.602; $p < 0.01$); THB (r = −0.453; $p < 0.01$)], suggesting that bacteria in water would do well and better at lower conductivity levels and that fecal matter present is not the source for high conductivity recorded from SW samples but occurring mutually exclusively mainly due to presence of salts and metalloids. For TDS, a strong statistical positive correlation was observed between TDS with [EC (r = 0.992; $p < 0.01$); Salinity (r = 0.676; $p < 0.01$)] while strong negative statistical significance was observed with all bacteriological stressors ($p < 0.001$).

Table 4. Correlation Matrix between measured parameters.

Correlation Matrix								
		Conductivity	**TDS**	**pH**	**Salinity**	**TC**	***E. coli***	**THB**
	Conductivity	1.000	0.992 **	0.044	0.676 **	−0.247 **	−0.602 **	−0.453 **
	TDS	0.992 **	1.000	0.054	0.676 **	−0.231 *	−0.582 **	−0.430 **
	pH	0.044	0.054	1.000	0.502 **	−0.024	−0.081	−0.190
Correlation	salinity	0.676 **	0.676 **	0.502 **	1.000	−0.263 *	−0.419 **	−0.454 **

Table 4. *Cont.*

	Correlation Matrix						
	Conductivity	**TDS**	**pH**	**Salinity**	**TC**	***E. coli***	**THB**
TC	−0.247 *	−0.231 *	−0.024	−0.263 *	1.000	0.571 **	0.467 **
E. coli	−0.602 **	−0.582 **	−0.081	−0.419 **	0.571 **	1.000	0.615 **
THB	−0.453 **	−0.430 **	−0.190	−0.454 **	0.467 **	0.615 **	1.000

Note: * p-value < 0.05; ** p-value < 0.01.

With the latter already described with the same perception as EC i.e., TDS to EC ($r = 0.992$; $p < 0.01$), the former suggests that Salinity to TDS relationship emanates from the same source. Therefore, this is likely to be from either the extraction process of water used for making SW containing rock salts or salt compounds added to reduce pH. Interestingly, the strong statistical correlation between pH and salinity ($r = 0.5025$; $p < 0.01$) could suggest that increased salinity was due to buffering pH as both increased directly. Additionally, an attempt was made to identify major pollution in the SW samples by conducting an exploratory statistical method using PCA/FA and communalities. PCA/FA is a dimension reduction that can be useful in whittling down variables to important factors depending on the subject at hand. It whittles this by grouping variables into factors or component loadings. The output of these component loading in a factor shows the relationship of the variables on each component. On the other hand, communalities show how well these variables explain the variability in a component and it is very useful as a performance evaluator of measured components extracted during PCA/FA procedures. However, both techniques are sensitive to the size of the dataset and produce spurious output when the variables used do not follow a normal curve distribution.

Before conducting the PCA/FA analysis, the variables were first tested for sampling adequacy using the KMO and Bartlett test. These tests were conducted for two main reasons. First, the KMO test revealed whether the data set to be used for PCA/FA statistical operations was suitable. Literature has suggested that KMO values greater than 0.5 are usually suitable but most will agree that higher values of KMO i.e., >0.5 are even better. Second, Bartlett's test hypothesizes that the data set to be extracted are typically uncorrelated which can be determined by the p-value. For our study as seen in Table 5, KMO was 0.653 which suggests that our dataset is suitable for extraction while Bartlett's test also showed that the dataset is uncorrelated ($p < 0.05$). Thereafter, normalization or transformation was carried out using the CLR method discussed elsewhere. The transformed dataset was then subjected to exploratory analysis and reported in Table 6.

Table 5. Kaiser-Meyer-Olkin Measure of Sampling Adequacy of sampled SW.

KMO and Bartlett's Test		
Kaiser-Meyer-Olkin Measure of Sampling Adequacy.		0.653
Bartlett's Test of Sphericity	Approx. Chi-Square	214.392
	df	21
	Sig.	0.000

Table 6. PCA and communality results of sampled SW.

	Rotated Component Matrix		ComVal
	Component		
	1	2	
Conductivity	0.802	−0.351	0.767
TDS	0.620	−0.073	0.389
Ph	0.662	0.064	0.442

Table 6. *Cont.*

Rotated Component Matrix			
	Component		ComVal
	1	2	
salinity	0.902	−0.251	0.876
TC	−0.088	0.828	0.693
E. coli	−0.021	0.795	0.633
THB	−0.248	0.696	0.546

The result from the PCA/FA was determined using two main rules (Table 5). The first rule is the Kaiser-Meyer-Olkin rule which suggests that meaningful components are usually extracted with Eigenvalue greater than 1 (Table 6). Next was the Elbow or Point of Inflexion shown on a Scree plot (Figure 3). From both rules, components were factored as 2. The major advantage of FA is that it helps in the grouping, which can be used in this context for source allocation. For example, in this study, component 1 had strong loadings on Salinity and Conductivity. While component 2 is correlated by *E. coli* and Total Coliform suggesting the presence of fecal contamination. This could only be possible as the distribution system accumulates bacterial films and sediments that do so well in retaining organisms in the water. Therefore, showing multiple lines of evidence that infrequent cleaning and lack of advanced treatment mechanisms are so lacking in the SW industry and require both the attention of consumers and relevant government agencies. The absence of these would only increase the level of water insecurity as new and old packaged water companies would show increased lackadaisical behavior towards producing packaged drinking water of better quality.

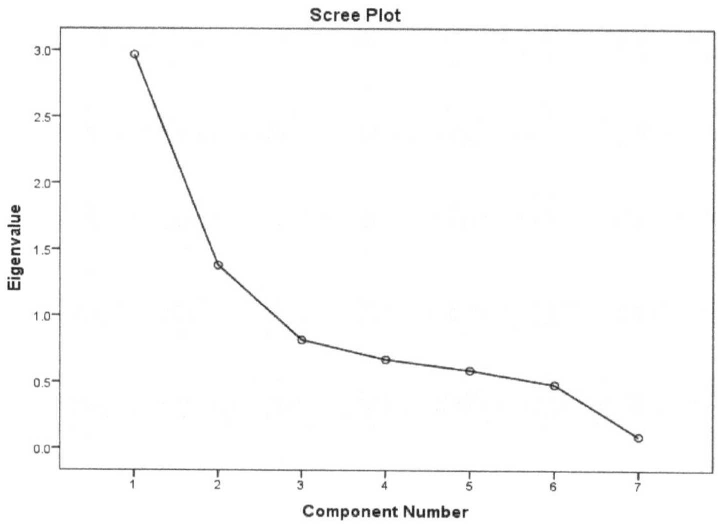

Figure 3. Scree plot showing the point of inflection for the PCA/FA analysis.

Communality analysis shown in Table 5 revealed that ionic contamination was well explained when compared to bacteriological contamination in the SW studied. Specifically, it revealed that 87.6% of salinity and 76.7% of conductivity were explained by factor analysis while 69.3% and 63.3% explained the variability of TC and *E. coli* in its Factor Component 2. An indication that shows or depicts a satisfactory performance suggests that lesser grouping was more effective than extended groupings.

4. Discussion

From this study, we investigated the presence of bacteriological contamination. Seemingly, the presence of fecal contamination was detected in 51 (≥81%) of SW investigated in the region. The contamination levels recorded were in amounts that require urgent attention. This is concerning because, before seven-fold serial dilution, the results initially recorded from the MPN test of SW samples revealed TNC observations of bacteriological contamination. This suggests that none of the SW is fit for consumption and that is very concerning. In the past, a study [36] investigated the presence of bacteria in 92 SW in Lagos, Nigeria. The findings from that study showed elevated *E. coli* values which were significantly higher than the threshold set by WHO. The values in this study revealed a much higher presence of bacterial contamination than those reported in earlier studies. Amongst all samples tested, only one sample was positive for the presence of staph. Although this study reports the presence of staph in the SW sample as statistically insignificant, it calls for a more proactive public health check to avoid any occurrences.

In the PCA/FA analysis, component 1 had strong loadings on Salinity and Conductivity. While component 2 is correlated by *E. coli* and Total Coliform suggesting the presence of fecal contamination. This could only be possible as the distribution system accumulates bacterial films and sediments that do so well in retaining organisms in the water. Therefore, showing multiple lines of evidence that infrequent cleaning and lack of advanced treatment mechanisms are so lacking in the SW industry and require both the attention of consumers and relevant government agencies to be properly addressed. The absence of these would only increase the level of water insecurity as new and old packaged water companies would show increased lackadaisical behavior towards producing packaged drinking water of better quality.

Staph infections have been reported to cause extremely serious or fatal infections in humans including bacteremia, sepsis (when bacteria spread to the bloodstream), pneumonia, endocarditis (infection of the heart valves), and osteomyelitis (bone infection) [37]. Another study [38] also investigated SW consumed by university students in Kumasi Ghana and found 50% of SW investigated contained TC & EC of similar contamination levels as this current study. However, their finding did not agree with findings from [14] that found no *E. coli* presence in SW samples investigated in Ghana. These studies suggest that there is need for a location-based continuous monitoring scheme to create more public awareness. Two SW brands in this study showed no presence of TC & EC. The first brand (Hebron) is a brand every SW consumer around the studied area would prefer to purchase but they see it beyond their reach as SW produced by this brand are bought off quickly before it spreads to the neighborhood, thus making it expensive to purchase compared to other brands in the region. The other brand is not well known and does not feature attractive quality packaging. Despite the good performance of the two SW brands after seven-fold serial dilution, it is opined that best management practices as stipulated by National Agency for Food and Drug Administration and Control (NAFDAC) should be unrelentingly followed. NAFDAC reported that persistent presence of pathogenic bacteria in SW is due to the use of untreated raw water as well as SW companies non-compliance for treatment of raw water. NAFDAC (2018) recommends that the treatment process in all SW production factories should follow the stated order (Figure 4). Previous studies have shown that microfiltration has the capacity to remove all bacteria from the water being treated. Hence, it can be deduced from the results obtained that SW manufacturers do not adhere to the recommended treatment guidelines prescribed by NAFDAC. Another reason for bacteria presence in packaged SW could be linked to the fouling of membranes in facilities where microfiltration systems are used [39–42]. Hence, a need for proper cleaning of the microfiltration membranes is also essential for maintaining water quality.

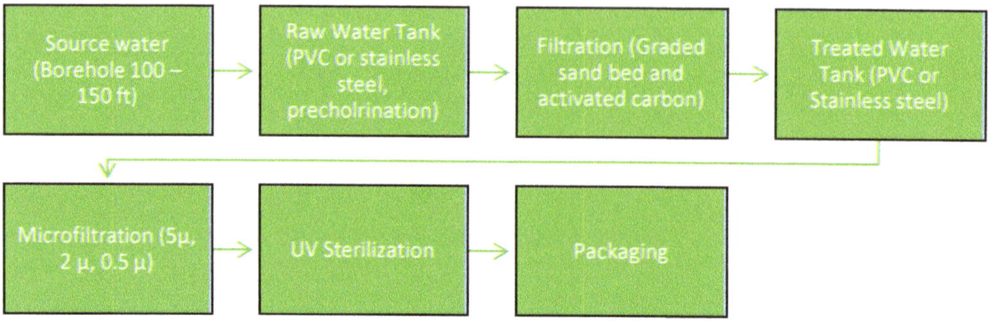

Figure 4. NAFDAC recommendations for sachet water production processes.

Poor hygiene practices in SW production facilities can be linked to the presence of staph in packaged water. Most SW factories visited during this study were in poor environmental conditions with waste polyethylene bags and papers littered all over. It is believed that this may lead to heightened microbial contamination due to the proximity of SW production facilities to waste disposal facilities, including septic tanks and small to medium dumpsters [43]. A deformation in the physical component of these facilities can lead to the escape of substances of public health concern into SW processes. [44,45] reported that *E. coli* and other microbes from fecal contamination can enter raw water sources when open defecation is practiced close to water bodies or when a leak exists in a septic tank. These contaminants can be transferred to packaged water in the case where proper treatment is not ensured, thus causing a public health crisis. This study, therefore, has revealed that there is a higher possibility of a range of non-treatment to inadequate treatment of water used for SW purposes even though many of the SW companies claim to involve in reverse osmosis, microfiltration process, and chlorination. This study embarked on gives new insight to the contamination level years after a similar investigation was conducted in a different region of West Africa. Our findings corroborate with the conditions of [46] that in high demand there is a possibility of the high occurrence of contamination levels because of limited or no maintenance of water supply distribution lines, as well as a non-compliance of SW manufacturers to good sanitation and hygiene measures. It is a great concern to note the high number of fecal indicators consistently found in SW across the country. Even though this is consistent with our study, being that it was conducted in a busy period where Christians from over the world gather en masse, the authors of this study see that the results might be the same irrespective of seasonality [11]. The reason for this conclusion was due to the environmental conditions of the SW factories during our study.

5. Conclusions

The bacteriological quality of 80 SW samples collected from three neighborhoods in Ota, Nigeria was investigated. The results obtained were used to conduct statistical analysis to determine the level of significance of the contamination. Additionally, bacterial dominance was correlated with a brand reputation to give an informed re-education of public perception as to what brands they should be consuming for the safety of their health. Initial bacteriological tests revealed that all SW are not fit for consumption. All the results obtained for 69 samples fall below the WHO limits for the bacteriological quality of drinking water. Although THB has no associated health risk, the WHO standard requires a maximum of 500 CFU/Ml in water. Additionally, a high concentration of THB is an indication of favorable conditions for the growth of more dangerous microbes such as *E. coli*, Legionella, and *S.* Typhi and these can cause serious harm to human health and in some cases, death. The results obtained in this study show THB counts over the recommended WHO limit which is a cause for concern.

Fecal contamination in SW has become a norm in SW at an alarming rate such that entrusting the monitoring of the SW production process to NAFDAC, which appears to have done less in enforcing regulations in this regard will be a grave mistake for the public. Frequent periodic monitoring programs and collaborations between SW manufacturers and researchers are encouraged to find a lasting solution to this crisis. It is also important to note that the problem of bacterial dominance in packaged water is location specific, and as such requires location-specific water treatment as opposed to a general treatment approach. Testing of source water quality should be mandated for all SW manufacturers and specific measures should be undertaken to remove the contaminants found in the source water. It is also encouraged that the public pays more attention to the SW they consume as it can impair their health negatively. Finally, properly cleaning water treatment facilities and distribution systems to remove sediments and clogged bacterial films can to a large extent help reduce the re-occurrence of bacteria into packaged SW after treatment. This study compared public perception of the quality of SW across various brands with the actual quality of SW based on bacteriological tests. Although previous studies have shown that storage and handling conditions have a significant impact on the quality of SW, this study did not consider these factors in our analysis. Despite the similarities in the quality of SW across many regions, literature suggests that conducting location-specific studies is crucial due to differences in source water quality, treatment methods, and handling techniques of SW. Thus, we recommend conducting further studies to investigate the impact of storage and handling conditions on SW in this study region. Additionally, as microplastics are a growing concern in products stored in polyethylene containers, we recommend exploring the possibility of microplastics in SW due to degradation during transportation and storage, as this poses a potential threat to human health. Finally, despite mentioning the lack of compliance of most if not all SW producers within the study area, this study was unable to determine whether NAFDAC policies were effective, lacked compliance or broken at worst. Future studies can conduct a panel data analysis on this subject across countries or regions where SW is ubiquitous.

Author Contributions: All authors contributed to the conception and design of the study. Material preparation, data collection, and spatial and statistical analysis were performed by I.T.T., E.O.B., C.C.E.-U., E.-E.E.E., N.M.O., C.V.E., O.A.D., O.S.O. and O.B.E. The first draft of the manuscript was written by I.T.T., and all the authors reviewed and commented on the previous versions of the manuscript. N.M.O., C.V.E. and O.B.E. contributed to the demographic and geographical information analysis. All authors have read and agreed to the published version of the manuscript.

Funding: This research received no external funding.

Institutional Review Board Statement: Not applicable.

Informed Consent Statement: Not applicable.

Data Availability Statement: All relevant data has been made available in the paper.

Conflicts of Interest: The authors declare no conflict of interest.

References

1. Turner, S.W.; Rice, J.S.; Nelson, K.D.; Vernon, C.R.; McManamay, R.; Dickson, K.; Marston, L. Comparison of potential drinking water source contamination across one hundred US cities. *Nat. Commun.* **2021**, *12*, 7254. [CrossRef] [PubMed]
2. Jiang, W.; Liu, H.; Sheng, Y.; Ma, Z.; Zhang, J.; Liu, F.; Chen, S.; Meng, Q.; Bai, Y. Distribution, source apportionment, and health risk assessment of heavy metals in groundwater in a multi-mineral resource area, North China. *Expo. Health* **2022**, *14*, 807–827. [CrossRef]
3. Li, P.; Karunanidhi, D.; Subramani, T.; Srinivasamoorthy, K. Sources and consequences of groundwater contamination. *Arch. Environ. Contam. Toxicol.* **2021**, *80*, 1–10. [CrossRef] [PubMed]
4. Chinedu, S.; Nwinyi, O. Assessment of Water Quality in Canaanland, Ota, Southwest Nigeria. *Agric. Biol. J. N. Am.* **2011**, *2*, 577–583. [CrossRef]
5. Olalekan, A.; Abubakar, B.; Abdul-Mumini, K. Physico-chemical characteristics of borehole water quality in Gassol, Taraba, State, Nigeria. *Afr. J. Environ. Sci. Technol.* **2015**, *9*, 143–154. [CrossRef]

6. Lin, L.; Yang, H.; Xu, X. Effects of water pollution on human health and disease heterogeneity: A review. *Front. Environ. Sci.* **2022**, *10*, 975. [CrossRef]
7. Vedachalam, S.; MacDonald, L.H.; Omoluabi, E.; Olaolorun, F.; Otupiri, E.; Schwab, K.J. The Role of Packaged Water in Meeting Global Targets on Improved Water Access. *J. Water Sanit. Hyg. Dev.* **2017**, *7*, 369–377. [CrossRef]
8. Tetteh, J.D.; Templeton, M.R.; Cavanaugh, A.; Bixby, H.; Owusu, G.; Yidana, S.M.; Moulds, S.; Robinson, B.; Baumgartner, J.; Annim, S.K.; et al. Spatial Heterogeneity in Drinking Water Sources in the Greater Accra Metropolitan Area (GAMA), Ghana. *Popul. Environ.* **2022**, *44*, 46–76. [CrossRef]
9. Ajala, O.J.; Ighalo, J.O.; Adeniyi, A.G.; Ogunniyi, S.; Adeyanju, C.A. Contamination Issues in Sachet and Bottled Water in Nigeria: A Mini-Review. *Sustain. Water Resour. Manag.* **2020**, *6*, 112. [CrossRef]
10. Angnunavuri, P.N.; Attiogbe, F.; Dansie, A.; Mensah, B. Evaluation of Plastic Packaged Water Quality Using Health Risk Indices: A Case Study of Sachet and Bottled Water in Accra, Ghana. *Sci. Total Environ.* **2022**, *832*, 155073. [CrossRef]
11. Emenike, P.G.C.; Tenebe, T.I.; Omeje, M.; Osinubi, D.S. Health Risk Assessment of Heavy Metal Variability in Sachet Water Sold in Ado-Odo Ota, South-Western Nigeria. *Environ. Monit. Assess.* **2017**, *189*, 1–16. [CrossRef] [PubMed]
12. Mosi, L.; Adadey, S.M.; Sowah, S.A.; Yeboah, C. Microbiological Assessment of Sachet Water "Pure Water" from Five Regions in Ghana. *AAS Open Res.* **2019**, *1*, 12. [CrossRef] [PubMed]
13. Stoler, J. Sachet Water Quality in Ghana: The Jury Remains Out. *Food Control* **2014**, *37*, 417–418. [CrossRef]
14. Stoler, J.; Tutu, R.A.; Ahmed, H.; Frimpong, L.A.; Bello, M. Sachet Water Quality and Brand Reputation in Two Low-Income Urban Communities in Greater Accra, Ghana. *Am. J. Trop. Med. Hyg.* **2014**, *90*, 272–278. [CrossRef]
15. Stoler, J.; Ahmed, H.; Asantewa Frimpong, L.; Bello, M. Presence of Pseudomonas Aeruginosa in Coliform-Free Sachet Drinking Water in Ghana. *Food Control* **2015**, *55*, 242–247. [CrossRef]
16. Omonigho, D. Bacteriological Analysis of Sachet Water Vended in Ugbor, Benin. *SAU Sci. Tech. J.* **2017**, *1*, 88–100.
17. Omoigberale, N.O.; Isibor, J.O.; Izegaegbe, J.I.; Iyamu, M.Y. Seasonal variation in the bacteriological quality of Ebutte River in ehor community Edo state, Nigeria. *Am. J. Res. Commun.* **2013**, *1*, 59–64.
18. Tenebe, I.T.; Emenike, P.C.; Nnaji, C.C.; Babatunde, E.O.; Ogarekpe, N.M.; Dede-Bamfo, N.; Omole, D.O. Bacterial Characterization and Quantification of Rainwater Harvested in a Rural Community in Nigeria. *Environ. Nanotechnol. Monit. Manag.* **2020**, *14*, 100370. [CrossRef]
19. Tenebe, I.; Emenike, P.C.; Babatunde, E.O.; Neris, J.B.; Fred-Ahmadu, O.H.; Dede-Bamfo, N.; Etu, E.E.; Ogarekpe, N.M.; Emakhu, J.; Nsikak, B.U. Assessing the State of Rainwater for Consumption in a Community in Dire Need of Clean Water: Human and Health Risk Using HERisk. *Water Pract. Technol.* **2022**, *17*, wpt2022109. [CrossRef]
20. Oluwafemi, F.; Oluwole, M. Microbiological Examination of Sachet Water Due to a Cholera Outbreak in Ibadan, Nigeria. *Open J. Med. Microbiol.* **2012**, *2*, 115–120. [CrossRef]
21. Adedire, O.M.; Atere, A.; Ogundipe, W.F.; Farinu, A.O. Effects of Direct and Indirect Sunlight on Polythene Packs, Sensory, Microbial and Chemical Properties of Sachet Water. *J. Adv. Biol. Biotechnol.* **2021**, *24*, 25–34. [CrossRef]
22. Onosakponome, O.R. The effects of sun radiation on the quality of sachet and bottled water. *Int. J. Eng. Appl. Sci. Technol.* **2017**, *6*, 192–208.
23. Samuel, O.A.; Tenebe, I.T.; Emenike, P.C.; Daniel, D.I.; Omole, D.O.; Maxwell, O.; Ben, N.U.; Kelechi, O.O.; Osagie, I. Preliminary Assessment of the Current Pollution Status of the River Atuwara, Nigeria, within an Industrial Site: A Bivariate Approach. *WIT Trans. Ecol. Environ.* **2019**, *234*, 209–219. [CrossRef]
24. Enyoh, C.E.; Enyoh, E.C.; Ohiagu, F.O. Evaluation of Some Groundwater Sources in Ota, Ogun State, Southwestern Nigeria. *World News Nat. Sci.* **2021**, *36*, 99–113.
25. Omole, D.O.; Ndambuki, J.M.; Balogun, K. Consumption of Sachet Water in Nigeria: Quality, Public Health, and Economic Perspectives. *Afr. J. Sci. Technol. Innov. Dev.* **2015**, *7*, 45–51. [CrossRef]
26. Omole, D.O.; Ogbiye, A.S.; Longe, E.O.; Adewumi, I.K.; Elemile, O.O.; Tenebe, T.I. Water quality checks on river Atuwara, South-West Nigeria. *WIT Trans Ecol Env.* **2018**, *228*, 165–173.
27. Emenike, P.C.; Neris, J.B.; Tenebe, I.T.; Nnaji, C.C.; Jarvis, P. Estimation of Some Trace Metal Pollutants in River Atuwara Southwestern Nigeria and Spatio-Temporal Human Health Risks Assessment. *Chemosphere* **2020**, *239*, 124770. [CrossRef]
28. Rice, E.W.; Geldreich, E.E.; Read, E.J. The Presence-Absence Coliform Test for Monitoring Drinking Water Quality. *Public Health Rep* **1989**, *104*, 54–58.
29. MacLeod, C.; Peletz, R.; Kere, F.; Baye, A.M.; Onyango, M.; Aw, S.; El Hadj Issabre, M.; Tung, R.; Khush, R. Are Presence/Absence Microbial Tests Appropriate for Monitoring Large Urban Water Supplies in Sub-Saharan Africa? *Water* **2019**, *11*, 491. [CrossRef]
30. Blodgett, R.; Food and Drug Administration. *Bacteriological Analytical Manual Appendix 2: Most Probable Number from Serial Dilutions*; Food and Drug Administration: Washington, DC, USA, 2010.
31. Wright, K.M.; Wright, P.J.; Holden, N.J. MacConkey Broth Purple Provides an Efficient MPN Estimation Method for Shigatoxigenic Escherichia Coli. *J. Microbiol. Methods* **2021**, *181*, 106132. [CrossRef]
32. Bolton, F.J.; Hinchliffe, P.M.; Coates, D.; Robertson, L. A Most Probable Number Method for Estimating Small Numbers of Campylobacters in Water. *J. Hyg.* **1982**, *89*, 185–190. [CrossRef] [PubMed]
33. Niu, C.; Zhang, Y.; Zhang, Y. Evaluation of a Most Probable Number Method for Detection and Quantification of Legionella pneumophila. *Pathogens* **2022**, *11*, 789. [CrossRef] [PubMed]

34. Thani, T.S.; Symekher, S.M.L.; Boga, H.; Oundo, J. Isolation and Characterization of Escherichia Coli Pathotypes and Factors Associated with Well and Boreholes Water Contamination in Mombasa County. *Pan Afr. Med. J.* **2016**, *23*, 12. [CrossRef] [PubMed]
35. Standard Methods Committee of the American Public Health Association; American Water Works Association; Water Environment Federation. 9221 Multiple-Tube Fermentation Technique for Members of the Coliform Group. In *Standard Methods For the Examination of Water and Wastewater*; Lipps, W.C., Baxter, T.E., Braun-Howland, E., Eds.; APHA Press: Washington, DC, USA, 2017. [CrossRef]
36. Olaoye, O.A.; Onilude, A.A. Assessment of Microbiological Quality of Sachet-Packaged Drinking Water in Western Nigeria and Its Public Health Significance. *Public Health* **2009**, *123*, 729–734. [CrossRef]
37. CDC. *Staphylococcus Aureus* in Healthcare Settings | HAI | CDC. Available online: https://www.cdc.gov/hai/organisms/staph.html (accessed on 17 January 2023).
38. Addo, B.E.; Amankwaa, G.; Gyasi, R.M. Physicochemical and Bacteriological Quality of Sachet Water Used by Ghanaian University Students: Implications for Public Health. *J. Water Sanit. Hyg. Dev.* **2019**, *9*, 56–63. [CrossRef]
39. Lay, H.T.; Wang, R.; Chew, J.W. Influence of Foulant Particle Shape on Membrane Fouling in Dead-End Microfiltration. *J. Membr. Sci.* **2022**, *647*, 120265. [CrossRef]
40. Poerio, T.; Denisi, T.; Mazzei, R.; Bazzarelli, F.; Piacentini, E.; Giorno, L.; Curcio, E. Identification of Fouling Mechanisms in Cross-Flow Microfiltration of Olive-Mills Wastewater. *J. Water Process Eng.* **2022**, *49*, 103058. [CrossRef]
41. Tanudjaja, H.J.; Anantharaman, A.; Ng, A.Q.Q.; Ma, Y.; Tanis-Kanbur, M.B.; Zydney, A.L.; Chew, J.W. A Review of Membrane Fouling by Proteins in Ultrafiltration and Microfiltration. *J. Water Process Eng.* **2022**, *50*, 103294. [CrossRef]
42. Yu, R.; Wang, H.; Wang, R.; Zhao, P.; Chen, Y.; Liu, G.; Liao, X. Polyphenol Modified Natural Collagen Fibrous Network towards Sustainable and Antibacterial Microfiltration Membrane for Efficient Water Disinfection. *Water Res.* **2022**, *218*, 118469. [CrossRef]
43. Nnaji, C.C.; Onuigbo, K.; Nnam, J.P. Assessment of Seasonal Variation of Nsukka Phreatic Aquifer Groundwater Quality. *Clim. Change* **2019**, *5*, 18.
44. Kabwama, S.N.; Bulage, L.; Nsubuga, F.; Pande, G.; Oguttu, D.W.; Mafigiri, R.; Kihembo, C.; Kwesiga, B.; Masiira, B.; Okullo, A.E.; et al. A Large and Persistent Outbreak of Typhoid Fever Caused by Consuming Contaminated Water and Street-Vended Beverages: Kampala, Uganda, January–June 2015. *BMC Public Health* **2017**, *17*, 23. [CrossRef]
45. Okullo, J.O.; Moturi, W.N.; Ogendi, G.M. Open Defaecation and Its Effects on the Bacteriological Quality of Drinking Water Sources in Isiolo County, Kenya. *Environ. Health Insights* **2017**, *11*, 1178630217735539. [CrossRef] [PubMed]
46. Stoler, J.; Tutu, R.A.; Winslow, K. Piped Water Flows but Sachet Consumption Grows: The Paradoxical Drinking Water Landscape of an Urban Slum in Ashaiman, Ghana. *Habitat Int.* **2015**, *47*, 52–60. [CrossRef]

water

MDPI

Article

Designing Water Inter-Plant Networks of Single and Multiple Contaminants through Mathematical Programming

Abeer M. Shoaib [1], Amr A. Atawia [1], Mohamed H. Hassanean [1], Abdelrahman G. Gadallah [2,3] and Ahmed A. Bhran [2,*]

1 Department of Petroleum Refining and Petrochemical Engineering, Faculty of Petroleum and Mining Engineering, Suez University, Suez P.O. Box 43221, Egypt; a.shoaib@suezuni.edu.eg (A.M.S.); amr.atawia@gmail.com (A.A.A.); mhassanean@yahoo.com (M.H.H.)
2 Chemical Engineering Department, College of Engineering, Imam Mohammad Ibn Saud Islamic University (IMSIU), Riyadh 11432, Saudi Arabia; agadallah@imamu.edu.sa
3 Chemical Engineering Department, National Research Center, Cairo 11241, Egypt
* Correspondence: aabahran@imamu.edu.sa; Tel.: +966-580772698

Abstract: Water is the meaning of life for humans, agricultural and industrial processes; controlling the distribution of water and wastewater between industrial processes is very vital for rationalizing water and preserving the environment. This paper addresses a mathematical approach to optimizing water inter-plant networks. The water network problem is formulated as a nonlinear program (NLP) that is solved by LINGO Software, version 14.0. A generalized two-step mathematical model is designed to be valid for solving networks containing large numbers of sources and sinks. The introduced model is proposed to be used for both single and multiple contaminant problems with up to six contaminants. Two mathematical models are presented to design water inter-plant networks efficiently. Firstly, the introduced model is solved by LINGO, in which the data given are applied; the obtained results are simultaneously sent to a second model (based on Excel Software 2019, v. 16.0), by which the obtained water networks are automatically drawn. The proposed approach has been applied in three case studies; the first case study contains five plants of single contaminants, the second case study contains three plants of single contaminants, and the third case study contains three plants of multiple contaminants. The results showed a noticeable reduction in the percentages of freshwater consumption in the investigated three case studies, which were 38.6, 4.74 and 8.64%, respectively, and the wastewater discharge of the three case studies were decreased by 38.1, 4.61 and 8.65%, respectively.

Keywords: inter-plant water network; nonlinear programming; freshwater consumption; mathematical approach; multiple contaminants

Citation: Shoaib, A.M.; Atawia, A.A.; Hassanean, M.H.; Gadallah, A.G.; Bhran, A.A. Designing Water Inter-Plant Networks of Single and Multiple Contaminants through Mathematical Programming. *Water* **2023**, *15*, 4315. https://doi.org/10.3390/w15244315

Academic Editor: Christos S. Akratos

Received: 15 October 2023
Revised: 10 December 2023
Accepted: 15 December 2023
Published: 18 December 2023

1. Introduction

The management of water in the inter-plant industrial process has been posed in the last decade since the consumption of global freshwater has increased continuously in industry.

Several processes in fertilizers, refineries and chemical companies use water in cooling systems, the scrubbing of gases, dilution and the adaptation of heat balance in heat exchangers.

Various methodologies have been presented in recent years for minimizing freshwater consumption and reducing the flowrate of wastewater discharge in the design of water inter-plant networks. A stochastic optimization model is proposed by Al-Redhwan et al. to minimize freshwater consumption and to produce a flexible wastewater network; they studied the distribution of wastewater in several processes in oil refinery plants. These processes include the atmospheric crude distillation unit, vacuum distillation unit, and the hydrocracker and kerosene desulfurization unit [1]. A genetic algorithm was presented by Ami et al. to manage the distribution of water in the contaminant sensor network to obtain the optimal system and multi-objective sensor model [2]. A pinch technique was proposed by Chew et al. for the reduction of freshwater and wastewater flowrates: a case study of an iron and steel mill was presented to show the effectiveness of their presented techniques; the processes contained mold cooling, slab cooling, rinsing and fume scrubbing [3]. The production of methanol from molasses was studied by Satyawali et al.: the effluent wastewater which is produced from methanol production included a high strength of pollution and the processes contained several equipment for maintaining temperature, such as cooling towers [4]. Iancu et al. introduced a mathematical model to design a regeneration wastewater network: they presented a case study of a petrochemical plant that contains one water source, six operation units, four contaminants and one regeneration unit to maximize the reuse of wastewater [5]. A case study of a steel plant was presented by Tian et al. to optimize the allocation of water and wastewater between several processes including the power plant, ore dressing, blast furnace, hot air furnace and rinsing residue; the chlorine concentration was presented as the limiting concentration in the design of water–wastewater networks [6]. A systematic methodology is presented by Kim et al. to minimize the cost estimation in the design of wastewater and heat exchange networks in oil refinery processes that contain multiple contaminants; in their work, a mixed integer non-linear programming formulation based on mass and heat balance between the processes is proposed; several processes such as hydrodesulfurization unit and an atmospheric distillation unit were introduced to show the effect of such processes on changing wastewater concentration [7]. An algorithm-based method is proposed by Chew et al. to minimize the flowrate of water resources in the single contaminant system of an inter-plant resource conservation network (IPRCN); they applied their algorithm to three water networks [8]. A different mathematical model is presented by Chen et al. to minimize the consumption of fresh water for the inter-plants; their work is applied to a case study of three plants with multiple contaminant systems [9]. Three wastewater treatment plants were studied by Julien et al. to manage the distribution of microbiological water in the Seine River [10]. An adaptive random search (ARS), which is an optimization approach introduced by Poplewski et al., is applied to several case studies of mixed integer nonlinear problems: a case study of a paper mill was presented with several processes which included pulping (dilution), a paper machine, a cylinder shower and felt showers [11]. An organic production plant was presented by Gopal et al., using treatment units to minimize the concentrations of contaminants and maximize the reuse of wastewater while minimizing the cost of treatment [12]. Yang et al. have proposed mathematical programming approaches that are based on mixed integer nonlinear programming to optimize reuse-recycle wastewater networks using treatment units; several methods, such as reverse osmosis, ion exchange, sedimentation, ultrafiltration and activated sludge, were used to decrease the concentration of contaminants [13]. A simultaneous optimization model was formulated to design a heat-integrated wastewater network based on mixed integer nonlinear programming to minimize the cost of freshwater consumption and the cost of wastewater treatment units [14]. A mixed integer nonlinear program was proposed using regeneration reuse and regeneration recycle in the hollow fiber reverse osmosis membrane to minimize the cost of freshwater and energy consumption; the presented model was applied to a refinery case study which included amine sweeting distillation, hydrotreating and desalting processes and considered the chemical oxygen demand and total dissolved solids to be limiting concentrations [15]. Bozkurt et al. have proposed a mathematical approach based on a framework to solve and optimize a multiple contaminant retrofitting problem; they studied the design of a wastewater treatment plant and the calculation of energy efficiency [16]. A reduction in the total annualized cost and wastewater discharge has been presented by Sueviriyapan et al. using a mixed integer nonlinear program; they applied their technique to a refinery plant and the results showed a decrease in the total annualized cost as well as the wastewater discharge [17]. A two-stage stochastic programming model has been presented to design an optimum water–wastewater network [18]: Naderi et al. studied the effect of hazards on environmental law. Hong et al. developed a strategy of multi-objective optimal control (MOOC) and multi-objective particle swarm optimization (MOPSO) to reduce the consumption of heat and increase the operational efficiency of the wastewater treatment plant [19]. A corn refinery case study was introduced to show the water management techniques presented by Mostafa et al.; several processes were presented to show the flexibility of the presented model, and these processes include gluten separation, starch separation, starch dewatering and glucose evaporation; chemical oxygen demands and total dissolved solids were presented as the limiting concentrations of contaminants in the allocation of freshwater and wastewater between processes [20]. Two techniques of centralized water header are proposed by Fadzil et al. to improve the reuse of wastewater in networks; they presented a case study of a single contaminant system that consists of five plants [21]. Lv et al. presented a step-by-step optimization method in the design of inter-plant water networks; a case study of a single contaminant in southern China was applied to show the applicability of their method [22]. A case study of inter-plant processes between an oil refinery plant and a petrochemical plant was presented by Reinaldo et al. to optimize the distribution of water and wastewater between several processes such as those of the cooling towers, condensers, coolers and boilers [23]. Robles et al. proposed model predictive control (MPC) and particle swarm optimization (PSO) to make a quality control of river basins in the presence of ammonium and nitrites [24]. A concentration potential concept was used by Wang et al. to design an optimal inter-plant water network; a case study of three plants and multiple contaminants was presented to show the effectiveness of their technique [25]. Fard et al. presented a Lagrangian relaxation-based model to make a control of water supply and wastewater collection; they studied the quality of the Azerbaijan province in Iran as a case study to minimize the water supply and wastewater discharge [26]. Mohammad and Chang studied the design of water–wastewater networks in the textile industry; according to the high temperature of the water, up to 60 °C, several contaminants were found in wastewater discharged streams such as chemical oxygen demand [27]. Kumawat et al. proposed a robust formulation in a continuous process to calculate the consumption of freshwater; their technique controlled the flowrates and qualities of the reused and recycled wastewater [28]. Three optimization models are proposed by Grzegorz and Dominic to design a flexible water network while minimizing the total length of the pipeline, the consumption of freshwater and the total annualized costs [29]. A textile industrial cluster was studied to manage the allocation of wastewater flowrate between sources and demands; zero liquid discharge was targeted in the design of a wastewater network that included a single contaminant TDS in several processes like the crystallizer, centrifuge and dilution processes [30]. A maximization of wastewater reuse in the textile dyeing industry was presented by Erkata et al.; several processes needed water in the dying industry such as the singeing, de-sizing, boiling, bleaching and printing processes [31]. A Bayesian optimization approach was proposed by Mariacrocetta et al. to manage the water quality of drainage systems [32]. A scrap tires-into-fuel processing facility was studied by Nessren et al. to design the wastewater network between several processes which include the condenser, decanter, separation, seal-pot and stripping processes; a graphical technique was used to optimize the distribution of wastewater between sinks to sources [33].

Reducing the consumption of freshwater usage and wastewater discharge in water inter-plant networks is a challenge in many plants, such as cement plants, polyethylene plants, oil refineries and fertilizers plants. Managing the distribution of water in inter-plant processes and the large amount of freshwater consumption in different industrial processes, such as those of the condensers, heat exchangers, vacuum systems, cooling and washing processes, refers to the need to minimize the freshwater consumption and wastewater discharge that are leading us to establish the proposed optimization program. Good management of water distribution between plants will consequently result in a considerable reduction in the cost of freshwater as well as wastewater treatment. To date, no generalized model has been introduced to help in designing inter-plant networks aiming to minimize the required freshwater consumption and wastewater discharge including a wide range and number of sources and sinks. In this paper, a generalized model, which is able to deal with up to five inter-plants having up to a hundred sources and a hundred sinks, is introduced. The introduced model could be applied to single contaminant networks as well as multiple contaminants networks. Additionally, the results of running the proposed model are presented simultaneously as a drawn network to facilitate the application of the proposed network construction. The proposed mathematical model is based on equations that are formulated as a nonlinear program with definite constraints and assumptions. After running the mathematical model, the obtained results are shown and sent to a designed Excel software which is able to achieve the water–wastewater inter-plant networks automatically. Three case studies are investigated, and their results are compared with the obtained results in the literature.

2. Methods

In this research, the minimization of freshwater consumption is presented as an objective function in the presence of a single or multi-contaminant system to design water–wastewater inter-plant networks. The present problem could be stated as follows:

- Given a set of sources, reaching up to one hundred sources, where each source (n) has a flowrate (F_{Rn}) in a multi-contaminant reach up to six contaminants (A, B, C, D, E and F), where the concentrations of contaminants in sources are X_{RnA}, X_{RnB}, X_{RnC}, X_{RnD}, X_{RnE} and X_{RnF}, the flowrate of each source has the probability to send to sinks by flowrate $g_{n\text{-}i}$ or send to waste by flowrate G_{n_waste}.
- Given a set of sinks, reaching up to one hundred sinks, where each sink (i) has a flowrate (G_i) with a limiting concentration of contaminants X_{giA}, X_{giB}, X_{giC}, X_{giD}, X_{giE} and X_{giF}, then:
- The freshwater flowrate (F_W) has the probability to feed each sink (i) with a concentration of contaminants X_A, X_B, X_C, X_D, X_E and X_F.
- The total wastewater flowrate is G_{waste} with a concentration of Xw_A, Xw_B, Xw_C, Xw_D, Xw_E and Xw_F.

As shown in Figure 1, the design of the water–wastewater network is illustrated in sequence procedures that started by applying an overall mass balance to each source (n), which has a flowrate (F_{Rn}) that has a probability to distribute to each sink (i) by flowrate $g_{n\text{-}i}$ and to waste by flowrate $G_{n-waste}$, which is shown in Equation (1).

$$F_{Rn} = \sum g_{n-i} + G_{n-Waste} \tag{1}$$

The overall mass balance is applied to each sink (i); the flowrate of each sink (G_i) has the probability to be fed by the freshwater flowrate (F_{wi}) and water flowrate from source to sink (g_{n-i}), as shown in Equation (2).

$$G_i = F_{wi} + \sum g_{n-i} \tag{2}$$

As shown in Equation (3), a component mass balance is applied on each sink having contaminant A: the product of the flowrate of each sink (G_i) by limiting the concentration of contaminant A (X_{giA}) is equal to the sum of the product of the freshwater flowrate (F_{wi}) and the concentration of the freshwater of contaminant A (X_A), and the product of the summation of the water flowrate from source to sink (g_{n-i}) and the concentrations of contaminant A in each source (X_{RnA}).

$$G_i * X_{giA} = F_{wi} * X_A + \sum g_{n-i} * X_{RnA} \tag{3}$$

A component mass balance of contaminant B is applied to each sink as shown in Equation (4): the product of (G_i) by the limiting concentration of contaminant B (X_{giB}) is equal to the sum of the product of F_{wi} and the concentration of the freshwater of contaminant B (X_B), and the product of g_{n-i} and the concentrations of contaminant B in each source (X_{RnB}).

$$G_i * X_{giB} = F_{wi} * X_B + \sum g_{n-i} * X_{RnB} \quad (4)$$

By applying a component mass balance of component C to each sink as shown in Equation (5), the result of the product of (G_i) and X_{giC} (the limiting concentration of contaminant C) is equal to the sum of the product of F_{wi} and X_C (the concentration of the freshwater of contaminant C) and the product of g_{n-i} and X_{RnC} (the concentrations of contaminant C in each source).

$$G_i * X_{giC} = F_{wi} * X_C + \sum g_{n-i} * X_{RnC} \quad (5)$$

As shown in Equations (6)–(8), a component mass balance is applied to each sink having contaminants (D, E and F), where X_D, X_E and X_F are the concentrations of contaminants D, E and F of the freshwater flowrate, respectively, and the concentrations of contaminants D, E and F are X_{RnD}, X_{RnE} and X_{RnF}, respectively.

$$G_i * X_{giD} = F_{wi} * X_D + \sum g_{n-i} * X_{RnD} \quad (6)$$

$$G_i * X_{giE} = F_{wi} * X_E + \sum g_{n-i} * X_{RnE} \quad (7)$$

$$G_i * X_{giF} = F_{wi} * X_F + \sum g_{n-i} * X_{RnF} \quad (8)$$

In Equation (9), the overall mass balance is applied to the waste discharge stream; each source has the probability of sending wastewater to waste by a flowrate G_{n_waste}, and the collected wastewater flowrate is G_{waste}.

$$G_{Waste} = \sum G_{n-Waste} \quad (9)$$

Furthermore, a component mass balance is applied to the wastewater discharge of six contaminants (A, B, C, D, E and F), as shown in Equations (10)–(15).

$$G_{Waste} * X_{wA} = \sum G_{n\_waste} * X_{RnA} \quad (10)$$

$$G_{Waste} * X_{wB} = \sum G_{n\_waste} * X_{RnB} \quad (11)$$

$$G_{Waste} * X_{wC} = \sum G_{n\_waste} * X_{RnC} \quad (12)$$

$$G_{Waste} * X_{wD} = \sum G_{n\_waste} * X_{RnD} \quad (13)$$

$$G_{Waste} * X_{wE} = \sum G_{n\_waste} * X_{RnE} \quad (14)$$

$$G_{Waste} * X_{wF} = \sum G_{n\_waste} * X_{RnF} \quad (15)$$

Each sink (i) has the probability of being fed by freshwater flowrate (F_{Wi}); the overall mass balance of the freshwater streams is shown in Equation (16).

$$F_w = \sum F_{wi} \quad (16)$$

LINGO Software, v. 14.0 is used in this work to get the optimum solution. LINGO Software is used to solve linear and nonlinear equations with definite constraints and assumptions; the mathematical approach is based on a nonlinear program (NLP) and the constraints and variables refer to the positive real number or zero values. After running the proposed mathematical model in LINGO Software, the obtained results are sent directly to the Excel software which has the ability to draw the water–wastewater inter-plant network automatically.

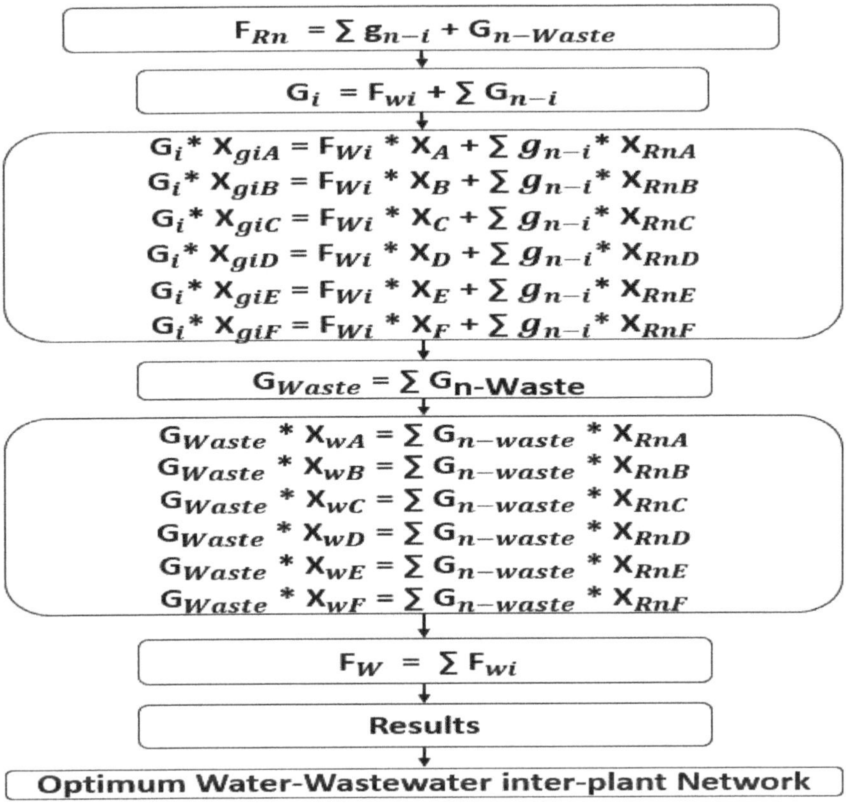

Figure 1. Procedure of optimum design for water–wastewater inter-plant network.

3. Case Studies

The proposed mathematical model was examined by applying it to three case studies that contain single and multi-contaminants to show its effectiveness in designing water–wastewater networks. The presented case studies include a different number of plants in each case study with different contaminants, including total suspended solids (TSS), chemical oxygen demand (COD), hydrocarbon, hydrogen sulfide (H_2S) and total dissolved solids (TDS); these contaminants should be controlled via a mathematical approach to avoid the fouling, cooling efficiency, hardness and corrosion problems in the plants. These case studies are described in the following subsections.

3.1. Case Study 1

Case study 1 contains a single contaminant, which is the total suspended solids (TSS); it was presented by Fadzil et al. [21]. This case study includes five plants; plant A has four sources and four sinks, plant B consists of four sources and four sinks, plant C contains five sources and five sinks, plant D has three sources and two sinks, and plant E contains five sources and five sinks, as shown in Table 1.

Table 1. Limiting flowrates and concentrations of sources and sinks in case study 1.

Plant	Sources and Sinks	Stream Number	Flow Rate (m^3/h)	TSS (ppm)	Plant	Sources and Sinks	Stream Number	Flow Rate (m^3/h)	TSS (ppm)
Plant A	Sources	S1	50	50	Plant B	Sources	S1	20	100
		S2	100	100			S2	100	100
		S3	70	150			S3	40	800
		S4	60	250			S4	10	800
	Sinks	K1	50	20		Sinks	K1	20	0
		K2	100	50			K2	100	50
		K3	80	100			K3	40	50
		K4	70	200			K4	10	400
Plant C	Sources	S1	105	17	Plant E	Sources	S1	40	200
		S2	182.35	44			S2	50	200
		S3	138.7	49			S3	30	400
		S4	92.55	83			S4	60	400
		S5	45.55	115			S5	40	600
	Sinks	K1	182.35	0		Sinks	K1	40	0
		K2	45.55	10			K2	50	100
		K3	138.7	10			K3	30	100
		K4	92.55	10			K4	60	300
		K5	105	87			K5	40	400
Plant D	Sources	S1	150	10	Plant D	Sinks	K1	200	20
		S2	60	50					
		S3	100	85			K2	80	75

3.2. Case Study 2

Case study 2, provided by Lv et al. [22], presents three plants (molasses treatment system (X), yeast production system (Y), and circulating cooling system (Z)) with a single contaminant, which is chemical oxygen demand (COD). Plant X contains five sources and five sinks, plant Y contains five sources and five sinks, while plant Z includes five sources and five sinks. The limiting flowrates and concentrations of contaminants of the sources and sinks are shown in Table 2.

Table 2. The limiting data of sources and sinks in plants X, Y and Z for case study 2.

Plant	Process	Stream Number	Flow Rate (m^3/h)	Limiting Concentration of Contaminant COD (ppm)
Molasses treatment system (X)	Sources	S1	20	100
		S2	66.67	80
		S3	100	100
		S4	41.67	800
		S5	10	800
	Sinks	K1	20	0
		K2	66.67	50
		K3	100	50
		K4	41.67	80
		K5	10	400

Table 2. *Cont.*

Plant	Process	Stream Number	Flow Rate (m^3/h)	Limiting Concentration of Contaminant COD (ppm)
Yeast production system (Y)	Sources	S1	20	100
		S2	66.67	80
		S3	15.63	400
		S4	42.86	800
		S5	6.67	1000
	Sinks	K1	20	0
		K2	66.67	50
		K3	15.63	80
		K4	42.86	100
		K5	6.67	400
Circulating cooling system (Z)	Sources	S1	20	100
		S2	80	50
		S3	50	125
		S4	40	800
		S5	300	150
	Sinks	K1	20	0
		K2	80	25
		K3	50	25
		K4	40	50
		K5	300	100

3.3. Case Study 3

The third case study of the current work was presented by Wang et al. [25]. This case study includes three plants with multiple contaminant systems including the contaminants hydrocarbon, hydrogen sulfide (H_2S) and total dissolved solids (TDS); plant 1 consists of eight sources and eight sinks, plant 2 contains seven sources and seven sinks, while plant 3 consists of three sources and three sinks as shown in Table 3.

Table 3. The limiting flowrates and concentrations of sources and sinks for case study 3.

Plant	Sources and Sinks	Flowrate (m^3/h)	Contaminant A (Hydrocarbon) (ppm)	Contaminant B (H_2S) (ppm)	Contaminant C (TDS) (ppm)
Plant 1	Source 1	30	100	90	50
	Source 2	16	50	70	70
	Source 3	75	150	80	70
	Source 4	21	160	100	90
	Source 5	29	210	200	120
	Source 6	65	80	70	80
	Source 7	61	300	290	170
	Source 8	57	210	170	100

Table 3. *Cont.*

Plant	Sources and Sinks	Flowrate (m^3/h)	Contaminant A (Hydrocarbon) (ppm)	Contaminant B (H_2S) (ppm)	Contaminant C (TDS) (ppm)
Plant 1	Sink 1	30	0	0	0
	Sink 2	16	0	0	0
	Sink 3	75	40	60	20
	Sink 4	21	30	40	70
	Sink 5	29	110	135	60
	Sink 6	65	0	0	0
	Sink 7	61	100	75	20
	Sink 8	57	90	50	34
Plant 2	Source 1	35	110	120	100
	Source 2	40	350	400	210
	Source 3	40	150	180	210
	Source 4	30	210	150	220
	Source 5	30	350	320	310
	Source 6	64	800	1100	1000
	Source 7	50	1500	2100	1800
	Sink 1	35	0	0	0
	Sink 2	40	200	170	150
	Sink 3	40	90	130	100
	Sink 4	30	110	80	150
	Sink 5	30	260	200	180
	Sink 6	64	340	350	400
	Sink 7	50	950	850	900
Plant 3	Source 1	30	900	4500	3000
	Source 2	34	120	12,500	180
	Source 3	56	220	45	9500
	Sink 1	30	150	700	800
	Sink 2	34	20	300	45
	Sink 3	56	120	20	200

4. Results and Discussions

The proposed approach for optimizing water–wastewater inter-plant networks in industrial inter-plants was applied to three case studies (with single and multiple contaminants) and the results are discussed in the following subsections.

4.1. Results and Discussions of Case Study 1

Controlling the limiting concentration of total suspended solids (TSS) in the industrial processes prevented them from causing plugging in the pipelines, cavitation in the pumps, erosion in the unit operation and accumulation which decreases the heat exchange, as shown in Julien et al. [10].

After introducing the data given for case study 1 into the LINGO program, the obtained results of the freshwater consumption flowrate, the flowrates from sources to demands and the flowrates from sources to waste are listed in Table 4 and shown in Figure 2.

Table 4. Freshwater flowrates to sinks, sources flowrates to sinks and to waste for case study 1.

Stream	Flowrate (t/h)	Stream	Flowrate (t/h)	Stream	Flowrate (t/h)	Stream	Flowrate (t/h)
F_w	412.3	$G_{4\text{-waste}}$	53	$G_{10\text{-}13}$	8.5	$G_{14\text{-}1}$	35.3
F_{w5}	20	$G_{5\text{-}3}$	1.9	$G_{10\text{-}14}$	55.8	$G_{14\text{-}11}$	39.1
F_{w9}	182.4	$G_{5\text{-}4}$	0.4	$G_{10\text{-}15}$	12.4	$G_{14\text{-}14}$	75.6
F_{w10}	35.2	$G_{5\text{-}6}$	0.8	$G_{10\text{-}17}$	3.5	$G_{15\text{-}2}$	4
F_{w11}	41	$G_{5\text{-}13}$	3	$G_{10\text{-}18}$	1.8	$G_{15\text{-}6}$	56
F_{w12}	71.5	$G_{5\text{-}15}$	6.7	$G_{10\text{-}19}$	0.7	$G_{16\text{-}4}$	4.2
F_{w14}	22.2	$G_{5\text{-}17}$	1.8	$G_{10\text{-waste}}$	7	$G_{16\text{-}6}$	1.2
F_{w16}	40	$G_{5\text{-}18}$	0.6	$G_{11\text{-}2}$	47.8	$G_{16\text{-}7}$	3.3
$G_{1\text{-}2}$	45	$G_{5\text{-}19}$	0.3	$G_{11\text{-}3}$	29.9	$G_{16\text{-}13}$	8.5
$G_{1\text{-}3}$	0.9	$G_{5\text{-waste}}$	4.5	$G_{11\text{-}4}$	1.5	$G_{16\text{-}15}$	50
$G_{1\text{-}6}$	1.5	$G_{6\text{-}4}$	5.8	$G_{11\text{-}6}$	22.7	$G_{16\text{-}17}$	6.5
$G_{1\text{-}7}$	2	$G_{6\text{-}7}$	1	$G_{11\text{-}7}$	7.1	$G_{16\text{-}18}$	5.9
$G_{1\text{-}17}$	0.4	$G_{6\text{-}18}$	3.3	$G_{11\text{-}13}$	12.3	$G_{16\text{-}19}$	17.8
$G_{1\text{-waste}}$	0.2	$G_{6\text{-}19}$	4.1	$G_{11\text{-}15}$	10.9	$G_{16\text{-}20}$	2.5
$G_{2\text{-}2}$	1.2	$G_{6\text{-}20}$	3	$G_{11\text{-}17}$	3.8	$G_{17\text{-}8}$	0.8
$G_{2\text{-}4}$	7.3	$G_{6\text{-waste}}$	82.8	$G_{11\text{-}18}$	1.9	$G_{17\text{-}19}$	1.1
$G_{2\text{-}13}$	63.6	$G_{7\text{-}8}$	1.9	$G_{11\text{-}19}$	0.9	$G_{17\text{-waste}}$	38.1
$G_{2\text{-}17}$	18	$G_{7\text{-}20}$	7	$G_{12\text{-}4}$	42.3	$G_{18\text{-}4}$	2.3
$G_{2\text{-}18}$	3.6	$G_{7\text{-waste}}$	31.1	$G_{12\text{-}6}$	1.3	$G_{18\text{-}8}$	1.6
$G_{2\text{-waste}}$	6.4	$G_{8\text{-waste}}$	10	$G_{12\text{-}8}$	2	$G_{18\text{-}18}$	2.5
$G_{3\text{-}3}$	5.3	$G_{9\text{-}11}$	58.6	$G_{12\text{-}13}$	8.4	$G_{18\text{-}20}$	14.1
$G_{3\text{-}4}$	4.8	$G_{9\text{-}14}$	46.4	$G_{12\text{-}17}$	6	$G_{18\text{-waste}}$	29.5
$G_{3\text{-}17}$	6.3	$G_{10\text{-}1}$	14.7	$G_{12\text{-}18}$	6.7	$G_{19\text{-waste}}$	30
$G_{3\text{-}18}$	2.7	$G_{10\text{-}2}$	2	$G_{12\text{-}20}$	2.3	$G_{20\text{-}8}$	2.2
$G_{3\text{-}19}$	5.3	$G_{10\text{-}4}$	1.4	$G_{12\text{-waste}}$	23.4	$G_{20\text{-}19}$	27
$G_{3\text{-waste}}$	45.6	$G_{10\text{-}6}$	16.6	$G_{13\text{-}3}$	42	$G_{20\text{-}20}$	8.7
$G_{4\text{-}17}$	1.7	$G_{10\text{-}7}$	26.5	$G_{13\text{-}13}$	0.7	$G_{20\text{-waste}}$	22.1
$G_{4\text{-}19}$	2.7	$G_{10\text{-}10}$	10.4	$G_{13\text{-}17}$	2	$G_{21\text{-}8}$	1.4
$G_{4\text{-}20}$	2.5	$G_{10\text{-}12}$	21	$G_{13\text{-}18}$	0.9	$G_{21\text{-waste}}$	38.6

According to the mass load of sources and sinks, the distribution of water and wastewater flowrates between sources and sinks is achieved. Regarding the obtained results, source 10 feeds thirteen sinks (K1, K2, K4, K6, K7, K10, K12, K13, K14, K15, K17, K18, K19) and waste by flowrates of 14.7, 2, 1.4, 16.6, 26.5, 10.4, 21, 8.5, 55.8, 12.4, 3.5, 1.8, 0.7 and 7 t/h, respectively. However, source 5 feeds seven sinks (K3, K4, K6, K13, K17, K18, K19) and waste by flowrates of 1.9, 0.4, 0.8, 3, 6.7, 1.8, 0.6, 0.3 and 4.5 t/h, respectively.

According to the low mass load of source 11, it does not supply any water to waste and its wastewater feeds ten sinks (K2, K3, K4, K6, K7, K13, K15, K17, K18 and K19) by flowrates of 47.8, 29.9, 1.5, 22.7, 7.1, 12.3, 10.9, 3.8, 1.9 and 0.9 t/h, respectively.

The obtained results show that the total freshwater consumption is 412.3 t/h, which is distributed to sinks K5, K9, K10, K11, K12, K14 and K16 by flowrates of 20, 182.4, 35.2, 41, 71.5, 22.2 and 40 t/h, respectively. However, source 1 supplies K2, K3, K6, K7, K17 and the waste by 45, 0.9, 1.5, 2, 0.4 and 0.2 t/h, respectively.

Regarding source 2, it feeds sinks K2, K4, K13, K17, K18 and waste by 1.2, 7.3, 63.6, 18, 3.6 and 6.4 t/h, respectively. Source 3 feeds five sinks (K3, K4, K17, K18, K19) and waste by flowrates of 5.3, 4.8, 6.3, 2.7, 5.3 and 45.6 t/h, respectively. Source 4 supplies its water to sinks K17, K19, K20 and waste by 1.7, 2.7, 2.5 and 53 t/h flowrates, respectively. Source 6 supplies the waste by 82.8 t/h and it supplies sinks K4, K7, K18, K19 and K20 by 5.8, 1, 3.3, 4.1 and 3 t/h, respectively. Source 7 is supplied to K8, K20 and waste by 1.9, 7 and 31.1 t/h, respectively. Source 8 sends all its water to waste with a flowrate of 10 t/h while source 9 feeds two sinks only (K11 and K14) with flowrates of 58.6 and 46.4 t/h. In addition, source 12 feeds waste by 23.4 t/h and it feeds sinks K4, K6, K8, K13, K17, K18 and K20 by flowrates of 42.3, 1.3, 2, 8.4, 6, 6.7 and 2.3 t/h, respectively.

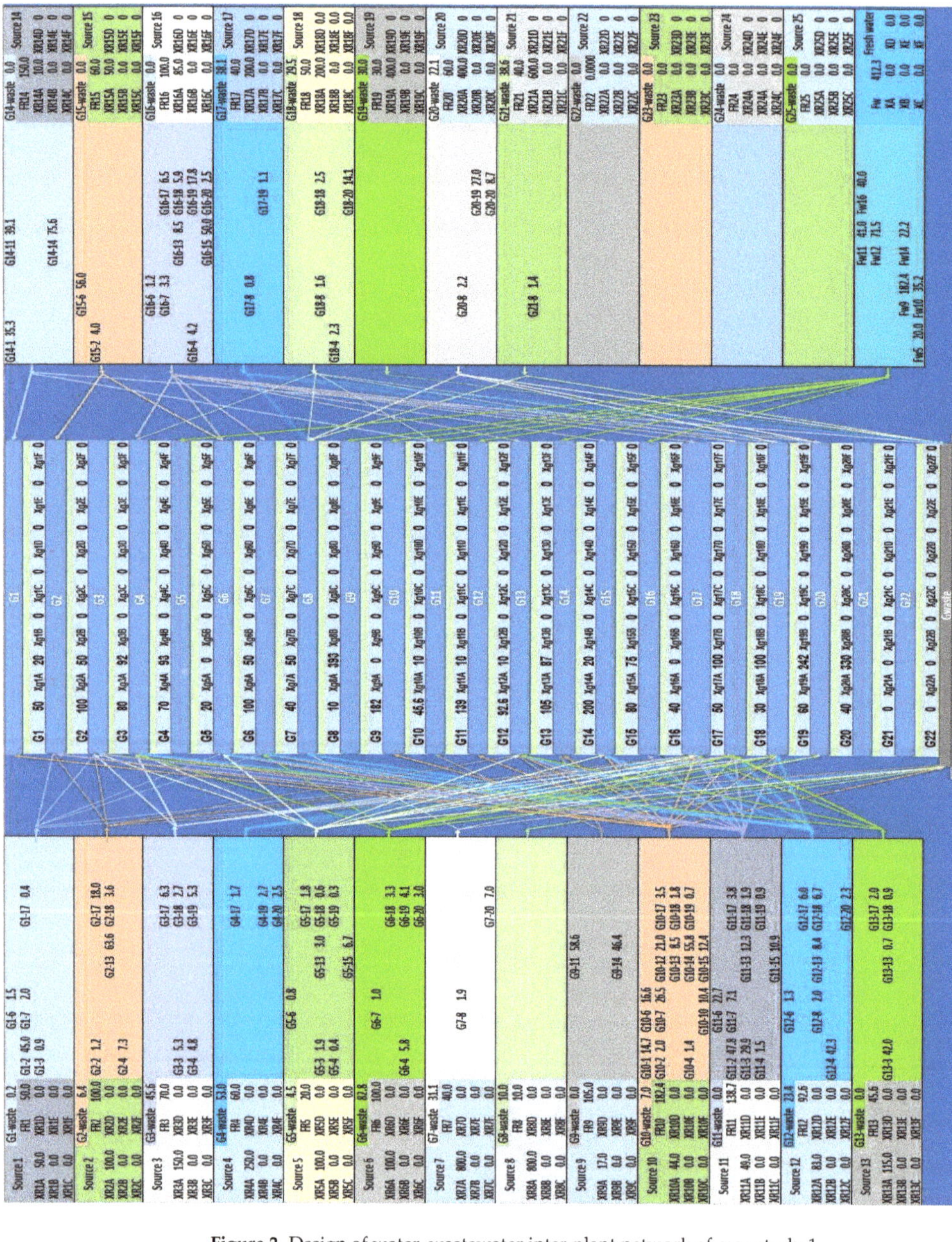

Figure 2. Design of water–wastewater inter-plant network of case study 1.

Source 13 is supplied to sinks K3, K13, K17 and K18 by 42, 0.7, 2 and 0.9 t/h, respectively. At the same time, source 14 feeds sinks K1, K11 and K14 by 35.3, 39.1 and 75.6 t/h flowrates, respectively. Source 15 supplies k2 and k6 by 4 and 56 t/h, respectively. Regarding source 16, it supplies sinks K4, K6, K7, K13, K15, K17, K18, K19 and K20 by 4.2, 1.2, 3.3, 8.5, 50, 6.5, 5.9, 17.8 and 2.5 t/h, respectively. Source 17 feeds sinks K8, K19 and waste by 0.8, 1.1 and 38.1 t/h, respectively. Also, source 18 feeds K4, K8, K18, K20 and waste by 2.3, 1.6, 2.5 and 14.1 t/h, respectively. All discharge water from source 19 is sent to waste by 30 t/h, while source 20 feeds K8, K19, K20 and waste by 2.2, 27, 8.7 and 22.1 t/h. In addition, source 21 feeds only sink 8 by 1.4 t/h and the remainder of its flowrate is supplied to waste by a flowrate of 38.6 t/h. Therefore, the total wastewater flowrate is equal to 422 t/h.

The LINGO results were applied to the introduced Excel program and the drawing of the water–wastewater inter-plant network was achieved automatically.

By comparing the results obtained by the proposed mathematical model with the results of the header design method of the original plants, it is clear that the freshwater consumption decreased from 671.7 to 412.3 t/h by a reduction percentage of 38.6%. Furthermore, the wastewater generated is reduced from 681.7 to 422 t/h by a reduction percentage of 38.1%. These results show the effectiveness of the introduced technique in designing water–wastewater networks by reducing the freshwater consumption as well as by decreasing the wastewater flowrate.

4.2. Results and Discussions of Case Study 2

Increasing the concentration of chemical oxygen demand (COD) leads to an increase in the fouling rate in the heat exchanger, a decrease in the cooling efficiency and blocking in the inner side of the pipelines, as shown in Mariacrocetta et al. [32].

After introducing the flowrates, concentrations of sources and sinks of the two plants to the proposed model, the results are obtained and shown in Table 5 and Figure 3. These results are sent to the prepared Excel software to show the final drawing of the water–wastewater inter-plant network.

Table 5. Freshwater flowrates to sinks, sources flowrates to sinks and to waste for case study 2.

Sinks	F_w (t/h)	Sources Flowrates (t/h)														
		S1	S2	S3	S4	S5	S6	S7	S8	S9	S10	S11	S12	S13	S14	S15
K1	20	0	0	0	0	0	0	0	0	0	0	0	0	0	0	0
K2	44.4	0	0	0	0	0	0	0	0	0	0	0	0	0	0	22.2
K3	47	20	0	0	0	0	0	25	0	0	0	0	0	8	0	0
K4	11.7	0	0	16.6	0	0	0	0	0	0	0	0	0	13.4	0	0
K5	0	0	0	0	0	2.8	0	0	2.7	0	0	0	0	0	0	4.5
K6	20	0	0	0	0	0	0	0	0	0	0	00	0	0	0	0
K7	25	0	0	0	0	0	0	41.7	0	0	0	0	0	0	0	0
K8	0	0	0	0	0	0	0	0	0	0	0	0	10.9	0	0	4.7
K9	0	0	0	0	0	0	0	0	0	0	0	0	14.3	28.6	0	0
K10	0	0	0	0	0	1.1	0	0	1.8	0	0	0	0	0	0.7	3
K11	20	0	0	0	0	0	0	0	0	0	0	0	0	0	0	0
K12	66.7	0	0	0	0	0	0	0	0	0	0	0	0	0	0	13.3
K13	31.8	0	0	0	0	0	0	0	0	0	0	0	14.8	0	0	3.4
K14	0	0	0	0	0	0	0	0	0	0	0	0	40	0	0	0
K15	27.7	0	66.7	83.4	0	0	20	0	0	0	0	20	0	0	0	82.1
waste	0	0	0	0	41.7	6.1	0	0	11.1	42.9	6.7	0	0	0	39.3	166.8

The obtained results from the LINGO Software showed that all wastewater of sources S4, S9 and S10 are sent to waste only by flowrates of 41.7, 42.9 and 6.7 t/h, respectively, which referred to the high mass load of sources rather than sinks.

The total consumption of freshwater flowrate is 314.36 t/h and is distributed to sinks K1, K2, K3, K4, K6, K7, K11, K12, K13 and K15 by 20, 44.4, 47, 11.7, 20, 25, 20, 66.7, 31.8 and 27.7 t/h, respectively.

The wastewater flowrate of source 1 is distributed to K3 by 20 t/h, while source 2 feeds K15 by a flowrate of 66.7 t/h. Source 3 distributed its water to K4 and K15 by flowrates of 16.6 and 83.4 t/h, respectively. Source 5 feeds two sinks (K5 and K10) and waste by flowrates 2.8, 1.1 and 6.1 t/h, respectively. Sources S6 and S11 feed only sink 15 by the same flowrates of 20 t/h, while source S7 feeds two sinks, K3 and K7, by 25 and 41.7 t/h, respectively.

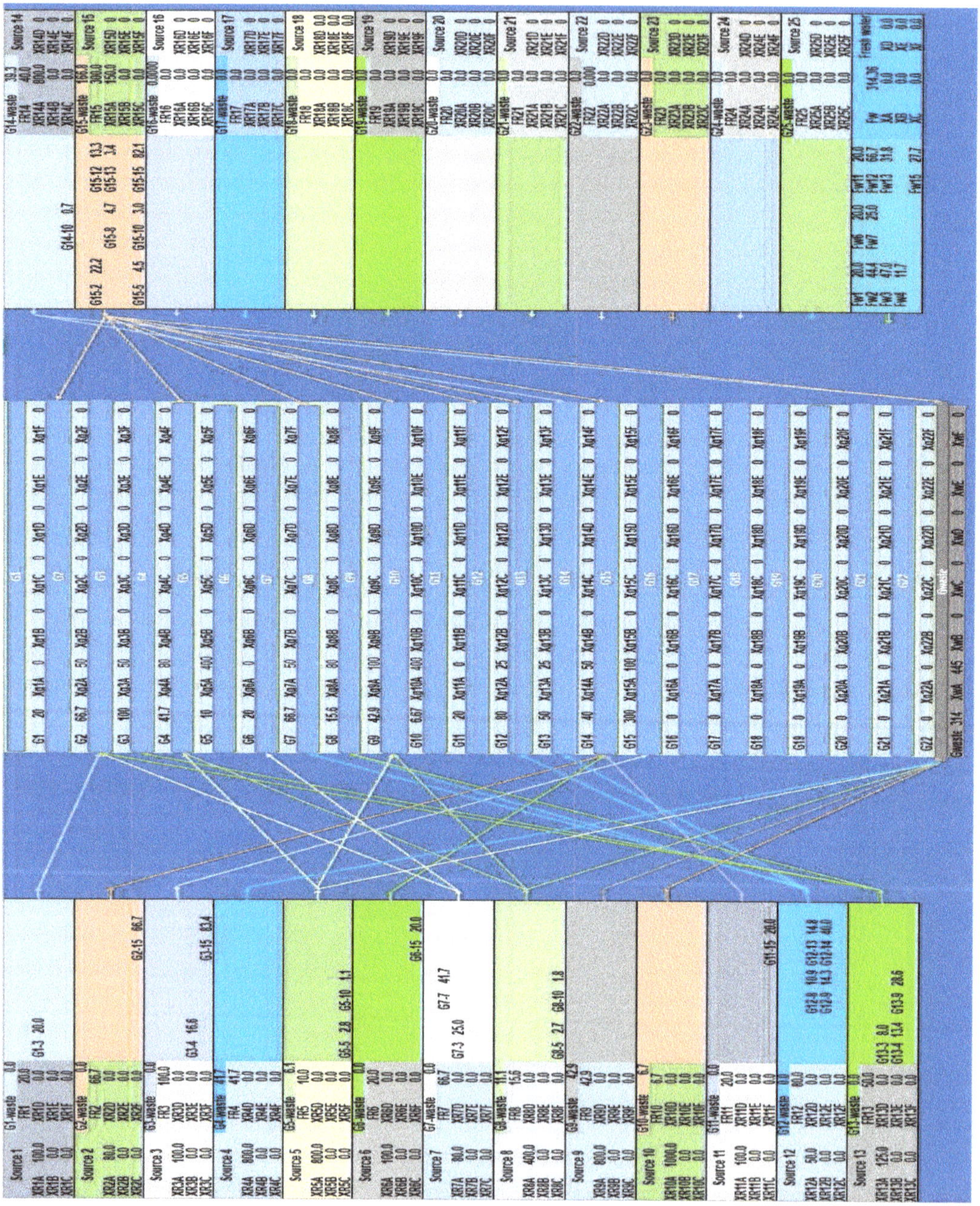

Figure 3. Design of water–wastewater inter-plant network of case study 2.

Source S8 supplied its wastewater to sinks K5, K10 and waste by flowrates of 2.7, 1.8, 11.1 t/h, while source S12 feeds four sinks, K8, K9, K13 and K14, by 10.9, 14.3, 14.8 and 40 t/h, respectively. Source S13 supplied sinks K3, K4 and K9 by 8, 13.4 and 28.6 t/h, respectively.

Source 14 feeds sink 10 and waste by 0.7 and 39.3 t/h, respectively, while source S15 feeds K2, K5, K8, K10, K12, K13, K15 and waste by 22.2, 4.5, 4.7, 3, 13.3, 3.4, 82.1 and 166.8 t/h, respectively.

As shown in Table 6, in the comparison between our technique, which is formulated as a nonlinear program (NLP), and the step-by-step optimization method (Lv et al. [22]), which is formulated as a linear programming model, the consumption of freshwater flowrate decreased from 330 to 314.36 t/h by a reduction percentage of 4.74%, and the wastewater discharge decreased from 329.54 to 314.36 t/h by a reduction percentage of 4.61%. In comparison with the optimization method (Chew et al. [3]) which is formulated by MINLP, the freshwater consumption decreased from 314.96 to 314.36 t/h by a reduction percentage of 0.19% and the wastewater discharge decreased from 538 to 314.6 t/h by a reduction percentage of 41.52%.

Table 6. Comparison between the introduced method and techniques of Chew et al. [3] and Lv et al. [22].

Integration Scheme	The Introduced Method	Optimization Method (Chew et al. [3])	Step-by-Step Optimization Method (Lv et al. [22])
Used Technique	Nonlinear Programming (NLP)	Mixed integer nonlinear programming (MINLP)	Linear Programming (LP)
Freshwater consumption (t/h)	314.36	314.96	330
Wastewater discharge (t/h)	314.36	538	329.54

4.3. Results and Discussions of Case Study 3

The data given in the third case study consist of three plants with multiple contaminants (hydrocarbon, hydrogen sulfide (H_2S) and total dissolved solids (TDS)). The effect of hydrocarbon appears in the increasing of organic matter in the water which increases the fouling rate in the pipelines of the heat exchanger, while the increase in hydrogen sulfide increases the acidity of the water, and consequently the rate of corrosion increases. On the other hand, the higher level of total dissolved solids results in an increase in the formation rate of scales as well as the hardness in the pipelines of plants, as shown in Buabeng et al. [15]. The obtained results of source flow rates to sinks and freshwater flowrates to sinks are shown in Table 7 after introducing these plants' data into the LINGO program. With passing these results to the Excel software, the design of the water–wastewater inter-plant network is achieved automatically, as shown in Figure 4.

Table 7. Freshwater flowrates to sinks, sources flowrates to sinks and to waste for case study 3.

Sources and Fresh Water	Sinks																		Waste
	K1	K2	K3	K4	K5	K6	K7	K8	K9	K10	K11	K12	K13	K14	K15	K16	K17	K18	
Fw	30	16	45	12.2	5	65	43.6	29.3	35	0	0	0	0	0	0	0	20.4	40.4	0
S1	0	0	30	0	0	0	0	0	0	0	0	0	0	0	0	0	0	0	0
S2	0	0	0	2.4	0	0	0	0	0	0	0	0	0	0	0	0	13.6	0	0
S3	0	0	0	0	18.1	0	17.4	27.7	0	0	0	6.5	0	0	0	0	0	5.2	0
S4	0	0	0	0	0	0	0	0	0	0	0	4.3	8.8	1.9	6	0	0	0	0
S5	0	0	0	0	00	0	0	0	0	1.3	0	0	0.2	1.9	1.9	0.2	0	0	0
S6	0	0	0	6.4	5.9	0	0	0	0	0	26.7	16.7	0	0	0	0	0	9.4	0
S7	0	0	0	0	0	0	0	0	0	0	0	0	0	42.7	18.3	0	0	0	0
S8	0	0	0	0	0	0	0	0	0	19	0	0	0.2	1.9	8.2	0.2	0	0	27.5
S9	0	0	0	0	0	0	0	0	0	19.4	13.3	2.2	0	0	0	0	0	0	0
S10	0	0	0	0	0	0	0	0	0	0	0	0	0.2	1.9	1.9	0	0	0	36
S11	0	0	0	0	0	0	0	0	0	0	0	0	0.2	1.9	1.9	28.1	0	0	7.8
S12	0	0	0	0	0	0	0	0	0	0	0	0	20.2	7.8	1.9	0	0	0	0
S13	0	0	0	0	0	0	0	0	0	0	0	0	0	1.9	1.9	0	0	0	26.2
S14	0	0	0	0	0	0	0	0	0	0	0	0	0	0	1.7	0	0	0	62
S15	0	0	0	0	0	0	0	0	0	0	0	0	0	0	0	0	0	0	50
S16	0	0	0	0	0	0	0	0	0	0	0	0	0	0	0	0	0	0	30
S17	0	0	0	0	0	0	0	0	0	0.8	0	0	0.2	0.5	2.5	1.3	0	0	30
S18	0	0	0	0	0	0	0	0	0	0.2	0	0.2	0	1.5	3.8	0.2	0	1.1	49

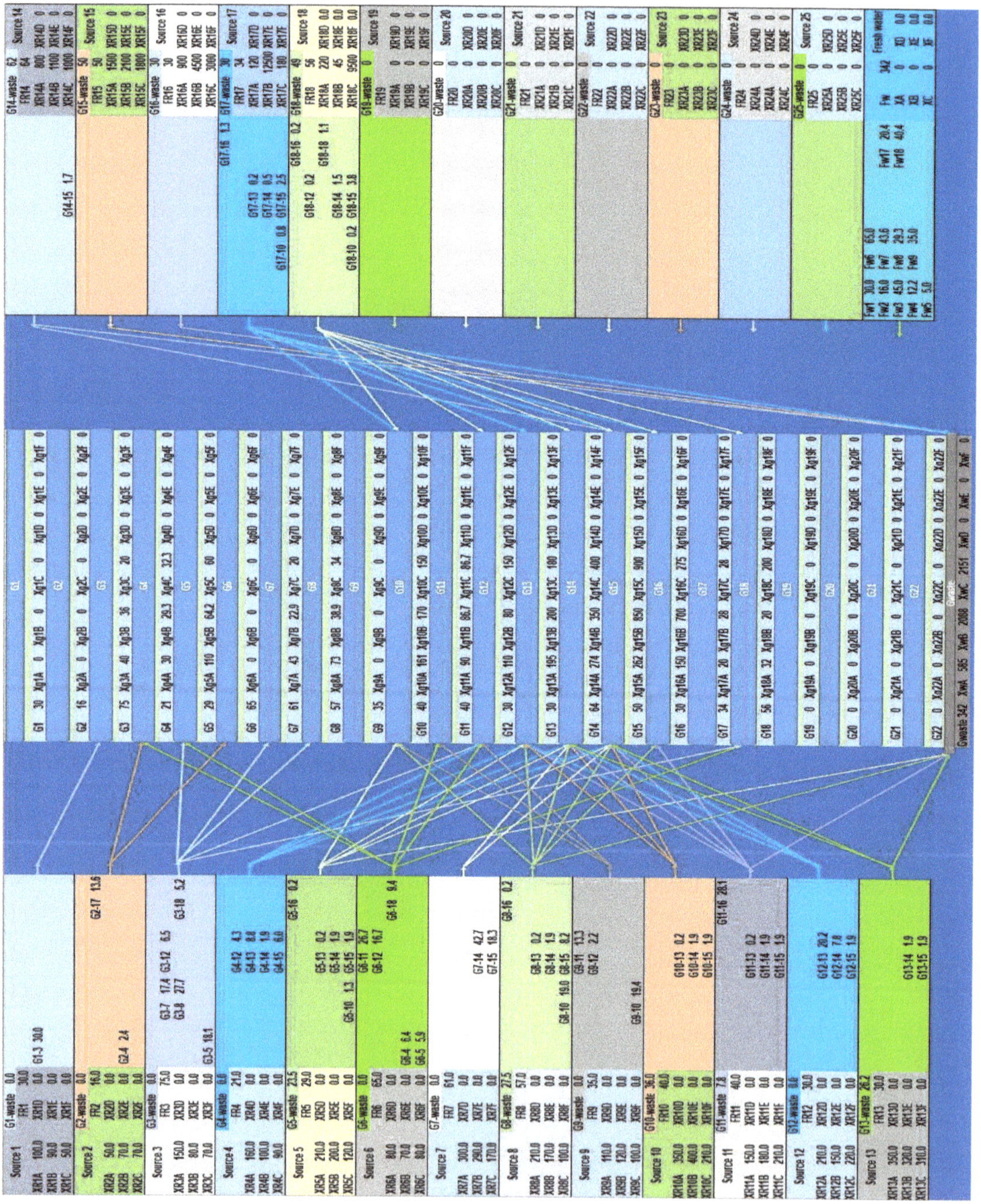

Figure 4. Design of water–wastewater inter-plant network of case study 3.

Regarding the obtained results, there was a decrease in the total consumption of freshwater flowrate from 374.3 t/h to 342 t/h by a reduction percentage 8.64% and the wastewater discharge decreased from 374.3 to 342 t/h by a reduction percentage 8.6%.

The waters of sources S15 and S16 are sent to waste directly because their mass loads are higher than the limiting mass loads of the sinks, but source 1 has a low mass load, so it feeds sink 3 only by 30 t/h.

Source 2 feeds sinks K4 and K17 by 2.4 and 13.6 t/h, respectively, while source S3 feeds K5, K7, K8, K12, and K18 by 18.1, 17.4, 27.7, 6.5 and 5.2 t/h, respectively. Source 4 supplies its wastewater to four sinks, K12, K13, K14 and K15, by flowrates of 4.3, 8.8, 1.9 and 6 t/h, respectively.

Source 5 supplies its wastewater to five sinks, K10, K13, K14, K15, K16, and waste by 1.3, 0.2, 1.9, 1.9 and 0.2 t/h, respectively, while source 6 feeds K4, K5, K11, K12 and K18 by 6.4, 5.9, 26.7, 16.7 and 9.4 t/h, respectively. Source 7 feeds two sinks, K14 and K15, by 42.7 and 18.3 t/h, while source 8 supplies its wastewater to K10, K13, K14, K15, K16 and waste by 19, 0.2, 1.9, 8.2, 0.2 and 27.5 t/h, respectively.

Source 9 feeds K10, K11 and K12 by flowrates of 19.4, 13.3 and 2.2 t/h, respectively, while source 10 supplies its wastewater to K13, K14, K15 and waste by flowrates of 0.2, 1.9, 1.9 and 36 t/h, respectively. Source 11 feeds four sinks K13, K14, K15, K16 and waste by 0.2, 1.9, 1.9, 28.1 and 7.8 t/h, respectively. Source 12 supplies its wastewater to K13, K14 and K15 by 20.2, 7.8 and 1.9 t/h, respectively.

The water of source 13 is sent to K14, K15 and waste at flowrates of 1.9, 1.9 and 26.2 t/h, respectively, while source 14 feeds K15 and waste by 1.7 and 62 t/h, respectively. Source 17 supplies sinks K10, K13, K14, K15, K16 and waste by 0.8, 0.2, 0.5, 2.5, 1.3 and 30 t/h, respectively while source 18 feeds K10, K12, K14, K15, K16, K18 and waste by flowrates of 0.2, 0.2, 1.5, 3.8, 0.2, 1.1 and 49 t/h, respectively.

5. Conclusions

This work is proposed to design water–wastewater inter-plant networks while minimizing the consumption of freshwater used in the plants' processes. A mathematical model is introduced to solve the equations that are formulated as a nonlinear program. Data given of sources and sinks (flowrates and limiting concentration) are introduced to the model and solved by the LINGO software. The obtained results are sent to the Excel software which is responsible for designing and drawing the water–wastewater inter-plant networks automatically. This mathematical approach has the ability to solve for a water system that contains single contaminant or multiple contaminants, with a reach of up to six contaminants. The proposed mathematical approach was applied to three case studies that contain single and multiple contaminants between several plants. The obtained results of the three case studies showed a reduction in the freshwater consumption by percentages of 38.6, 4.74 and 8.64% while the wastewater discharge decreased by percentages of 38.1, 4.61 and 8.6% for case study 1, 2 and 3, respectively. The introduced mathematical model is easy to use and understand because it is required only to enter the flowrates and concentrations of the sources and sinks into the LINGO software and the obtained results will be sent directly to the Excel software which is able to generate and draw the water–wastewater inter-plant network design automatically. This advantage makes this proposed technique beneficial for several industrial plants in the designing of their optimum water inter-plant networks with single and/or multiple contaminants.

Author Contributions: Conceptualization, A.M.S., M.H.H. and A.A.B.; methodology, A.M.S. and M.H.H.; software, A.M.S. and A.A.A.; validation, A.A.A., A.M.S. and A.A.B.; formal analysis, A.A.A. and M.H.H.; investigation, A.A.A. and A.G.G.; resources, A.M.S. and A.A.B.; data curation, A.A.A. and A.G.G.; writing—original draft preparation, A.A.A.; writing—review and editing, A.M.S., A.G.G. and A.A.B.; visualization, A.M.S., M.H.H. and A.G.G.; supervision, A.M.S., M.H.H. and A.A.B.; funding acquisition, A.G.G. and A.A.B. All authors have read and agreed to the published version of the manuscript.

Funding: This work was supported and funded by the Deanship of Scientific Research at Imam Mohammad Ibn Saud Islamic University (IMSIU) (grant number IMSIU-RG23020).

Data Availability Statement: Data are available upon request through the corresponding author.

Conflicts of Interest: The authors declare no conflict of interest.

References

1. Al-Redhwan, S.A.; Crittenden, B.D.; Lababidi, H.M.S. Wastewater minimization under uncertain operational conditions. *Comput. Chem. Eng.* **2005**, *29*, 1009–1021. [CrossRef]
2. Ami, P.; Avi, O.; Asce, M. Multiobjective contaminant sensor network design for water distribution system. *J. Water Resour. Plann. Manag.* **2008**, *134*, 366–377. [CrossRef]
3. Chew, I.M.L.; Foo, D.C.Y. Automated targeting for inter-plant water integration. *Chem. Eng. J.* **2009**, *153*, 23–36. [CrossRef]
4. Satyawali, Y.; Balakrishnan, M. Wastewater treatment in molasses-based alcohol distilleries for COD and color removal: A review. *J. Environ. Manag.* **2008**, *86*, 481–497. [CrossRef] [PubMed]
5. Iancu, P.; Plesu, V.; Lavric, V. Regeneration of internal streams as an effective tool for wastewater network optimization. *Comput. Chem. Eng.* **2009**, *33*, 731–742. [CrossRef]
6. Tian, J.R.; Zhou, P.J.; Lv, B. A process integration approach to industrial water conservation: A case study for a Chinese steel plant. *J. Environ. Manag.* **2008**, *86*, 682–687. [CrossRef] [PubMed]
7. Kim, J.; Kim, J.; Kim, J.; Yoo, C.; Moon, I. A simultaneous optimization approach for the design of wastewater and heat exchange networks based on cost estimation. *J. Clean. Prod.* **2009**, *17*, 162–171. [CrossRef]
8. Chew, I.M.L.; Foo, D.C.Y.; Tan, R.R. Flowrate targeting algorithm for interplant resource conservation network. Part 2: Assisted integration scheme. *Ind. Eng. Chem. Res.* **2010**, *49*, 6456–6468. [CrossRef]
9. Chen, C.L.; Hung, S.W.; Lee, J.Y. Design of inter-plant water network with central and decentralized water mains. *Comput. Chem. Eng.* **2010**, *34*, 1522–1532. [CrossRef]
10. Julien, P.; Nouho, K.; Jean, M.; Vincent, R.; Pierre, S. Impact of an intense combined sewer overflow event on the microbiological water quality of the Seine River. *Water Res.* **2011**, *45*, 893–903. [CrossRef]
11. Poplewski, G.; Jezowski, J.M.; Jezowska, A. Water network design with stochastic optimization approach. *Chem. Eng. Res. Des.* **2011**, *89*, 2085–2101. [CrossRef]
12. Sahu, G.; Garg, A.; Majozi, T.; Bandyopadhyay, S. Optimum Design of Waste Water Treatment Network. *Am. Chem. Soc.* **2013**, *52*, 5161–5171. [CrossRef]
13. Yang, L.; Diaz, R.S.; Grossmann, I.E. Water network optimization with wastewater regeneration models. *Ind. Eng. Chem. Res.* **2014**, *53*, 17680–17695. [CrossRef]
14. Ahmetovic, E.; Ibric, N.; Kravanja, Z. Optimal design for heat-integrated water-using and wastewater treatment networks. *Appl. Energy* **2014**, *4*, 63–81. [CrossRef]
15. Buabeng, E.; Majozi, T. Effective Synthesis and Optimization Framework for Integrated Water and Membrane Networks: A Focus on Reverse Osmosis Membranes. *Ind. Eng. Chem. Res.* **2015**, *54*, 9394–9406. [CrossRef]
16. Bozkurt, H.; Van Loosdrecht, M.C.M.; Gernaey, K.V.; Sin, G. Optimal WWTP process selection for treatment of domestic wastewater—A realistic full-scale retrofitting study. *Chem. Eng. J.* **2015**, *286*, 447–458. [CrossRef]
17. Sueviriyapan, N.; Suriyaprapadilok, U.; Siemanond, K.; Quaglia, A.; Gani, R. Industrial wastewater treatment network based on recycling and rerouting strategies for retrofit design schemes. *J. Clean. Prod.* **2015**, *111*, 231–252. [CrossRef]
18. Naderi, M.J.; Pishvaee, M.S.A. A stochastic programming approach to integrated water supply and wastewater collection network design problem. *Comput. Chem. Eng.* **2017**, *104*, 107–127. [CrossRef]
19. Mostafa, B.; Mohammad, H.S.; Hashem, A.; Hossein, N.F. Water management methods in food industry: Corn refinery as a case study. *J. Food Eng.* **2018**, *74*, 262–283.
20. Han, H.G.; Zhang, L.; Liu, H.X.; Qiao, J.F. Multiobjective design of fuzzy neural network controller for wastewater treatment process. *Appl. Soft Comput.* **2018**, *67*, 467–478. [CrossRef]
21. Fadzil, A.F.A.; Alwi, S.R.W.; Manan, Z.A.; Klemes, J.J. Maximizing total site water reuse via a two-way centralized water header. *ACS Sustain. Chem. Eng.* **2018**, *6*, 2563–2573. [CrossRef]
22. Lv, Z.; Song, Y.; Chen, C.; Jiang, B.; Sun, H.; Lyu, Z. A novel step-by-step optimization method for interplant water. *J. Environ. Manag.* **2018**, *213*, 255–270. [CrossRef] [PubMed]
23. Reinaldo, C.M.; Fernando, L.P.; Silvio, A.B. Industrial wastewater reuse in inter-plant water netwoks. *Braz. J. Dev.* **2019**, *5*, 8366–8374.
24. Robles, C.E.; Ben-Ayed, A.; Bernier, J.; Rocher, V.; Dochain, D. Management of an integrated network of wastewater treatment plants for improving water quality in a river basin. *IFAC-Pap. Online* **2019**, *52*, 358–363. [CrossRef]
25. Wang, X.; Fan, X.; Liu, Z.Y. Design of interplant water network of multiple contaminants with an interplant water main. *Chem. Eng. Trans.* **2019**, *72*, 295–300. [CrossRef]
26. Fard, A.M.F.; Keshteli, M.H.; Tian, G.; Li, Z. An Adaptive Lagrangian Relaxation-based algorithm for a coordinated water supply and wastewater collection network design problem. *Inf. Sci.* **2019**, *512*, 1335–1359. [CrossRef]
27. Mohammad, J.D.; Chang, K.Y. Modeling and extensive analysis of the energy and economics of cooling, heat, and power trigeneration (CCHP) from textile wastewater for industrial low-grade heat recovery. *Energy Convers. Manag.* **2020**, *205*, 112451. [CrossRef]

28. Kumawat, P.K.; Chaturvedi, N.D. Robust targeting of resource requirement in a continuous water network. *Chem. Eng. Trans.* **2020**, *81*, 1003–1008. [CrossRef]
29. Poplewski, G.; Foo, D.C.Y. An extended corner point method for the synthesis of flexible water network. *Process Saf. Environ. Prot.* **2020**, *148*, 210–224. [CrossRef]
30. Bhoomi, S.; Deepak, C.; Ajit, P.S. Wastewater allocation and pricing model for the efficient functioning of CETP serving a textile industrial cluster. *Adv. Energy Environ.* **2021**, *142*, 247–263.
31. Erkata, Y.; Rinaldi, I.; Roy, H.S.; Carolus, B.R.; Satriyo, K.W.; Rangga, K.M.; Juris, B.; Herry, S. Reducing energy and water consumption in textile dyeing industry with cleaner production by inlet-outlet modification to reuse wastewater. *Phys. Comput. Sci.* **2021**, *58*, 49–58.
32. Mariacrocetta, S.; Gabriele, F. Strategies for improving optimal positioning of quality sensors in urban drainage systems for non-conservative contaminants. *Water* **2021**, *13*, 934. [CrossRef]
33. Nessren, M.F.; Dina, A.K.; Ayat, O.G.; Mamdouh, A.G.; Mai, K.F. Graphical design and analysis of mass exchange networks using composition driving forces. *S. Afr. J. Chem. Eng.* **2021**, *36*, 94–104. [CrossRef]

water

MDPI

Article

Multi-Dimensional Surface Water Quality Analyses in the Manawatu River Catchment, New Zealand

Imokhai T. Tenebe [1,*], Jason P. Julian [2], PraiseGod C. Emenike [3], Nathaniel Dede-Bamfo [4], Omeje Maxwell [5], Samuel E. Sanni [6], Eunice O. Babatunde [7] and Darlan D. Alves [8]

1 Mineta Transportation Institute, San Jose State University, San Jose, CA 95192, USA
2 Department of Geography and Environmental Studies, Texas State University, San Marcos, TX 78666, USA
3 Department of Civil Engineering, Covenant University, Ota 112104, Nigeria
4 Albert B. Alkek Library, Texas State University, San Marcos, TX 78666, USA
5 Department of Physics, Covenant University, Ota 112104, Nigeria
6 Department of Chemical Engineering, Covenant University, Ota 112104, Nigeria
7 Civil Engineering Program, Ingram School of Engineering, Texas State University, San Marcos, TX 78666, USA
8 Politechnical School, Unisinos University, Sao Leopoldo 93022-750, Brazil
* Correspondence: yoshearer@gmail.com

Abstract: Land Use and Land Cover (LULC) properties give vital information about pollution signatures in rivers, and they help develop best management practices (BMPs) for effective water resource management. This work employs multivariate statistical methods, receptor modeling, connectivity analysis, and univariate trend analysis to investigate pollution sources across spatiotemporal scales in the Manawatu River, New Zealand. A positive matrix factorization (PMF) method was applied to interpret possible contamination sources. A 25-year dataset (1989–2014) comprising 12 water quality variables from three sites was used. Runoff connectivity analyses identified high-producing grassland (HG) as the most dominant pollution class in all sub-catchments. Univariate analyses revealed that nutrients and sediments were higher than in the initial monitoring years. The PMF analysis found possible pollutants causing impairment, which required attention from waste managers. PMF also showed that point, natural, and agricultural sources significantly contributed to pollution downstream of the river. In the midstream, the erosion, point, and agricultural sources were significant contributing factors. Agricultural pollution and soil erosion were the main contributors to the upstream sub-catchment area. This study suggests that BMPs with a high retention capacity are needed in specific locations in the catchment area to filter high concentrations of pollutants generated.

Keywords: pollution status; river water quality; source apportionment; agricultural pollution; LULC

Citation: Tenebe, I.T.; Julian, J.P.; Emenike, P.C.; Dede-Bamfo, N.; Maxwell, O.; Sanni, S.E.; Babatunde, E.O.; Alves, D.D. Multi-Dimensional Surface Water Quality Analyses in the Manawatu River Catchment, New Zealand. *Water* **2023**, *15*, 2939. https://doi.org/10.3390/w15162939

Academic Editor: Christos S. Akratos

Received: 19 June 2023
Revised: 4 August 2023
Accepted: 9 August 2023
Published: 15 August 2023

1. Introduction

Clean water is vital for many daily activities and good health [1,2]. However, with advances in agricultural, industrial, and urban development, maintaining water quality at preferred standards has been difficult in recent times [3–6]. The quality of water sources, particularly surface water, is the sum of the components of the surrounding watershed, with land-use activities identified as being responsible for the impairment of water quality [7,8]. One of the main reasons is that land use decisions are made without considering the watershed's assimilatory capacity [9,10]. The authors in a recent publication [11] found that an excess influx of nutrients beyond the river's assimilation capacity results in poor water quality.

When water quality is adversely affected, ecosystem function is disrupted. The consequences are eutrophication or sedimentation [12]. For example, phosphorus and nitrogen from fertilizer in agricultural areas have been major causes of pollution, leading to algal blooms in rivers and lakes. These impairments can reduce the aesthetic quality of the river due to a reduction in clarity [13], and they can reduce oxygen levels when dead zones are created. Changes in the levels of other variables, such as dissolved oxygen, temperature, pH, and total suspended solids, can affect ecosystem performance, especially when they fall out of the recommended ranges. Thus, it is crucial to monitor water quality, reduce soil erosion, and prevent runoff entering water bodies to ensure they are maintained at an acceptable standard. The implementation of a wide range of watershed best management practices (BMPs) has been suggested in the literature. These include waste stabilization ponds, wetlands, animal fencing, and riparian restoration [14]. Riparian restoration has proven to be the preferred method of policy makers, as it can be a cost-effective option when installed in an appropriate location within a river catchment.

In the absence of riparian restoration, a consistent water quality assessment is required to determine the state of the water bodies. This assessment will assist watershed managers in developing optimal policies, decisions, and management practices. Water quality assessments can be performed using routinely monitored stations. These stations enable relevant authorities to gather data that can be analyzed using univariate and multivariate statistical methods. These methods have been applied collectively to detect potential sources of pollution and the relationships between variables in clusters. However, performing these analyses in isolation may lead to incorrect conclusions. Several studies in the literature have utilized these techniques to propose cause-and-effect relationships between environmental pollution associated with groundwater [15–18] and surface water [19].

New Zealand (NZ) has been identified as a nation that experiences surface water quality problems that are associated with intensive land use [20–22]. Certainly, New Zealand stands as a significant global exporter of sheep products, powdered milk, and butter. This surface water quality issue can be linked to the nation's heightened level of agricultural productivity [23]. This dominance in agriculture is likely to stress river water quality, as more fertilizers are required to increase agricultural production. NZ began consistently collecting and monitoring water quality data on a national scale in 1989 [24]. As reported in Davies-Colley et al. [25], their National River Water Quality Network (NRWQN) has been in operation for three decades, and it has consistently monitored water quality variables. This robust dataset encompasses the spatial and temporal variability of water quality. Seventy-seven sites were monitored throughout thirty-five rivers across NZ, with each site close to a hydrometric station. Although several studies have used this dataset to reveal water quality issues on a national scale [21,26–30], very few have used it to identify sources of pollution on regional and catchment scales.

As water quality issues are location-dependent, they are influenced by natural processes and anthropogenic activities in the watershed. This study applied multivariate statistical analyses to determine the seasonal and spatial patterns of water quality impairments in the Manawatu River. Additionally, this study assessed the overall state of the watershed in terms of pollutant concentrations by apportioning sources using a receptor modeling technique. Finally, we compared our water quality findings with runoff connectivity and land use/land cover (LULC) data to identify specific landscape connections and contributions. This multi-dimensional analysis can be used to inform water-management strategies.

2. Materials and Methods

2.1. Study Area Physical Geography

The 5879 km^2 Manawatu catchment is situated on the southern tip of the North Island of NZ (Figure 1). Within this watershed, three NRWQN monitoring stations have been used to collect water quality variables since 1989. The first station in the upstream section was mainly surrounded by grassland and pasture, whereas the midstream station was adjacent to large built-up areas. The last station was situated near the catchment outlet. Larned et al. [27] and Abott et al. [31] reported that the Manawatu River is affected by sediment primarily because of intensive land use on moderate to steep slopes comprising erodible soils. The southern and eastern regions of the Manawatu catchment are mountainous and hilly, respectively, and they are covered by natural forests and shrublands, some of which have been converted to pastures for beef cattle, dairy, and sheep farming [14]. It is situated on soft sedimentary rocks that are characterized by mudstone and sandstones. The mountains contain hard, dark grey-brown soils that produce fine deposits when eroded, whereas the hills are characterized by tertiary-aged mudstone or sandstone [14]. Substantial amounts of sediment from this catchment (proliferated by land clearing) are generated in the ocean at a rate of Ca 3.74 Mt yr^{-1} [32].

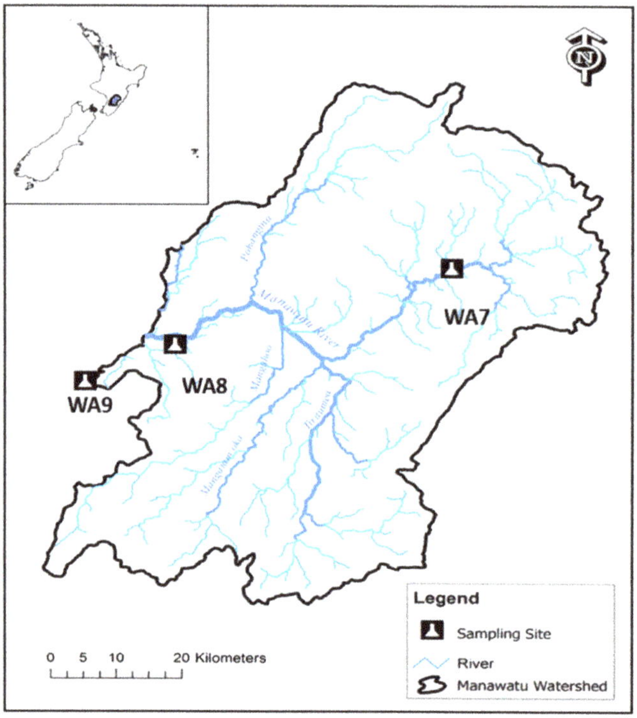

Figure 1. Manawatu River catchment showing major river networks and sampling sites.

In accordance with Land Air Water Aotearoa [33], the Manawatu River is 235 km long. It has several large tributaries, such as the Mangahoo, Mangatain oka, Oroua, Pohangina, and Tiraumea rivers, which are 86, 71, 131, 71, and 69 km, respectively. The river starts in the eastern part of NZ and gradually moves into the Tasman Sea. Land use in the Manawatu catchment predominantly consists of agricultural activities (70%). Half of the agricultural land is used for sheep and beef farming [33]. Within the catchment, Palmerston North, where the WA8 sampling site is situated, had the largest urban settlement. In addition, there are numerous other small communities within the catchment. The settlers in these communities practice intensive agriculture and have done so for years, which suggests that the primary consumptive use of the river within the catchment is for irrigation purposes.

2.2. Water Quality

The Manawatu catchment is located within the Manawatu–Wanganui region, which has some distinctive topography, including the Ruahine, Tararua, and Puketoi Ranges. These ranges exceed 1000 m in terms of elevation, but they drop to form a ridge as they approach the Manawatu gorge. Water quality data were collected monthly by the National Institute for Water and Atmospheric Research (NIWA) from three monitoring sites within the Manawatu catchment for 25 consecutive years (1989–2014). These three stations comprised a longitudinal study, cutting across the upland river with an altitude of >150 m, and they were assumed to be less polluted (WA7) than the lowland (lower elevation) river with an altitude of <150 m. These lowland rivers will likely witness higher background concentrations as the accumulated diffuse pollution flows downstream (WA8 and WA9). A summary of different site characteristics is presented in Table 1. More specifically, WA7 (Manawatu at Weber Road) is in the upstream section of the Manawatu catchment, with a median temperature of 13 °C and a catchment area of 705 km^2. WA8 (Manawatu at Teachers College) was located at an intermediate point.

Table 1. Characteristics and location of monitoring sites in the Manawatu Catchment.

Site Code	Catchment Area (km^2)	Site Description
WA7	705.28	Located upstream
WA8	3897.37	Major river mid-stream located in the center of Palmerston North
WA9	4222.22	Located downstream

The catchment area of WA8 encompasses 3897 km^2, and it is located in the center of Palmerston North. It has experienced significant soil erosion, which has affected the river's water quality owing to its low elevation, as compared with WA7. It consists of major rivers with significant pastoral development, and it has a median temperature of 13.8 °C. WA9 (Manawatu at Opiki Bridge) is located in the downstream section of the river. This catchment has a median temperature of 14.4 °C, and it receives treatment for industrial and sewage discharges.

The dataset used for this study included data collected from the inception of the water quality monitoring program, which was established by the NRWQN between Jan 1989 and Dec 2014, considering 12 variables (Table 2). Of the 12 parameters measured, five were measured in situ, and seven others were measured in the laboratory. The water quality variables measured in the field were Discharge (Q), Dissolved Oxygen (DO), Water Clarity (CLAR), turbidity (TURB), and Water Temperature (Tw). Moreover, pH, Total Phosphorus (TP), Total Nitrogen (TN), Dissolved Reactive Phosphorus (DRP), Oxidized Nitrogen (NOx^-), Colored Dissolved Organic Matter (CDOM), and Conductivity (COND) were measured in the laboratory [24]. Smith and McBride [24] and Davies-Colley et al. [34] provided detailed accounts of how the water samples were collected. At the beginning of the sampling years, and up until 2004, samples taken for laboratory testing were simultaneously collected in 2 L High-Density polyethylene (HDPE) bottles at each site. In 2005, sampling with 2 L HDPB was replaced with a single 500 mL HDPB of the same quality, and it was produced by the same manufacturer. Water samples were stored in an insulated bin filled with slush ice, and they were immediately transported to a water quality center in Hamilton, New Zealand. These samples were transported immediately to ensure that both chemical and biological tests were conducted within 24 h, in accordance with laboratory standards.

2.3. Statistical Analysis

To determine the quality of the monitored river over time, and to measure average changes in water quality, we grouped the dataset into 5 year periods from 1989 to 2014, except for 1994, which was exempted for all variables due to contamination with ammoniacal nitrogen (NH_4^+) [34]. This study used median values, rather than mean values, to compare and monitor water quality changes; this is because the latter is sensitive to spread and to extreme values. Median values for each 5 year period were calculated and compared in order to trigger values prepared by Australia and New Zealand for water quality assessments; these values varied between the lowlands and uplands (distinguished in accordance with the 150 m elevation threshold) (Table 2) depending on where the sampling stations were situated (Australian and New Zealand Environment and Conservation Council; ANZECC [35]). These triggered values do not necessarily suggest an immediate threat, but they are a warning sign for future risk if not curtailed. Statistical testing procedures were performed using the SPSS package (V.23) and an R package (R Core Team, 2020), whereas receptor modeling was performed using EPA PMF 5.0 [36]. Before statistical testing and modeling were conducted, missing data were excluded from the analyses. After that, a normality test was performed using the Kolmogorov–Smirnov (K–S) test to determine the distribution of the dataset. This normality test was crucial for identifying the appropriate statistical test to implement the forward movement. For example, a normality test showing a non-normal distribution is usually analyzed using non-parametric statistical methods. In this study, the normality test showed that the dataset was not normally distributed ($p < 0.05$), except for one or two parameters at some sites. Seasonal pollution patterns were investigated for each site by stratifying the datasets into seasons to determine which season might be an apparent contributor to river water quality compared with the overall pollution status, especially the trigger values.

Table 2. Description of Water Quality Data Used for Analysis of the Manawatu River in NZ from 1989–2014.

	Unit	Abbreviation	Missing Data (%)	Trigger Values (Lowland/Upland †)
Flow rate	m^3/s	Flow	1.4	
Water Temperature	°C	Tw	0.3	
Electrical conductivity	μScm^{-1}	EC	-	
Dissolved Oxygen	g/m^3	DO	1.3	6
Dissolved Oxygen (%)	%	DO%	1.3	98/99
pH	-	pH	0.2	7.2/7.3
Turbidity	NTU	Turb	-	5.6/4.1
Total Phosphorus	g/m^3	TP	0.3	33/26
Total Nitrogen	g/m^3	TN	0.6	614/295
Visual clarity	m^{-1}	Clar	-	0.8/0.6
Dissolved Reactive Phosphorus	g/m^3	DRP	-	10/9
Ammoniacal nitrogen	g/m^3	NH_4-N	-	21/10
Nitrate	g/m^3	NO_3-N	-	444/167

Note: DO % and pH: lower limits were considered. † Lowland and Upland are distinguished by the 150-m elevation threshold.

According to Abott et al. [31], the Manawatu Catchment has two different seasons (summer—November to May; and winter—June to October). Summer, winter, overall, and trigger comparison observations were conducted for both seasons to identify the contribution of seasonality to the water quality variables measured at different sites. A two-tailed, non-parametric Spearman's correlation test was applied to determine the relationship between the variables ($\alpha = 0.05$ and 0.01). Spearman correlation coefficient values were reported using the raw data because there was no change in statistical significance central to this study. This study focused on direction and statistically, and the significant difference developed by the variable rather than the coefficient, which was consistent when the transformed Pearson's correlation coefficient was used [37–40]. In addition, there were several significant relationships at the 95% and 99% confidence intervals, although many seemed weak. To reduce the number of relationships observed, only Spearman correlation values of (r) > 0.75 were reported.

The positive matrix factorization (PMF) method is used on environmental datasets to identify potential sources and to apportion possible weights in percentages of pollution parameters. This outcome was achieved by decomposing a large temporal dataset into single quantified weights in the form of factor contributions, factor profiles, or factor fingerprints. These factor profiles were sub-divided into concentrations of species (pollutants and their respective percentages). The percentages of these factor profiles were then interpreted as the pollutant prevalence at the sites under investigation. The PMF model can be expressed in the following general form:

$$X = GF + E \tag{1}$$

The X matrix was decomposed into G and F matrices, where G represents the factor contribution, and F represents the factor profiles. Matrix E is the residual error, which should be minimized. Minimizing these errors in each variable, such that they tend toward zero, would almost guarantee that the dataset or variable was from a normal distribution. The PMF model generates covariance and correlation matrices to be decomposed, with one of its strengths being the generation of non-negative factors [41]. For this study, the dataset was cleaned to ensure no missing data. In the event of missing data, it was replaced with the median value of a specific variable. The file that is subjected to PMF modeling is called the concentration file. After that, an uncertainty concentration file is created. This refers to the minimum values that a measuring device can record; below this value, no readings can be obtained. The use of an uncertainty dataset in modeling is essential in pollution studies and risk assessment because of the presence of unknown processes or activities, such as experimental precision, instrumental errors, environmental instability effects (climate nonstationarity), and seasonal variability. This phenomenon may affect the output, as well as the corresponding decisions made when neglected, as the model results might have been significantly underestimated. The process of mathematically handling uncertain data files using the PMF approach has been well documented elsewhere [36].

The Uncertainty concentration is obtained, as follows:

$$= \sqrt{((\text{Error fraction} \times \text{concentration})^2 + (0.5 \times \text{MDL})^2)} \quad (2)$$

where MDL is the minimum detection limit.

When performing the PMF analysis for this study, all datasets entered showed a strong S/N ratio. This was attributed to the consistency of the data recorded, fewer missing values, and the large dataset or sample size used for the analysis (Table 2). Despite this, some variables were excluded from the analysis because the study focused on variables directly involved in pollution, without including strongly correlated variables that could replace one another. Therefore, temperature, DO, CLAR, and pH were classified as "bad weight", and the program neglected those variables during computation. However, a "normal", "bad", "weak", or "strong" weight was given to datasets with large, average, and minimum missing values to alert the software program. In this study, all the variables were allocated or described as strong by default. Regression plots showing the corresponding R-squared values for each water quality variable were obtained.

2.4. Land Use Land Cover (LULC) Mapping

A map depicting the existing LULC in the Manawatu catchment was developed to determine the contribution of LULC activity to water quality. Land cover specifies the identifiable features of the land, including the presence of crops, forest plantations, or scrub-grassland covers [11]. Identifying various LULCs is essential as it helps to provide some insight into the sources of diffused pollution within a catchment. LULC data obtained from the land cover database (LCDB v4.1, 2015) were used for this purpose. Thirty-five land use classes were obtained. Moreover, they showed identical and conflicting classifications with the land use and carbon analysis system (LUCAS) data operated by the NZ Ministry for the Environment. Julian et al. [21] detailed these differences and reclassified the LULC to be suitable for effectively establishing a water quality impairment relationship. This study followed the same procedure as [21] to ensure consistency in terms of reporting in accordance with the authors' detailed output2.5. Landscape Connectivity Analysis

This study used an existing watershed connectivity model, developed by [42], to connect the river to pollution source areas. This map showed that the LULC was directly related to the river via surface runoff. This model was carefully developed following the detailed procedure reported in [43]. The stream channels headed for the Manawatu catchment were identified using 0.5 m rural aerial photos obtained from Land Information, New Zealand, using two-year period data (2010–2012) as the reference point. The watershed was then delineated from the headwaters of the catchment to establish the flow direction, from upstream to downstream. Each 15 m pixel on the digital elevation model, with a greater than 5° slope, along the flow direction, as well as pixels adjacent to the river, were categorized as "connected". With the landscape connectivity map (Figure 2) developed, it was clipped, combined, and reclassified to select catchments connected to floodplains in the Manawatu catchment area.

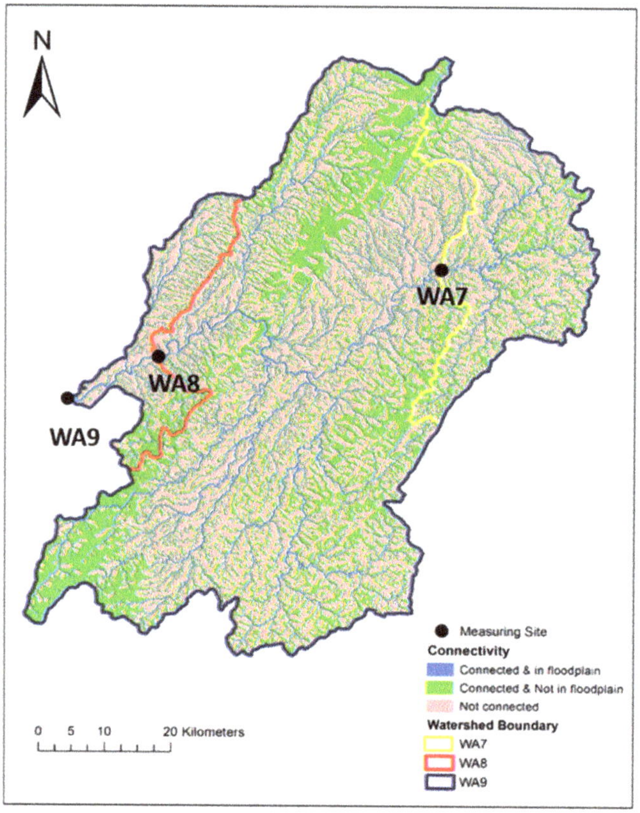

Figure 2. Landscape connectivity map for the Manawatu River Catchment.

3. Results

3.1. Land Use Map and Analysis

From LULC mapping (Figure 3), different LULC categories were observed at the three sites within the Manawatu catchment. Eight classes were present in the upstream area, as follows (WA7): shrub/grassland (SG), urban (UR), non-plantation forest (NF), plantation forest (PF), vegetated wetland (VW), high-producing grassland (HG), open water (OW), and barren/other (BO). These categories revealed that perennial and annual croplands did not exist upstream. The catchment of WA7 is dominated by HG (88.5%), NF (3.4%), and PF (3%). For the intermediate site (WA8), all ten LULC categories were obtained. This catchment is dominated by HG (74.1%), SG (14.2%), and NF (8%). The entire Manawatu catchment (WA9; most downstream station) was dominated by HG (74.5%), SG (10.8%), and NF (7.8%).

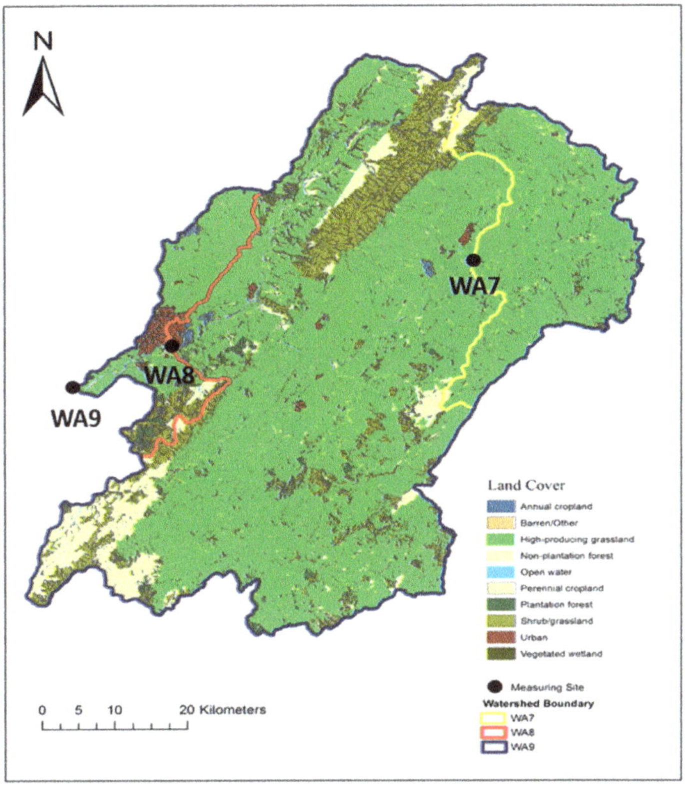

Figure 3. Land Use Land Cover (LULC) map for Manawatu River Catchment.

3.2. Landscape Connectivity

Landscape connectivity analyses (Table 3) revealed that 83.6 km^2 of the WA7 catchment (or 11.9%) was directly connected to streams via surface runoff and floodplains. Another 196 km^2 (27.8%) was related to streams via surface runoff, but not floodplains. That is, these were steep hillslopes that directly contributed to the surface runoff to streams. In total, For WA7′s catchment, 39.7% of the area was directly connected to streams via surface runoff. Approximately 426 km^2, more than half of the catchment area, is not directly connected, and it contributes surface runoff to the Manawatu River. HG (88%) was mainly related to a floodplain, which suggests a significant pollution source from the upstream section of the Manawatu River. For the intermediate site (WA8), landscape connectivity results suggest that ~2191 km^2 of a total of 3897 km^2 of the WA8 catchment (or 56.2%) was not directly connected to the river (Table 3). However, 11.3% of the total area in the WA8 catchment (or 440.9 km^2) was directly related to the stream via surface runoff and located in floodplains. Moreover, 32.5% of WA8′s catchment (or 1265.3 km^2) was directly connected to a stream via surface runoff and not located in a floodplain. Of note, HG (75.3%) dominated a substantial portion of the area connected to the floodplain, and ~64% of HG (HG area connected and HG not connected to floodplains but connected to catchment) contributed to the Manawatu River. Therefore, for WA8, 43.8% of the catchment area was connected to streams via surface runoff. Regarding the downstream sub-catchment (WA9), 483.9 km^2 of the catchment (or 11.5%) was directly connected to the stream via surface runoff and were in floodplains. Another 1343.6 km^2 of WA9′s catchment area (or 31.8%) was directly connected to the stream via surface runoff, but not located in floodplains, and 2395.29 km^2

(or 56.7%) was not directly connected to either floodplains or the stream via surface runoff (Table 3). Similarly, HG (66%) was the dominant LC in all three sub-catchments.

Table 3. (**a**) Reclassified watershed properties for WA7. (**b**) Reclassified Watershed Properties for Site WA8. (**c**) Reclassified Watershed Properties for Site WA9.

Land Use/ Land Cover Category	Connected Area in Floodplain (km^2)	Connected Area Not in Floodplain (km^2)	Area Not Connected (km^2)	Total (km^2)
		(a)		
Shrub/grassland	3.39	14.20	15.67	33.26
Urban	0.12	0.12	0.36	0.60
Non-plantation forest	2.69	10.92	10.99	24.60
Plantation forest	3.36	6.77	11.11	21.24
Vegetated wetland	0.00	0.01	0.03	0.04
High-producing grassland	73.47	163.76	386.83	624.06
Open water	0.49	0.28	0.49	1.26
Barren/other	0.05	0.08	0.09	0.22
Perennial cropland	-	-	-	-
Annual cropland	-	-	-	-
Total (km^2)	83.57	196.14	425.57	705.28
		(b)		
Shrub/grassland	47.51	294.89	207.40	549.80
Urban	1.93	2.45	17.53	21.91
Non-plantation forest	35.27	174.98	100.38	310.63
Plantation forest	11.78	28.27	50.55	90.60
Vegetated wetland	0.1	0.18	0.58	0.86
High-producing grassland	332.21	756.14	1799.32	2887.67
Open water	6.86	3.34	4.26	14.28
Barren/other	3.75	3.77	3.19	10.71
Perennial cropland	0.07	0.07	0.31	0.45
Annual cropland	1.43	1.19	7.66	10.28
Total (km^2)	440.91	1265.28	2191.18	3897.37
		(c)		
Shrub/grassland	52.08	319.23	225.73	597.04
Urban	5.30	5.73	39.29	50.32
Non-plantation forest	37.51	181.05	105.76	324.32
Plantation forest	14.09	35.27	60.82	110.18
Vegetated wetland	0.10	0.18	0.58	0.86
High-producing grassland	360.67	792.61	1944.12	3097.40
Open water	8.01	3.82	5.05	16.88
Barren/other	4.22	3.92	3.45	11.59
Perennial cropland	0.08	0.09	0.41	0.58
Annual cropland	1.85	1.70	10.08	13.63
Total (km^2)	483.91	1343.60	2395.29	4222.22

A comparison was made with their corresponding trigger values to determine the extent to which the water quality parameters were affected by watershed activities, as stipulated by ANZECC [35].

3.3. Seasonality and Trends in Water Quality in the Manawatu Catchment

The TP values (Figure 4a) for WA7 recorded an increasing trend for the first 15 years of sampling and a reduction for the last ten years during the monitoring period. However, the values recorded during this period exceeded the trigger values of 33 g/m^3, except for the first and last five years. The lowest median value measured in the first five years of sampling was initially lower than the trigger value, but it later increased between 2000 and

2004. For WA8 (Figure 4a), a similar trend was observed, with a minor difference observed in the first and last five years. The first five years had slightly higher median values than the trigger values, whereas the values during the previous five-year period were lower than the trigger values. This difference may be due to non-point source pollution at WA7 (upland river), which flowed into lowland rivers (WA8 and WA9), as well as the increase in connected pasture areas connected to the floodplain in WA9, compared with WA7. However, this was not the case for WA9. Until 2014, the TP values recorded were higher than the trigger values (Figure 4a). The higher TP values recorded may result from the accumulation of pollutants as they flow downstream. The seasonality comparison for all sites revealed similar trends for WA7 and WA8 (Figure 5a). The TP values in WA7 and WA8 were higher in winter than in summer, whereas the overall median values for both sites were higher than those of the stipulated trigger values. For WA9, which was not the case in summer and winter, the overall median values were higher than the trigger values (Figure 5a).

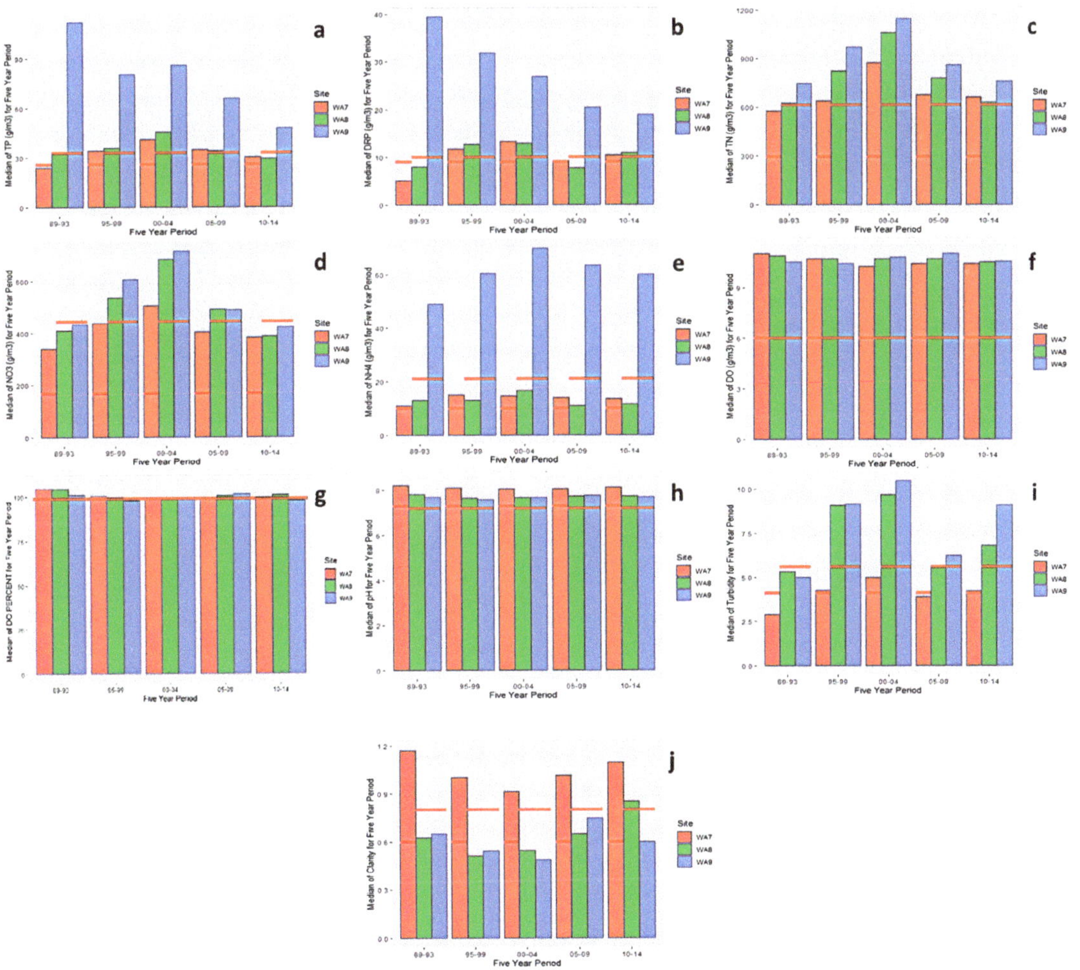

Figure 4. (**a–j**): Water Quality patterns for 25 years, at 5-year intervals, at the three stations. The red lines represent the ANZECC trigger values for the respective WQ variables.

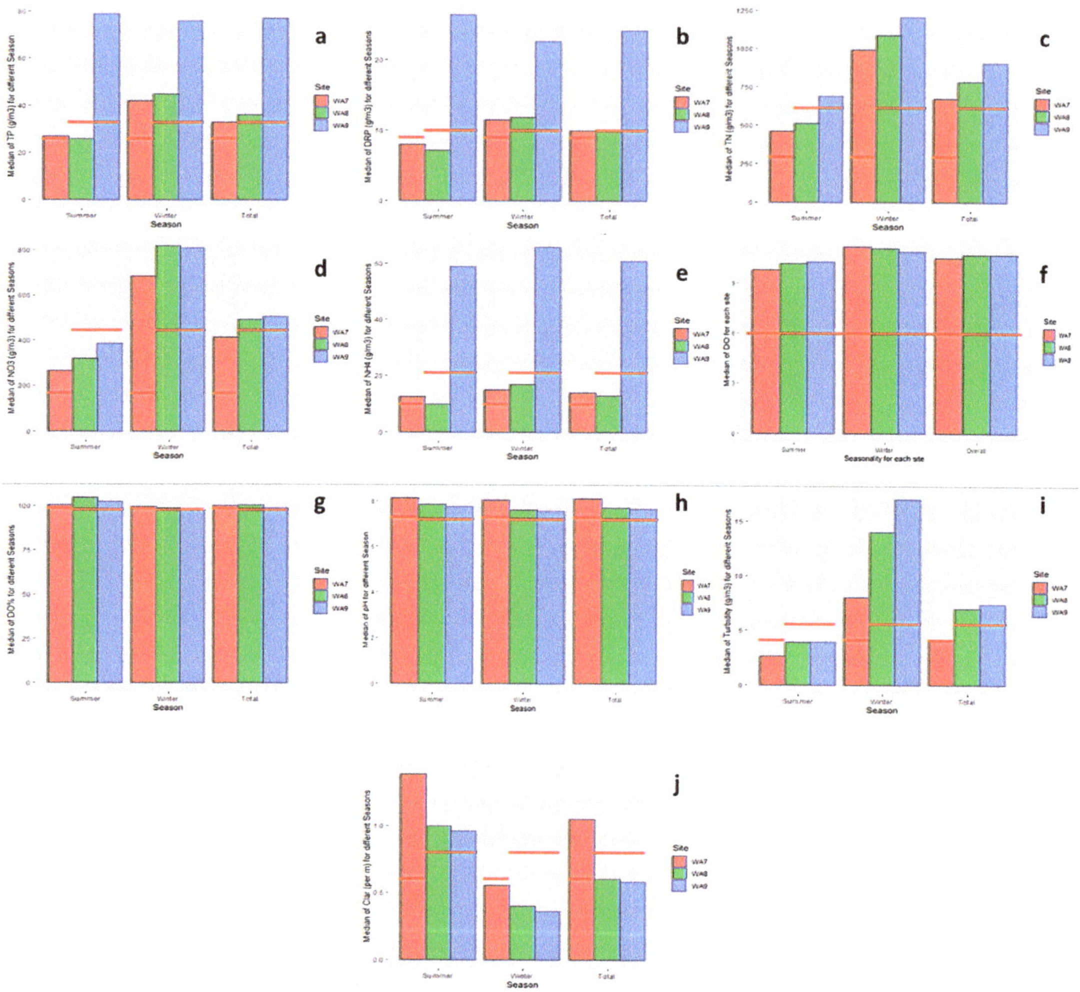

Figure 5. (**a**–**j**): Water Quality patterns for 25 years, at the three stations, in different seasons. The red lines represent the ANZECC trigger values for the respective WQ variables.

Throughout the monitoring period, the DRP values were higher than the trigger values for WA9, measuring 9 g/m^3 (Figure 4b). Between 1989 and 1993, and 2005 and 2009, there was a reduction in DRP; their levels were lower than the trigger values in WA7 and WA8. These values increased in 2014 and were higher than trigger values. Seasonality comparisons for all sites showed that seasonal values were higher than trigger values for WA9 (Figure 5b). At the same time, the same was only observed for winter and overall median values in WA7 and WA8. As for TN, all three sites (WA7, WA8, and WA9) (Figure 4c) showed similar distribution patterns to the patterns mentioned above from the beginning. The TN values recorded were above the trigger value of 295 g/m^3 until 2014. Interestingly, these values were significantly higher between 2000 and 2004. Seasonality across sites revealed that the summer, winter, and overall TN values exceeded the trigger values for all sites (Figure 5c). For NO_3^-, the values recorded throughout all periods were higher than the trigger values for all three sites (Figure 4d). High values above the trigger value of 167 g/m^3 were measured during the winter and summer periods (Figure 5d). For NH_4^+, all the site concentration values monitored over the 25 year period showed that the trigger values of 10 g/m^3 were exceeded to different extents (Figure 4e). More specifically, the NH_4^+ values recorded in WA9 were much higher than those at the other sites. Seasonality values also showed similar observations, with winter, summer, and overall median values exceeding the stipulated trigger values (Figure 5e). All sites (WA7, WA8, and WA9) had DO values below the threshold. Throughout all seasons, the data revealed that despite the number of pollutants entering the river, or that were already present in the river, the DO values were significantly above the trigger values of 6 g/m^3 (Figure 4f). The seasonal comparison also corroborates this finding. The functional reaeration capacity of the river allows for more oxygen to be re-introduced when used up, owing to the bathymetry of the river (Figure 5f). Additionally, observation of DO% revealed that all measured values were above the trigger values of 99% for the different sites during all study periods (Figure 4g). The same was observed for the median values, as seasonality was also observed (Figure 5g). The PH values were expected to be above the stipulated trigger value of 7.3. The values recorded during the monitoring periods for the three sites revealed that the median pH values were below the trigger values (Figure 4h). The same was observed during seasonality comparisons, thus corroborating a significant concern regarding high pH values (Figure 5h). For turbidity, WA8 and WA9 (Figure 4i) exhibited similar trends. Initially, the recorded values were below the trigger values, but these values increased over time to be above the trigger value. Nevertheless, in WA7 (Figure 4i), a slightly different pattern was observed. Slightly elevated values were recently recorded, whereas elevated values were measured between 1994 and 1999, and 2000 and 2004, with the highest value recorded between 2000 and 2004. Seasonality analysis revealed that the overall median and winter turbidity values exceeded the trigger values (Figure 5i). WA7 had good overall water clarity compared with WA8 and WA9 (Figure 4j). The water clarity values measured in WA7 exceeded the expectations of trigger values, whereas WA8 and WA9 had poor water clarity and did not meet expectations, except for the values measured in WA8 between 2010 and 2014. The seasonality comparison showed that most of the exceeded values were measured during the summer (Figure 5j), whereas the winter period had poor water quality for all three sites.

3.4. Correlation Matrix for the Different Sites

WA7, situated upstream of the Manawatu River, revealed that the flow had a strong, positive, and significant association with nutrient pollutants (NO_3^-, TN, and TP at the 1% level) (Figure 6). Turbidity had a strong but negative correlation with other physical parameters, such as clarity and EC, at the 1% level. This relationship suggests that nutrients are introduced into the river from the surrounding land via runoff or mass weathering, and they are likely to remain undisturbed. Clarity showed an inverse relationship with TN, TP, and turbidity; however, the opposite was true for EC. These relationships suggest that a prolonged influx of pollutants into the river can obstruct river water clarity. Turbidity showed a strong positive correlation with TP and TN.

Soil matter has been reported as a receptor for pollutants on land, and it enters the stream network through sediments. Therefore, this study agrees with other findings that suggest a significantly high affinity between phosphorus and sediments. In addition, a strong positive relationship was observed between DRP and TP, as well as TN and NO_3^-, suggesting that the pairs emanate from the same non-point sources. At the same time, with increasing temperature, DO% decreased at a rate that reflected the relationship observed between the two. For site WA8 (Figure 7), DO% showed a positive and robust correlation with pH and a negative correlation with turbidity and TP at the 1% significance level. Flow showed a consistent relationship with TP, TN, EC, turbidity, and clarity, as reported in the WA7 site. Again, clarity was inversely correlated with turbidity and TP, but proportional to EC. DO% showed strong negative relationships with turbidity and TP but was positively correlated with pH. The downstream site (WA9) showed several strong statistical relationships based on the research cut-off points. Turbidity increased with flow, whereas EC and clarity decreased under high-flow conditions (Figure 8). Similarly, the turbidity increase was associated with a decrease in clarity and EC, suggesting that the elevated EC measurements are likely to be from metallic salts instead of organic pollutants. This revealed that $NH_4^+$and TN possibly entered the river from the same source, showing strong positive statistically significant relationships at the 1% level.

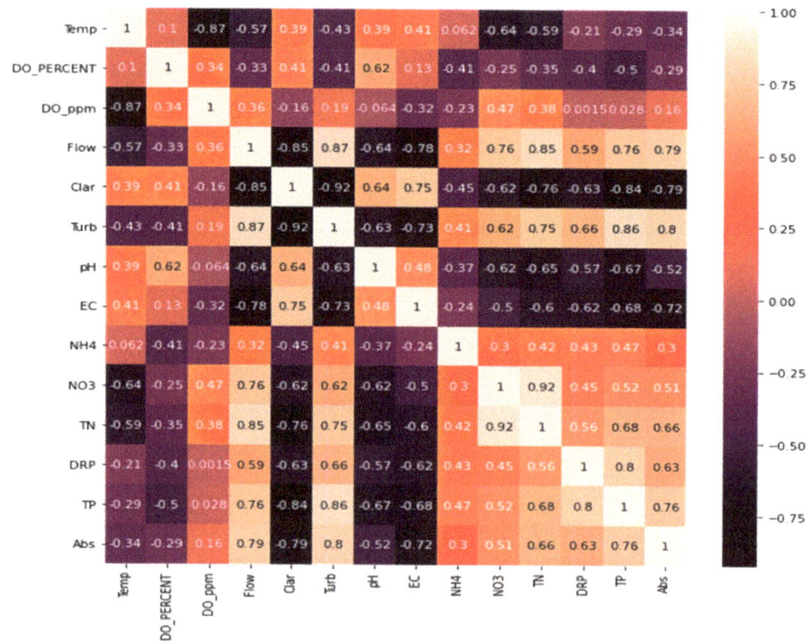

Figure 6. Spearman's correlation matrix for Site WA7.

3.5. *Potential Pollution Sources Using the Positive Matrix Factorization Method*

The PMF method was applied to interpret potential pollution sources using the EPA-PMF 5.0 software package. The receptor modeling technique was performed in accordance with the expected output from a combination of base model displacement, base model bootstrapping, and base model BS-DISP methods [36]. These techniques were applied using trial and error, usually by selecting several factors in sequence, and ensuring that all modeling conditions were met with minimum errors occurring; the optimal R2 was set manually. Figures 9–11 show the result from the extraction and modeling process, called the factor profile, and it is measured as a percentage (small red boxes). Four factors were generated for site WA7 (Figure 9), as follows: DRP (100%), NH_4^+ (79.1%), and CDOM (45.3%) in Factor 1; TN (59.3%) and NO_3^- (84%) in Factor 2; turbidity (83.9%) and TP (58.9%) in Factor 3; and DO (59.2%) and EC (73.1%) in Factor 4. For WA8 (Figure 10), Factor 1 was dominated by NH_4^+ (85.6%); factor 2 selected DRP (78.8%) and CDOM (36.1%); Factor 3 selected EC (72.7%) and DO (62.1%); Factor 4 selected NO_3^- (73.5%) and TN (48.9%); and Factor 5 selected turbidity (84.2%). Finally, six factors were extracted for WA9 (Figure 11). Factor 1 selected NH_4^+ (86.6%); Factor 2 selected DO (61.4%) and CDOM (42.4%); Factor 3 selected DRP (80.4%); Factor 4 selected turbidity (85.5%) and TP (39.4%); Factor 5 selected NO_3^- (75.4%) and TN (46.4%); and Factor 6 selected TP (32.2%).

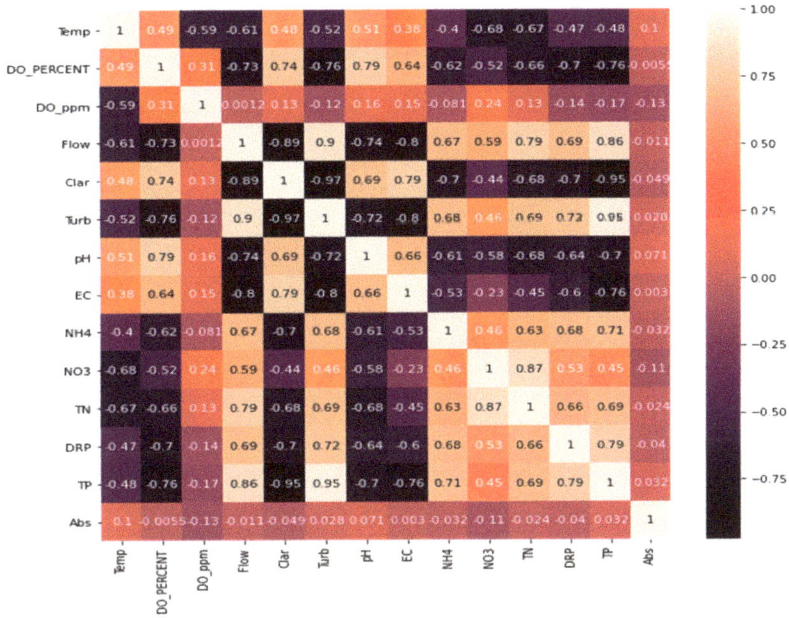

Figure 7. Spearman's correlation matrix for Site WA8.

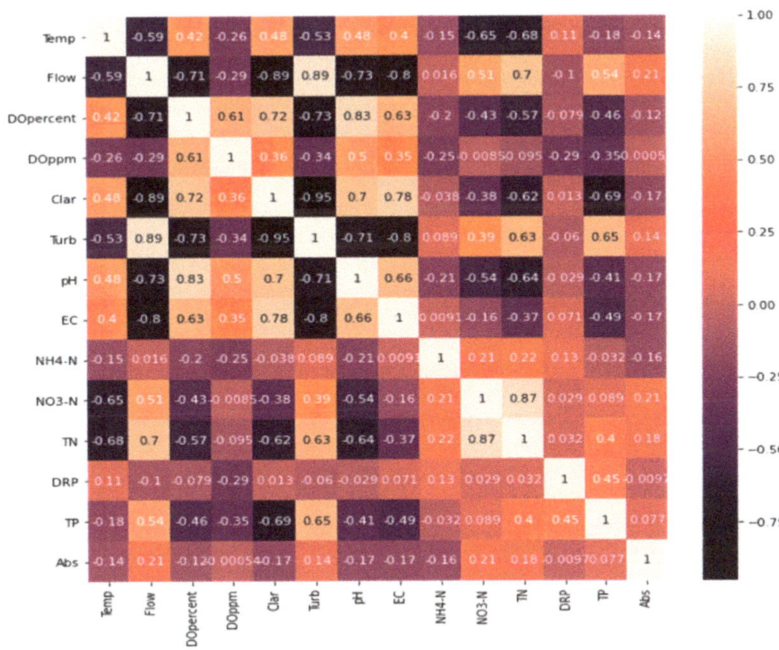

Figure 8. Spearman's correlation matrix for Site WA9.

3.6. Model Performance of PMF for Manawatu Catchment

The PMF model output showed outstanding performance in terms of modeling the trends and explaining the variability of each variable. The R^2 produced showed that except for DO, NH_4^+, and CDOM, other variables were well predicted for WA7 in the following order: DO (0.15) < CDOM (0.45) < NH_4^+ (0.62) < EC (0.76) < DRP (0.84) < turbidity (0.96) < TP (0.98) < TN (0.98) < NO_3^- (0.99). For WA8, R2 values for each variable were as follows: DO (0.23) < aCDOM (0.54) < EC (0.84) < TP (0.94) < DRP (0.99) < NO_3-N (0.99) < NH_4-N (0.99) < TN (0.98) < turbidity (0.93). For the WA9 site, the R2 values were in the following order: CDOM (0.05) < DO (0.78) < EC (0.80) < TN (0.97) < TP (0.98) < turbidity (0.98) < DRP (0.99) < NO_3-N (0.99) < NH_4-N (0.99).

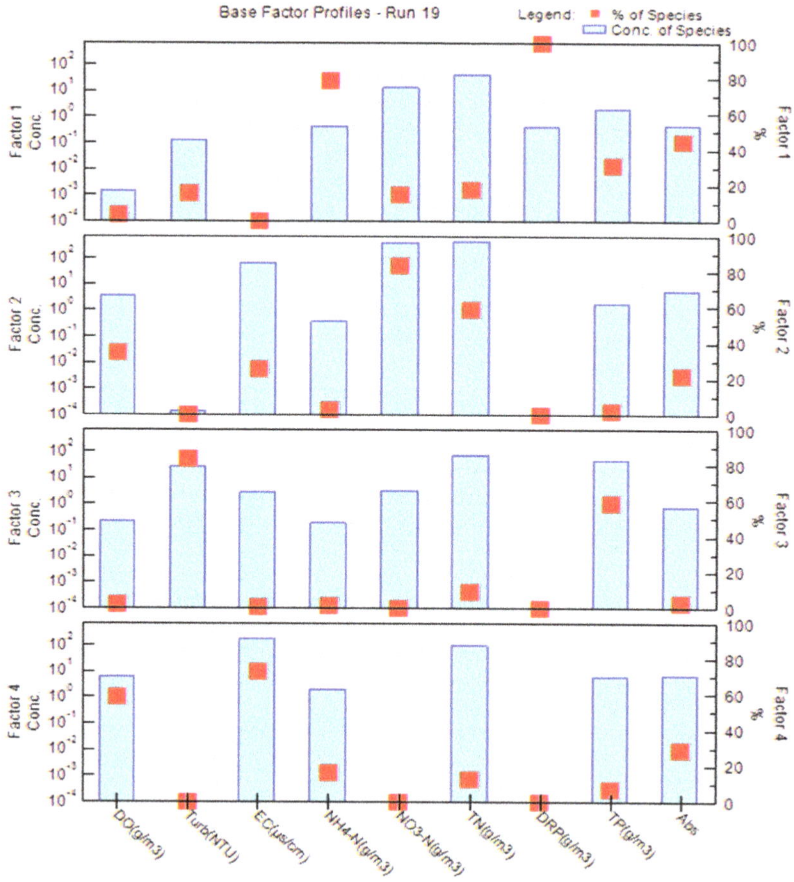

Figure 9. Profile concentrations for WA7 using PMF.

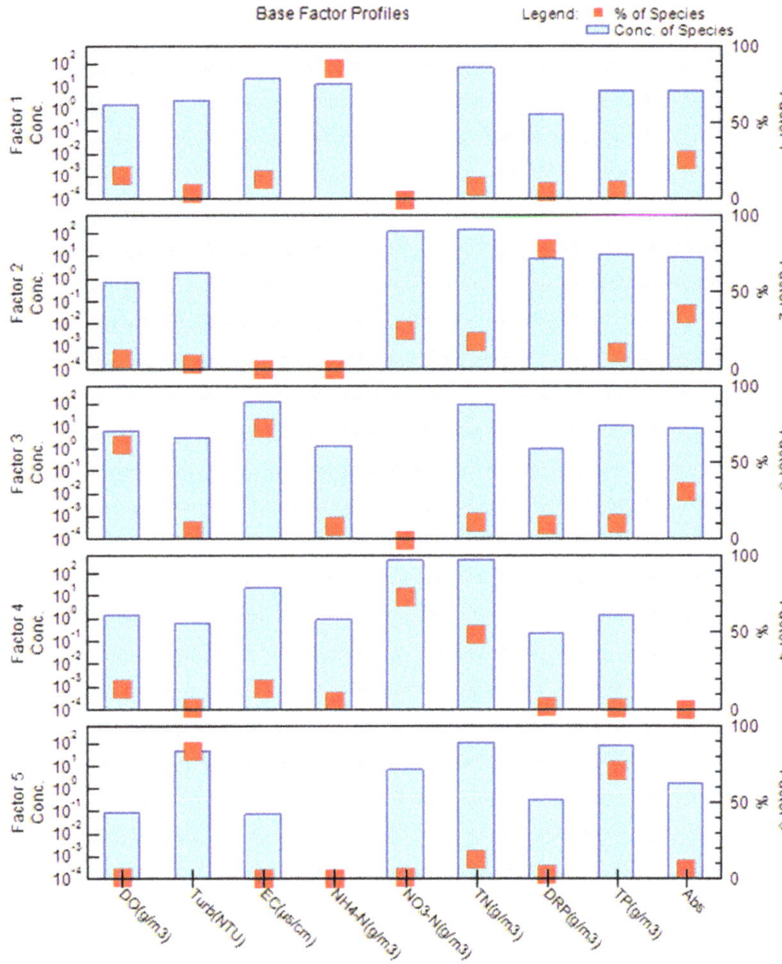

Figure 10. Profile concentrations for WA8 using PMF.

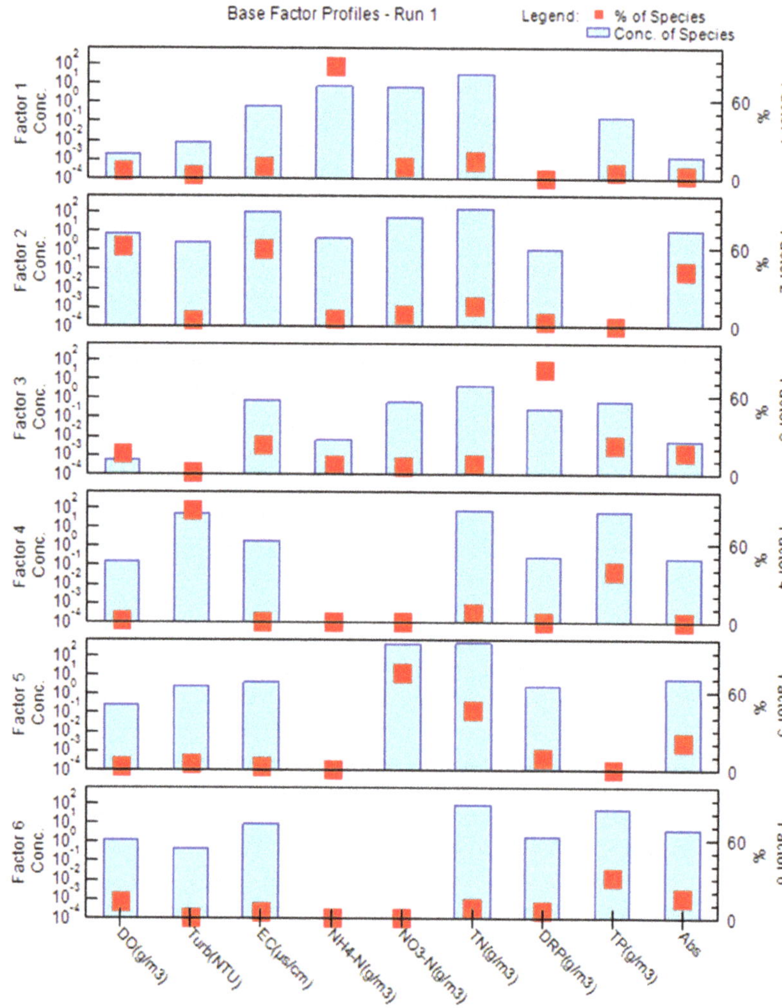

Figure 11. Profile concentrations for WA9 using PMF.

4. Discussion

4.1. Land Connectivity and Effects on Water Quality

The rivers within the Manawatu catchment have poor water quality, with median nutrient values above the ANZECC trigger values. For TP, all three stations had 5 year median values above the ANZECC threshold, with the highest TP values measured at WA9. Thus, the baseline phosphorus concentrations reveal a legacy of catchment sources that go beyond the inputs from storm events. The rainfall plots show that periods of high rainfall did not correspond with high TP values. Although high TP values were observed at the three stations in 2004, the overall rainfall pattern suggests that in-channel erosion was more likely to be the source of TP pollution. From the land-use analyses (Figure 3), it was apparent that high-producing grasslands (66%) were the dominant land-use class. However, connectivity analyses showed that a large amount of TP was introduced through the watershed from upstream (WA7).

Connectivity analyses revealed a large area of high-producing grassland connected to the floodplain. Several studies have shown that high-producing grasslands are a significant source of TP, among other nutrients [11,21,44]. The seasonality effect showed that TP values were higher than the standard in both summer and winter for all three stations, except WA9 in the summer. This also underscores that there is no direct relationship between rainfall patterns and recorded TP. The TP values may have been continuously and gradually introduced from sources other than those directly from the watershed. The gradual release of TP may have been influenced by in-channel erosion or the presence of riparian buffers. The presence of riparian buffers can limit the runoff rate and reduce the concentration of pollutants in rivers. A similar pattern was observed for DRP, NH_4^+, turbidity, TN, and NO_3^-. The observed trends suggest that these pollutants were released after a storm event. In the rainfall graphs, spikes were observed in 1995, 2004, and 2011.

The storm events in 2004 and 2011 were similar in magnitude, but they did not release the same number of pollutants. This pattern may be attributed to the presence of riparian buffers, as well as scaled-back fertilizer application over time. This finding corroborates the findings in the study by Abbott et al. [31]. They found that despite landslides occurring in the Manawatu catchment, minimal suspended solids (SS) are transported into the river because only a small fraction of upland LU/LC is connected. Similar increasing trends in NH_4^+, NO_3^-, and TN were observed for all three sites; however, a different pattern was observed for NH_4^+. The similarity between WA7 and WA8 for NO_3^- and TN suggests that both sites received considerable NO_3^- and TN pollution. However, a disproportionate amount of NH_4^+ was observed at different sites. Higher values of NH_4^+ were observed in WA9 than in the other two sites (WA7 or WA8). The reason for this is likely the presence of urban settlements that produce substantial amounts of waste rich in NH_4^+.

4.2. Water Quality Assessment

Analyses of 25 years of consistent river water quality data provided remarkable insights into the pollution status of the Manawatu catchment. The results for all sites clearly show that pollution varied over time and were mostly above ANZECC trigger values. High pollution (sediment and nutrients) was observed between 2000 and 2004 at all three sites. This was a response to a major storm and increased rainfall compared with other years.

These findings were revealed by post hoc tests of medians, which showed a statistical difference in nutrient concentrations. The increase in nutrients has been attributed to extreme rainfall events in the catchment between 2000 and 2004 (Figure 12). In accordance with Dymond et al. [14], rainfall was significantly elevated in February 2004. It was responsible for flushing a large amount of sediment and nutrients into the river, particularly in 2004 and 2011. Records show that the lower North Island experienced a large storm, with over 20 h of rainfall during that period. Abott et al. [31] reported that a significant storm produced an exceptional amount of sediment, especially in the loose and hilly terrain of the Manawatu watershed. The substantial increase in nutrients and sediments during the 2000–2004 period proves that a large amount of the eroded soil transporting these pollutants was deposited at that time. Consequently, these findings are corroborated with the findings in the studies of Larned et al. [27] and Kamarinas et al. [43], as they revealed that water quality in NZ is poor and will continue to degrade due to a cyclic influx of nutrients that have accumulated in the soil for up to 50 years.

Turbidity, electrical conductivity, total phosphorus, and absorbance are significant pollutants. Previous studies have shown that turbidity results from soil erosion and runoff processes [45–48]. However, it is more likely that these pollutants originated from in-channel sources. In the Manawatu River catchment, in-channel sources emanate from floodplains that trap sediments and nutrients over time and release them during rainfall. Electrical conductivity can also be a marker for the influence of mass weathering effects on water quality [49], whereas high loadings of TP stem from fertilizer application in intensive agricultural areas [46,50]. Finally, high loadings of CDOM represent the presence of dissolved organic matter comprising humic substances [51]. Cruz et al. [50] reported that TP and NO_3-N entered the Siriri River in Brazil, but the sources of this pollution were different. The correlation matrix in our study showed that TP was correlated with turbidity and TN, suggesting that they emanate from the same source. As established earlier, turbidity likely originated from soil erosion. Therefore, NO_3^- and TN are nutrients deposited around the river's soil.

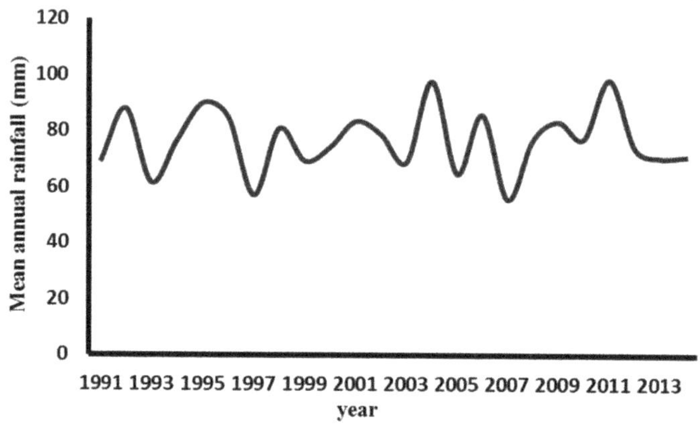

Figure 12. Mean annual rainfall for Palmerston North in the Manawatu Catchment.

Woldeab et al. [12] reported that TN, DRP, and NO_3^- were prevalent and significantly higher in vegetated and agricultural areas. NH_3-N (NH_4^+) can be attributed to dissolved livestock manure within the watershed. Therefore, because DRP is correlated with TP, with DRP as a constituent of TP, it can serve as a possible indicator of a natural pollution source from soil erosion. Dymond et al. [14] reported that the soil material in Manawatu is rich in phosphorus. Table 3a shows that in WA7, large areas of high-producing grassland (73.5 km^2) are connected to the river. This makes it highly likely that agricultural pollution enters the floodplain quickly and significantly. Similarly, Kamarinas et al. [43] revealed that high-producing grasslands and plantation forests produce substantial sediments.

A significant increase in pollution was observed in WA8, as the proportion of plantation forests and high-producing grasslands connected to WA8 increased. This finding is supported by the study by Julian et al. [21]. They reported that the amount of NOx leaking into most rivers in NZ has increased since 1989, despite the reduction in fertilizer application since the early eighties, and that it is consistent with an increase in turbidity and nutrients observed in lowland rivers [47]. The NO_3^- and TN at site WA9 may be due to the runoff of domestic sewage from nearby wastewater treatment plants and urban areas. One of the attributes of this site is that it has a wastewater treatment plant installed to treat waste from both industrial and domestic sources. This finding is supported by Alves et al. [52], who reported that the presence of organic matter in water could be attributed in no small amount to domestic sewage and industrial wastewater that could have been treated using traditional methods. Nutrient pollution seems likely because the downstream site had a wastewater treatment plant installed. Therefore, based on the large variability in NO_3^- and TN within this site, it is safe to suggest that NH_4^+originated from aerobic pollution caused by domestic or livestock waste. Moreover, NH_4^+, DRP, and TP are attributed to the effects of agricultural land use, specifically fertilizer application and animal waste [53].

Factor 2 showed high loadings of NO_3^-, TN, and DO (Figure 9). This can be attributed to the presence of sufficient oxygen, which can play a significant role in the oxidation of nutrients entering the river from fertilizer application, causing degradation. Therefore, it is not surprising to find NH_4^+ in Factor 1, which could be an oxidation byproduct of $NO_3^-$and TN. Factor 3 selected turbidity and TP, and this factor can be related to soil erosion. Factor 4 has been described as a physiochemical source. For WA8, high loadings of TP and turbidity may likely be attributed to soil erosion or mass weathering, as represented in Factor 1 (Figure 10). Factor 3 may be classified as agricultural pollution. Shrestha and Kazama [48] reported that nitrogen compounds were found in the Fuji River because of the application of nitrogenous fertilizers on agricultural lands around the river. Kazama and Yoneyama [54] reported similar findings. Factor 2 was likely a physiochemical source because of the presence of EC and DO. Factor 4 can be described as pollution from livestock and agricultural waste. In WA9, NH_4^+ was dominant, and it was likely released from domestic sewage (Figure 11). As reported in the literature, studies have shown that NH_4^+ can be released from several sources. These sources depend on typical land use or anthropogenic activities associated with pollution sites. However, a high percentage of NH_4^+ can be attributed to runoff from industrial wastewater treatment plants or domestic sewage. These findings align with the findings of Haji et al. [55], which proved that large loadings of ammoniacal-N emanated from an industrial or domestic source.

The high presence of DRP in Factor 2 in WA9 indicates the likely presence of nutrients from agricultural catchments. It is important to note that WA9 is a larger watershed encompassing WA8 and WA7. WA8 and WA7 empty into WA9, with higher agricultural practices occurring in WA7 and WA8, which could be the likely reason for the high DRP. The high presence of TP and turbidity in Factor 3 can be attributed to mass weathering pollution, one of the characteristics of lowland catchments. EC and DO in Factor 4 represent physicochemical source pollution. In contrast, factors 5 and 6 represent organic and agricultural pollution sources based on the presence of CDOM and nutrients. The results of the watershed connectivity model supported these findings. Table 3 reveals less coverage of plantation forests and high-producing grasslands within this watershed, which is a conduit for transporting pollutants into the river. Julian et al. [21] reported that high-producing grasslands increased nutrients significantly in NZ rivers, and it will likely continue to increase from legacy nutrient stores throughout the catchment.

The observations from this study are likely to be skewed in this regard, as elevated concentrations of nutrients (TN, NH_4 -N, TP, and DRP) and sediment loads were recorded across New Zealand. This resulted from the presence of cattle, deer, and dairy cattle [25,28,56]. The findings of Julian et al. [21] further support these findings, as it has been reported that the number of sheep stocks in the uplands, where steep slopes are present, is higher than in the lowlands in New Zealand. As more livestock gather, intensive grazing and the movement of livestock can expose the soil to erosion. Additionally, between the early 1990s and 2012, the number of dairy cattle in NZ increased by a factor of two, which resulted in a significant increase in the application of P and N fertilizers to meet feed demands. When lactating dairy cows graze on pastures treated with P- and N-based fertilizers, approximately 0.8 and 0.6 of the total amounts of P and N fertilizer used, respectively, is deposited on the soil as animal waste [57]. In accordance with Ledgard [58], these values remain underestimated because different dairy pastures are likely to incorporate extra atmospheric N through pasture grazing, strip grazing, and cropping harvest. Pasture grazing is predominant in NZ and has been identified as the root cause of soil catchment exposure to erosion [21]. These findings support the findings from our study as more nutrients, and higher sediment loads, were characteristic of sites within the Manawatu Catchment, wherein large areas of plantation forest (PF) and high-producing grassland (HG) were connected to the floodplain.

The broader impacts of our study will be relevant for receiving waters such as lakes and bays. A study by Abell et al. [59] concerning 101 national lakes showed that high-producing grasslands increased mean TP and TN concentrations. A similar study was carried out by Özkundakci et al. [60], which revealed that high-producing grasslands also increased nutrients in 25 national lakes. The Manawatu River drains into an open estuary, and farther out to the South Taranaki Bight, neither of which is monitored as comprehensively as the river. However, we do know that there are frequent warnings following rainfall in the catchment that discourage contact recreation in the estuary due to the water quality issues we have outlined in this article.

5. Conclusions

This study aimed to identify pollutants and their potential sources within the Manawatu River catchment using multivariate statistical methods to assess the relationships between LULC, watershed connectivity, and water quality. High-producing grassland was interpreted to be the dominant pollution source at all sites. At the same time, greater urban coverage in the western part of the catchment increased pollution in the downstream section (WA9), especially for NH_4^+. Connectivity analyses revealed that 73.4% of the entire catchment was dominated by high-producing grassland, and 43.3% was directly connected to streams via runoff. Domestic pollution sources were mostly found in the downstream section of the Manawatu catchment. Connectivity studies also revealed the role of LULC in water quality, as high-producing grassland areas contributed to increased pollution at a higher rate than the other LULCs combined. Furthermore, this study revealed that nutrients such as TP and TN showed declining trends in concentration at all three sites; however, all median values remained above the ANZECC trigger values. NH_4^+ also decreased significantly in WA8, but it remained elevated in WA7 and WA9. NO_3^- declined to numbers below the trigger values in lowland rivers over time, and it remained of inferior quality in the upland sub-catchment.

The extremely low or insignificant rates at which pollutant concentrations decline could be a cause of concern. The introduction or improvement of retention capacities in the wetlands and riparian buffers, in location-specific areas, would be a viable solution for the Manawatu catchment area. In general, the PMF revealed that point, natural, and agricultural sources were responsible for the pollution in the downstream section of the river. In the intermediate sub-catchment, soil/bank erosion and agricultural sources were the major contributors. Agricultural pollution and soil erosion were likely responsible for the pollution in the upstream section of the catchment. Future work within the Manawatu River catchment should include the development of reaeration models to provide insight into its assimilatory capacity. Risk analyses are necessary to determine the health risks associated with using the Manawatu River in agricultural and pastoral farming. Lastly, this study showed the need for continuous and consistent water quality monitoring to evaluate water quality variables and the effectiveness of current wetlands or riparian buffers already in place. Finally, effectively tracking pollution sources, to determine which portion of a watershed should have a riparian buffer installed, remains a major challenge that water resource managers face worldwide.

Author Contributions: All authors contributed to the study conception and design. Material preparation, data collection, spatial and statistical analysis were performed by J.P.J., P.C.E., N.D.-B., O.M., S.E.S., E.O.B. and D.D.A. The first draft of the manuscript was written by I.T.T. and all authors commented on previous versions of the manuscript. All authors have read and agreed to the published version of the manuscript.

Funding: The authors declare that no funds, grants, or other support were received during the preparation of this manuscript.

Data Availability Statement: The authors have no affiliations with or involvement in any organization or entity with any financial interest or non-financial interest in the subject matter or material discussed in this manuscript. The dataset referenced in this manuscript is available at: https://hydrowebportal.niwa.co.nz/ accessed on 22 April 2021.

Conflicts of Interest: The authors declare that they have no known competing financial interests or personal relationships that could have appeared to influence the work reported in this paper.

References

1. Jéquier, E.; Constant, F. Water as an Essential Nutrient: The Physiological Basis of Hydration. *Eur. J. Clin. Nutr.* **2010**, *64*, 115–123. [CrossRef]
2. Mandal, P.; Upadhyay, R.; Hasan, A. Seasonal and Spatial Variation of Yamuna River Water Quality in Delhi, India. *Environ. Monit. Assess.* **2010**, *170*, 661–670. [CrossRef]

3. Almeida, C.A.; Quintar, S.; González, P.; Mallea, M.A. Influence of Urbanization and Tourist Activities on the Water Quality of the Potrero de Los Funes River (San Luis–Argentina). *Environ. Monit. Assess.* **2007**, *133*, 459–465. [CrossRef]
4. Chelsea Nagy, R.; Graeme Lockaby, B.; Kalin, L.; Anderson, C. Effects of Urbanization on Stream Hydrology and Water Quality: The Florida Gulf Coast. *Hydrol. Process.* **2012**, *26*, 2019–2030. [CrossRef]
5. Peters, N.E. Effects of Urbanization on Stream Water Quality in the City of Atlanta, Georgia, USA. *Hydrol. Process.* **2009**, *23*, 2860–2878. [CrossRef]
6. Tenebe, I.T.; Emenike, C.P.; Daniel Chukwuka, C. Prevalence of Heavy Metals and Computation of Its Associated Risk in Surface Water Consumed in Ado-Odo Ota, South-West Nigeria. *Hum. Ecol. Risk Assess. Int. J.* **2019**, *25*, 882–904. [CrossRef]
7. ul Hassan, Z.; Shah, J.A.; Kanth, T.A.; Pandit, A.K. Influence of Land Use/Land Cover on the Water Chemistry of Wular Lake in Kashmir Himalaya (India). *Ecol. Process.* **2015**, *4*, 9. [CrossRef]
8. Huang, J.; Zhan, J.; Yan, H.; Wu, F.; Deng, X. Evaluation of the Impacts of Land Use on Water Quality: A Case Study in The Chaohu Lake Basin. *Sci. World J.* **2013**, *2013*, e329187. [CrossRef]
9. Tenebe, I.T.; Ogbiye, A.S.; Omole, D.O.; Emenike, P.C. Estimation of Longitudinal Dispersion Co-Efficient: A Review. *Cogent Eng.* **2016**, *3*, 1216244. [CrossRef]
10. Tenebe, I.T.; Ogbiye, A.S.; Omole, D.O.; Emenike, P.C. Parametric Evaluation of the Euler-Lagrangian Approach for Tracer Studies. *Desalin. Water Treat.* **2018**, *109*, 344–349. [CrossRef]
11. Larned, S.T.; Moores, J.; Gadd, J.; Baillie, B.; Schallenberg, M. Evidence for the Effects of Land Use on Freshwater Ecosystems in New Zealand. *N. Z. J. Mar. Freshw. Res.* **2020**, *54*, 551–591. [CrossRef]
12. Woldeab, B.; Ambelu, A.; Mereta, S.T.; Beyene, A. Effect of Watershed Land Use on Tributaries' Water Quality in the East African Highland. *Environ. Monit. Assess.* **2018**, *191*, 36. [CrossRef] [PubMed]
13. Julian, J.P.; Gardner, R.H. Land Cover Effects on Runoff Patterns in Eastern Piedmont (USA) Watersheds. *Hydrol. Process.* **2014**, *28*, 1525–1538. [CrossRef]
14. Dymond, J.R.; Serezat, D.; Ausseil, A.-G.E.; Muirhead, R.W. Mapping of Escherichia Coli Sources Connected to Waterways in the Ruamahanga Catchment, New Zealand. *Environ. Sci. Technol.* **2016**, *50*, 1897–1905. [CrossRef]
15. Taiwo, A.M. Source Identification and Apportionment of Pollution Sources to Groundwater Quality in Major Cities in Southwest, Nigeria/Identifikacija Izvora Oneciscenja Podzemnih Voda u Vecim Gradovima Jugozapadne Nigerije. *Geofizika* **2012**, *29*, 157–175.
16. Emenike, C.P.; Tenebe, I.T.; Omole, D.O.; Ngene, B.U.; Oniemayin, B.I.; Maxwell, O.; Onoka, B.I. Accessing Safe Drinking Water in Sub-Saharan Africa: Issues and Challenges in South–West Nigeria. *Sustain. Cities Soc.* **2017**, *30*, 263–272. [CrossRef]
17. Emenike, C.P.; Tenebe, I.T.; Jarvis, P. Fluoride Contamination in Groundwater Sources in Southwestern Nigeria: Assessment Using Multivariate Statistical Approach and Human Health Risk. *Ecotoxicol. Environ. Saf.* **2018**, *156*, 391–402. [CrossRef]
18. Gulgundi, M.S.; Shetty, A. Identification and Apportionment of Pollution Sources to Groundwater Quality. *Environ. Process.* **2016**, *3*, 451–461. [CrossRef]
19. Chounlamany, V.; Tanchuling, M.A.; Inoue, T. Spatial and Temporal Variation of Water Quality of a Segment of Marikina River Using Multivariate Statistical Methods. *Water Sci. Technol.* **2017**, *76*, 1510–1522. [CrossRef]
20. Bruesewitz, D.A.; Hamilton, D.P.; Schipper, L.A. Denitrification Potential in Lake Sediment Increases Across a Gradient of Catchment Agriculture. *Ecosystems* **2011**, *14*, 341–352. [CrossRef]
21. Julian, J.P.; de Beurs, K.M.; Owsley, B.; Davies-Colley, R.J.; Ausseil, A.-G.E. River Water Quality Changes in New Zealand over 26 Years: Response to Land Use Intensity. *Hydrol. Earth Syst. Sci.* **2017**, *21*, 1149–1171. [CrossRef]
22. Schallenberg, M. Determining Reference Conditions for New Zealand Lakes. *Sci. Conserv.* **2019**, *334*, 1.
23. OECD/FAO. *Aglink-Cosimo Model Documentation—A Partial Equilibrium Model of World Agricultural Markets*; FAO: Rome, Italy, 2022; ISBN 978-92-5-136969-2.
24. Smith, D.G.; McBride, G.B. New Zealand's National Water Quality Monitoring Network—Design and First Year's Operation1. *JAWRA J. Am. Water Resour. Assoc.* **1990**, *26*, 767–775. [CrossRef]
25. Davies-Colley, R.J.; Nagels, J.W.; Smith, R.A.; Young, R.G.; Phillips, C.J. Water Quality Impact of a Dairy Cow Herd Crossing a Stream. *N. Z. J. Mar. Freshw. Res.* **2004**, *38*, 569–576. [CrossRef]
26. Ballantine, D.J.; Davies-Colley, R.J. Water Quality Trends in New Zealand Rivers: 1989-2009. *Environ. Monit. Assess.* **2014**, *186*, 1939–1950. [CrossRef]
27. Larned, S.; Snelder, T.; Unwin, M.; McBride, G. Water Quality in New Zealand Rivers: Current State and Trends. *N. Z. J. Mar. Freshw. Res.* **2016**, *50*, 389–417. [CrossRef]
28. Mcdowell, R.W.; Larned, S.T.; Houlbrooke, D.J. Nitrogen and Phosphorus in New Zealand Streams and Rivers: Control and Impact of Eutrophication and the Influence of Land Management. *N. Z. J. Mar. Freshw. Res.* **2009**, *43*, 985–995. [CrossRef]
29. Reff, A.; Eberly, S.I.; Bhave, P.V. Receptor Modeling of Ambient Particulate Matter Data Using Positive Matrix Factorization: Review of Existing Methods. *J. Air Waste Manag. Assoc.* **2007**, *57*, 146–154. [CrossRef]
30. Smith, D.G.; McBride, G.B.; Bryers, G.G.; Wisse, J.; Mink, D.F.J. Trends in New Zealand's National River Water Quality Network. *N. Z. J. Mar. Freshw. Res.* **1996**, *30*, 485–500. [CrossRef]
31. Abbott, S.; Julian, J.P.; Kamarinas, I.; Meitzen, K.M.; Fuller, I.C.; McColl, S.T.; Dymond, J.R. State-Shifting at the Edge of Resilience: River Suspended Sediment Responses to Land Use Change and Extreme Storms. *Geomorphology* **2018**, *305*, 49–60. [CrossRef]
32. Hicks, D.M.; Shankar, U.; McKerchar, A.I.; Basher, L.; Lynn, I.; Page, M.; Jessen, M. Suspended Sediment Yields from New Zealand Rivers. *J. Hydrol.* **2011**, *50*, 81–142.

33. Land Air Water Aotearoa. Available online: https://www.lawa.org.nz/download-data/ (accessed on 14 June 2023).
34. Davies-Colley, R.J.; Smith, D.G.; Ward, R.C.; Bryers, G.G.; McBride, G.B.; Quinn, J.M.; Scarsbrook, M.R. Twenty Years of New Zealand's National Rivers Water Quality Network: Benefits of Careful Design and Consistent Operation1. *JAWRA J. Am. Water Resour. Assoc.* **2011**, *47*, 750–771. [CrossRef]
35. ANZECC & ARMCANZ (2000) Guidelines. Available online: https://www.waterquality.gov.au/anz-guidelines/resources/previous-guidelines/anzecc-armcanz-2000 (accessed on 8 June 2023).
36. US EPA. Positive Matrix Factorization 5.0 Fundamentals and User Guide. Available online: https://www.epa.gov/air-research/epa-positive-matrix-factorization-50-fundamentals-and-user-guide (accessed on 14 June 2023).
37. Kaiser, H.F. An Index of Factorial Simplicity. *Psychometrika* **1974**, *39*, 31–36. [CrossRef]
38. Aitchison, J. Measures of Location of Compositional Data Sets. *Math. Geol.* **1989**, *21*, 787–790. [CrossRef]
39. Blake, S.; Henry, T.; Murray, J.; Flood, R.; Muller, M.R.; Jones, A.G.; Rath, V. Compositional Multivariate Statistical Analysis of Thermal Groundwater Provenance: A Hydrogeochemical Case Study from Ireland. *Appl. Geochem.* **2016**, *75*, 171–188. [CrossRef]
40. Emenike, P.C.; Tenebe, I.; Ogarekpe, N.; Omole, D.; Nnaji, C. Probabilistic Risk Assessment and Spatial Distribution of Potentially Toxic Elements in Groundwater Sources in Southwestern Nigeria. *Sci. Rep.* **2019**, *9*, 15920. [CrossRef]
41. Manousakas, M.; Papaefthymiou, H.; Diapouli, E.; Migliori, A.; Karydas, A.G.; Bogdanovic-Radovic, I.; Eleftheriadis, K. Assessment of PM2.5 Sources and Their Corresponding Level of Uncertainty in a Coastal Urban Area Using EPA PMF 5.0 Enhanced Diagnostics. *Sci. Total. Environ.* **2017**, *574*, 155–164. [CrossRef] [PubMed]
42. Kamarinas, I. Geospatial Analyses of Terrestrial-Aquatic Connections Across New Zealand and Their Influence on River Water Quality. Ph.D. Thesis, Texas State University, San Marcos, TX, USA, August 2018.
43. Kamarinas, I.; Julian, J.P.; Hughes, A.O.; Owsley, B.C.; De Beurs, K.M. Nonlinear Changes in Land Cover and Sediment Runoff in a New Zealand Catchment Dominated by Plantation Forestry and Livestock Grazing. *Water* **2016**, *8*, 436. [CrossRef]
44. Snelder, T.H.; McDowell, R.W.; Fraser, C.E. Estimation of Catchment Nutrient Loads in New Zealand Using Monthly Water Quality Monitoring Data. *JAWRA J. Am. Water Resour. Assoc.* **2017**, *53*, 158–178. [CrossRef]
45. Kemker, C. Turbidity, Total Suspended Solids & Water Clarity. Available online: https://www.fondriest.com/environmental-measurements/parameters/water-quality/turbidity-total-suspended-solids-water-clarity/ (accessed on 14 June 2023).
46. Salim, I.; Sajjad, R.U.; Paule-Mercado, M.C.; Memon, S.A.; Lee, B.-Y.; Sukhbaatar, C.; Lee, C.-H. Comparison of Two Receptor Models PCA-MLR and PMF for Source Identification and Apportionment of Pollution Carried by Runoff from Catchment and Sub-Watershed Areas with Mixed Land Cover in South Korea. *Sci. Total Environ.* **2019**, *663*, 764–775. [CrossRef]
47. Memon, S.; Paule, M.C.; Lee, B.-Y.; Umer, R.; Sukhbaatar, C.; Lee, C.-H. Investigation of Turbidity and Suspended Solids Behavior in Storm Water Run-off from Different Land-Use Sites in South Korea. *Desalin. Water Treat.* **2015**, *53*, 3088–3095. [CrossRef]
48. Shrestha, S.; Kazama, F. Assessment of Surface Water Quality Using Multivariate Statistical Techniques: A Case Study of the Fuji River Basin, Japan. *Environ. Model. Softw.* **2007**, *22*, 464–475. [CrossRef]
49. Ogwueleka, T.C. Use of Multivariate Statistical Techniques for the Evaluation of Temporal and Spatial Variations in Water Quality of the Kaduna River, Nigeria. *Environ. Monit. Assess.* **2015**, *187*, 137. [CrossRef] [PubMed]
50. Cruz, M.A.S.; Gonçalves, A.d.A.; de Aragão, R.; de Amorim, J.R.A.; da Mota, P.V.M.; Srinivasan, V.S.; Garcia, C.A.B.; de Figueiredo, E.E. Spatial and Seasonal Variability of the Water Quality Characteristics of a River in Northeast Brazil. *Environ. Earth Sci.* **2019**, *78*, 68. [CrossRef]
51. Reynolds, D.M. The Differentiation of Biodegradable and Non-Biodegradable Dissolved Organic Matter in Wastewaters Using Fluorescence Spectroscopy. *J. Chem. Technol. Biotechnol.* **2002**, *77*, 965–972. [CrossRef]
52. Alves, D.D.; Riegel, R.P.; de Quevedo, D.M.; Osório, D.M.M.; da Costa, G.M.; do Nascimento, C.A.; Telöken, F. Seasonal Assessment and Apportionment of Surface Water Pollution Using Multivariate Statistical Methods: Sinos River, Southern Brazil. *Environ. Monit. Assess.* **2018**, *190*, 384. [CrossRef]
53. Dils, R.M.; Heathwaite, A.L. The Controversial Role of Tile Drainage in Phosphorus Export from Agricultural Land. *Water Sci. Technol.* **1999**, *39*, 55–61. [CrossRef]
54. Kazama, F.; Yoneyama, M. Nitrogen Generation in the Yamanashi Prefecture and Its Effects on the Groundwater Pollution. *Environ. Sci.* **2002**, *15*, 293–298. [CrossRef]
55. Haji Gholizadeh, M.; Melesse, A.M.; Reddi, L. Water Quality Assessment and Apportionment of Pollution Sources Using APCS-MLR and PMF Receptor Modeling Techniques in Three Major Rivers of South Florida. *Sci. Total Environ.* **2016**, *566–567*, 1552–1567. [CrossRef]
56. Buck, O.; Niyogi, D.K.; Townsend, C.R. Scale-Dependence of Land Use Effects on Water Quality of Streams in Agricultural Catchments. *Environ. Pollut.* **2004**, *130*, 287–299. [CrossRef]
57. Monaghan, R.M.; Hedley, M.J.; Di, H.J.; McDowell, R.W.; Cameron, K.C.; Ledgard, S.F. Nutrient Management in New Zealand Pastures—Recent Developments and Future Issues. *N. Z. J. Agric. Res.* **2007**, *50*, 181–201. [CrossRef]
58. Ledgard, S.F. Nitrogen Cycling in Low Input Legume-Based Agriculture, with Emphasis on Legume/Grass Pastures. *Plant Soil* **2001**, *228*, 43–59. [CrossRef]

59. Abell, J.M.; Özkundakci, D.; Hamilton, D.P. Nitrogen and Phosphorus Limitation of Phytoplankton Growth in New Zealand Lakes: Implications for Eutrophication Control. *Ecosystems* **2010**, *13*, 966–977. [CrossRef]
60. Özkundakci, D.; Hamilton, D.P.; Kelly, D.; Schallenberg, M.; de Winton, M.; Verburg, P.; Trolle, D. Ecological Integrity of Deep Lakes in New Zealand across Anthropogenic Pressure Gradients. *Ecol. Indic.* **2014**, *37*, 45–57. [CrossRef]

water

MDPI

Article

Analysis of the Spatial Distribution Characteristics of Emerging Pollutants in China

Man Zhang [1,2], Yong Sun [3], Bin Xun [1,2,*] and Baoyin Liu [4,*]

1 College of Urban and Environmental Sciences, Northwest University, Xi'an 710127, China; 202121533@stumail.nwu.edu.cn
2 Shaanxi Key Laboratory of Earth Surface System and Environmental Carrying Capacity, Xi'an 710127, China
3 School of Public Administration, Guangzhou University, Guangzhou 510006, China; sunyong@gzhu.edu.cn
4 Institutes of Science and Development, Chinese Academy of Sciences, Beijing 100190, China
* Correspondence: xunbin@nwu.edu.cn (B.X.); liubaoyin@casisd.cn (B.L.)

Abstract: Pollutant types are increasing along with the rapid development of society and economy. Some emerging pollutants from chemicals have begun to appear and endanger public and ecosystem health. However, the research and development of emerging pollutant monitoring technology is still in its infancy, with no complete monitoring system in place. This makes it impossible to access and identify the spatial pattern of emerging pollutants. Therefore, this paper reviews the existing quantitative research results on four common emerging pollutants in China's water environment—namely, endocrine disruptors, brominated flame retardants, perfluorinated compounds, and microplastics—extracts the quantitative monitoring results of emerging pollutants in the case studies, and outlines the spatial distribution characteristics of emerging pollutants in the water environment. The results show that the emerging pollutants have a large distribution area that has covered most of China. The level of pollution from emerging pollutants correlates with the level of economic development and the pollution level in economically developed regions such as the Yangtze River Delta, the Pearl River Delta, and the Beijing–Tianjin–Hebei region is significantly higher than in other regions. This study provides a reference for the prevention and control of emerging pollutants in China.

Keywords: emerging pollutants; spatial distribution; endocrine disruptors; brominated flame retardants; perfluorinated compounds; microplastics

Citation: Zhang, M.; Sun, Y.; Xun, B.; Liu, B. Analysis of the Spatial Distribution Characteristics of Emerging Pollutants in China. *Water* **2023**, *15*, 3782. https://doi.org/10.3390/w15213782

Academic Editor: Imokhai Theophilus Tenebe

Received: 25 September 2023
Revised: 20 October 2023
Accepted: 24 October 2023
Published: 29 October 2023

1. Introduction

In recent years, with the improvement of living standards, people have paid increasing attention to the quality of the ecological environment. Currently, a relatively comprehensive water quality monitoring system has been established for traditional pollutants such as nutrients [1] (nitrogen, phosphorus, etc.), organics [2] (xylene, PCBs, etc.), and heavy metals [3] (mercury, nickel, etc.). However, some emerging pollutants have surfaced and made certain impacts on the ecological environment and human health [4–8]. Some research found that emerging pollutants such as endocrine disruptors (EDCs), perfluorinated compounds (PFCs), brominated flame retardants (BFRs), and microplastics (MPs) are widely distributed in water bodies and have significant impacts on ecosystems and human health [9–13]. For example, the frequently detected EDCs in the environment and human samples, such as organophosphates (OPEs), can disrupt thyroid hormone levels and liver receptors in the human body, causing thyroid disruption and DNA damage, and affecting liver metabolism [14,15]. Perfluorooctane sulfonic acid (PFOS) and perfluorooctanoic acid (PFOA), which are widely present in tap water and surface water, can induce oxidative stress and acute toxicity in human macrophages and zebrafish embryos [16,17]. BFRs are present in almost all chemical products, and tetrabromobisphenol A (TBBPA) is one of the most common types. It can cause hemolysis in human red blood cells and mitochondrial oxidative phosphorylation in rats. It also reduces the survival rate of young fish [18,19]. Microplastics are distributed throughout the entire aquatic ecosystem and can be ingested by various animals and plants. They can be toxic to human lung cells, liver cells, and brain cells, as well as cause damage to plant root systems, affecting their ability to absorb water and nutrients [20,21]. Studies have shown that these emerging pollutants may disrupt the normal functioning of the human endocrine system, leading to reproductive issues, neurological problems, immune system disorders, metabolic issues, and cancers [22–24]. They also have the potential to cause damage to the environmental ecosystem and can negatively impact the stability and diversity of aquatic and terrestrial organisms, as well as disrupt food chains, ecological balance, and biodiversity [25,26]. Additionally, emerging pollutants are characterized by their persistence and bioaccumulation. They are difficult to degrade in the environment and can accumulate within living organisms, and this accumulation can lead to a gradual increase in concentration within the food chain and ultimately enter the human body [27–29]. Therefore, it is crucial to conduct comprehensive monitoring for emerging pollutants.

Research has been conducted on the monitoring remediation of emerging pollutants in small-scale areas due to their impact on the ecological environment and human health. For instance, the bioaccumulation of specific emerging pollutants has been monitored through the collection of samples from animals, plants, and human tissues using biomonitoring techniques [30]. Biological indicators such as mussels and fish have also been used for monitoring the concentration of emerging pollutants in field and laboratory settings [31]. In addition to monitoring, there have been initial studies on the remediation of emerging pollutants in water bodies. Nanobiochar [32,33] and bioelectrochemical systems [34] are among the methods used for the removal of emerging pollutants from water. However, there are concerns regarding the accumulation of pollutants within the biochar and the potential re-release of these pollutants into the water [35]. Additionally, bioelectrochemical systems have complex operation requirements. As a result, the methods and technologies for the remediation of emerging pollutants are not yet perfect. The existing monitoring and remediation measures have been implemented only in small-scale areas, providing valuable data for these specific regions [36–39]. However, the lack of quantitative monitoring techniques hinders the large-scale implementation of these methods [25,26]. The incomplete development of technology for establishing a comprehensive water quality monitoring system for emerging pollutants has also limited the availability of widespread monitoring data [40–43]. The complexity of geographical environments further hampers the extrapolation of small-scale data to larger areas, preventing the comprehensive mapping of emerging pollutant distribution. Therefore, this study aims to extract and analyze China's regional concentration monitoring data based on the literature to obtain spatial distribution patterns of emerging pollutants in the area.

China is currently facing a serious pollution problem of emerging contaminants in its water environment. These pollutants mainly come from industrial emissions [28], agricultural activities [29], atmospheric deposition, and other sources, posing a significant threat to the environment and human health [44,45]. In 2021, Fan et al. [46] conducted quantitative monitoring of several highly concerned EDCs in water bodies in Jiangsu Province and found that the average concentration of these endocrine disruptors was higher than 300 ng/L. In 2013, Wang et al. [47] quantitatively monitored the concentration of PFCs in the surface water of the Han River in Wuhan, and the study showed that the maximum concentration of total PFCs was 568 ng/L. In 2021, Xia et al. [48] conducted quantitative monitoring of MPs in the surface water of Sangou Bay, China and found that the average concentration of MPs was 20.06 item/L. Although there has been some progress in quantitative monitoring of emerging contaminants at a small scale, there is a lack of monitoring data at a large scale. This makes it difficult to accurately assess the distribution characteristics and the extent of the impact of emerging contaminants in China, thereby causing a challenge in formulating effective pollution control strategies. It is essential to obtain large-scale information on emerging pollutants in China. Therefore, this paper chooses the Chinese region as the research area, systematically reviews the relevant literature on emerging pollutants in China in the past decade, extracts the quantitative monitoring results of pollutants in the case study area, and constructs a spatial distribution map of emerging pollutants in Chinese water bodies. This spatial distribution map can help decision makers understand the distribution of emerging pollutants and provide a scientific basis for formulating pollution control measures.

2. Definition and Characterization of Emerging Pollutants

2.1. Definition of Emerging Pollutants

The definition of emerging pollutants is not yet clear. Different scholars have elaborated on the definition of emerging pollutants from different perspectives (Table 1). The initial definition of emerging pollutants only applied to newly emerging substances that cause environmental problems [49–51]. Later, it gradually extended from substances causing environmental problems to substances with potential environmental risks [49–52]. As more is known about emerging pollutants, people are realizing that not only emerging substances, but also substances that have previously existed but have recently posed environmental risks, and substances that have not yet been regulated by laws and standards can also be considered as emerging pollutants [49–55].

Table 1. Definition of emerging pollutants.

Year	Author	The Definition of Emerging Pollutants
2006	Field [51]	Emerging pollutants refer to those caused by human activities which are currently known to exist but are not yet regulated or inadequately regulated by laws and standards, and pose a threat to living and ecological environments. These pollutants are produced in all aspects of production, construction, or other activities.
2008	Farre [49]	Newly emerged pollutants are defined as compounds currently not covered by existing water quality regulations, previously unstudied, and considered to be potential threats to environmental ecosystems as well as human health and safety.
2010	Houtman [50]	The first category consists of compounds newly introduced into the environment, such as industrial compounds that have only recently been developed; the second category of compounds may have been present for a long time but have only recently been detected in the environment; the third category of emerging pollutants consists of compounds that may have been known for a long time but have only recently been considered potentially harmful to ecosystems or humans.
2011	Bell [56]	The term "emerging pollutants" mainly refers to pollutants for which there are currently no regulations requiring monitoring or public reporting of their presence in our country's water supply or wastewater discharge.
2011	Deblonde [52]	Emerging pollutants refer to new products or chemicals without regulatory status, of which the impact on the environment and human health is unknown.
2012	Thomaidis [57]	The term "emerging pollutants" refers to substances released into the environment for which there are currently no regulations in place for their environmental monitoring.
2014	Sauve [54]	Naturally occurring, manufactured, or artificially created chemicals or materials that are now found or suspected to be present in various environmental compartments, the toxicity or persistence of which is likely to significantly alter the metabolism of organisms.
2015	Geissen [55]	Emerging pollutants (EPs) are defined as synthetic or naturally occurring chemicals that are not commonly monitored in the environment but have the potential to enter the environment and cause known or suspected adverse ecological and/or human health effects.
2021	Li [53]	Newly discovered or focused pollutants that pose a threat to the ecological environment or human health which have not yet been included in management or the existing management measures are insufficient to effectively prevent and control their risks.

Based on the above understanding of emerging pollutants, we can summarize the definition of emerging pollutants as pollutants produced in production, construction, or other activities that are caused by human activities which clearly exist but have not been regulated or are poorly regulated by laws and regulations and are harmful or potentially harmful to the living environment or the ecological environment.

2.2. Characteristics of Emerging Pollutants

Compared to traditional pollutants such as sulfur dioxide and nitrogen oxides, most of the emerging pollutants are more persistent, accumulative, and migratory. They can persist in the environment and are far more difficult to manage than traditional pollutants. The characteristics are specifically manifested as follows: (1) The chemical properties are very stable and not easy to degrade. For example, perfluorinated compounds refer to hydrocarbon compounds and their derivatives which are formed after all hydrogen atoms are replaced by fluorine atoms. Since fluorine is the most electronegative element, it makes the carbon–fluorine bond highly polar; therefore, perfluorinated compounds have chemical stability, surface activity, excellent temperature resistance, and so on [58–60], and are not easily decomposed in the environment. (2) The sources of emerging pollutants are widespread, as their superior chemical properties are used in all aspects of life. For example, brominated flame retardants, with their low cost, low addition amount, and good flame retardant performance, are widely used in products such as electrical appliances, textiles, automobiles, and building materials [61–63]. (3) Emerging pollutants are widely distributed. First, they are versatile. Due to the large demand in life, various new pollutants are used in various industries and products. For example, endocrine disruptors are widely used in the production of personal care products, detergents, thermal paper, and plastics [64,65]. Second, the distribution area is wide. Due to their strong migratory nature, they are found all over the world. For example, González-Pleiter [66] detected the presence of microplastics in freshwater in the Antarctic protected area. Third, they have a wide presence in various carriers. Emerging pollutants have been detected in various carriers such as air, water, and soil [67–69]. (4) Emerging pollutants possess various types of biological toxicity, including organ toxicity, neurotoxicity, and reproductive and developmental toxicity. They also have a certain degree of accumulation in organisms, posing certain harm or potential harm to the ecological environment or human health [70–73]. (5) Emerging pollutants have a certain incubation period, with pollution being lagging and remediation being long-term. Since pollutants need to accumulate to a certain concentration to cause environmental changes [74,75], the detection of pollution takes a certain amount of time. Moreover, the current imperfection of monitoring technology for new pollutants is more likely to cause lagging pollution.

Based on the definition and pollution characteristics of emerging pollutants, we select microplastics (MPs), brominated flame retardants (BFRs), perfluorinated compounds (PFCs), and endocrine disrupting chemicals (EDCs) as the most common emerging pollutants. On the basis of the literature review, the spatial distribution characteristics of these four emerging pollutants in China were studied.

3. Methodology

3.1. Literature Situation

We searched for keywords such as "emerging pollutants", "endocrine disruptors", "perfluorinated compounds", "brominated flame retardants", "microplastics", "MPs", "EDCs", "PFCs", "BFRs", etc., and retrieved a total of 3158 related articles. As can be seen from Figure 1, from 2000 to 2022, the total number of articles on the study of emerging pollutants shows an upward trend; in particular, after 2010, the number of articles has risen sharply. This is mainly because around 2010, the harm of emerging pollutants to the public and the environment became increasingly obvious, and the academic community has paid more attention to the research of emerging pollutants since [52,53]. Although the number of quantitative literature is also increasing, there are less than 300 articles on the quantitative analysis of pollution. This reflects the underdeveloped monitoring technology of emerging pollutants in the world, which still needs to be improved.

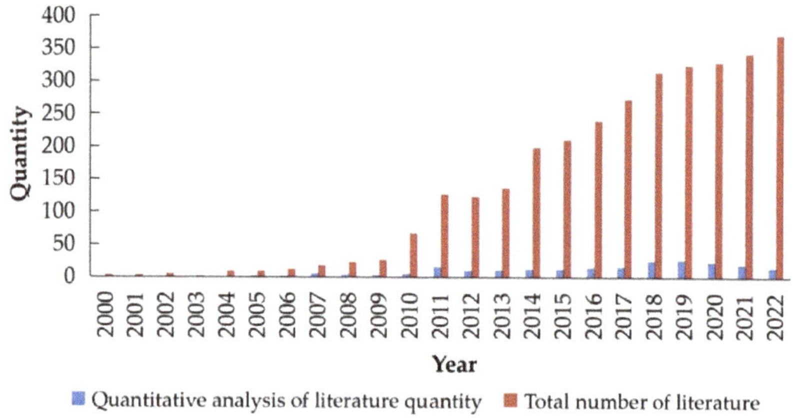

Figure 1. Annual number of articles on emerging pollutants.

3.2. Method of Extracting Pollutant Concentration Data

Due to the numerous carriers of emerging pollutants, e.g., plants, water bodies, atmosphere, sediments, etc., and the different measurement units used for different carriers (e.g., the concentration of EDCs in water bodies is usually expressed in ng/L, while in sediments, it is ng/Kg), it is difficult to convert the concentration values of pollutants between different carriers. Therefore, this study focuses on selecting the literature which studies water bodies and extracts the concentration values of pollutants between different water bodies. There are four methods for sampling and extracting the concentration of emerging pollutants from monitored water bodies, namely, surface water sampling [76,77], solid-phase extraction (SPE) sampling [78–80], vertical profiling sampling [81], and automatic water quality monitoring systems [82]. Although these different methods can monitor the concentration of emerging pollutants in water, the operational differences among these methods theoretically result in minimal variations in the obtained results. Therefore, in this study, we will not provide specific evaluations of these methods and only extract the final measured concentration values as the research results. Additionally, due to the existence of different pollution concentration values derived from the same water body in different literature, this study adheres to the following principles when extracting pollution concentration values. (1) Emerging pollutants are divided into four categories for extraction: EDCs, BFRs, PFCs, and MPs. These four categories all contain subcategories. If there are inconsistent values in the subcategories, the value with the greatest concentration in the subcategories is selected as the extraction result of this category. For example, Gong et al. [83] detected in 2011 that the concentration ranges of five types of EDCs in the Pearl River Delta rivers, namely OP (Octylphenol), NP (Nonylphenol), BPA (Bisphenol A), E1 (Estrone), and E2 (17a-ethynylestradiol), were 1.6–577 ng/L, 0.28–14.9 ng/L, 87–639 ng/L, 1.5–11.5 ng/L, and 1.1–1.7 ng/L, respectively. The maximum pollution concentration value of BPA (Bisphenol A) (87–639 ng/L) was selected as the pollution concentration of EDCs in the Pearl River Delta rivers. (2) If there are multiple quantitative monitoring results for the same body of water, the most recent monitoring result is extracted. For example, both Gang et al. [84] in 2016 and Guo et al. [85] in 2011 monitored the pollution concentration of PFCs in Tai Lake. This study chose the more recent pollution concentration data in 2016 as the extraction result. (3) If the years are also the same, then the larger concentration value is selected as the extraction result.

4. Result

4.1. The Characteristics of EDCs Distribution

From the distribution map of EDCs pollution concentration in the water body, it can be found that the concentration of EDCs pollution decreases from southeast to northwest. The distribution of quantitative monitoring results is most intensive in the southeast coastal area, while there are no quantitative monitoring results in the western region (Figure 2). However, on the whole, EDCs are widely distributed in China, such as Inner Mongolia, which is vast and sparsely populated, but a certain concentration of EDCs was also detected [86]. The Yangtze River is the longest river in China with a vast drainage area, which may lead to the accumulation of EDCs in the water, resulting in the highest detected concentration of EDCs is found in the Yangtze River region, reaching 144,000 ng/L [87]. There is a certain correlation between the level of economic development and the concentration of emerging pollutants. To some extent, the higher the level of economic development, the higher the concentration of pollutants detected. The Yangtze River Delta and the Pearl River Delta are economically developed regions in China, with abundant water resources and dense river networks. In terms of industrial level, these areas have large-scale industrial enterprises, petrochemical plants, and manufacturing clusters, involving sectors such as electronics, chemicals, textiles, machinery, etc., which produce a large amount of chemical products, petroleum products, and electricity. The high industrial level also brings about serious water pollution issues. Therefore, in China's eastern coastal areas such as the Yangtze River Delta and southern coastal areas such as the Pearl River Delta, the economy is developed, monitoring points are densely distributed, and the concentration values are at a higher level.

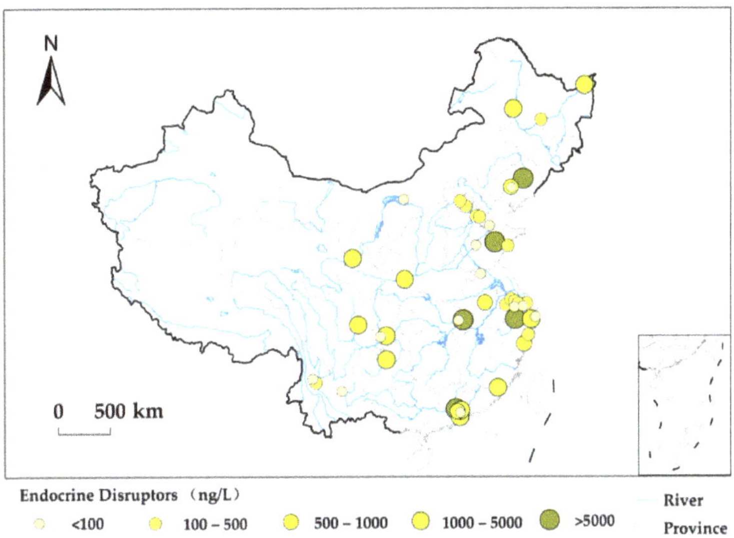

Figure 2. Distribution of Endocrine Disruptors in China based on Literature Research.

4.2. *The Characteristics of BFRs Distribution*

At present, there are few quantitative monitoring studies on the distribution of BFRs in waterbodies. The sites in Figure 3 are sparse and dispersed. These sites are mainly distributed in the eastern region, especially in the Bohai Bay area, and the pollution concentration is high. In the western, northern, and central regions of China, quantitative monitoring of BFRs has been limited. The Bohai Bay region is an important economic area in Northeast China, with a high level of industrial development. It is home to numerous industrial parks, ports, and cities, covering industries such as petrochemicals, steel, energy, and equipment manufacturing. Located between the Bohai Sea and the Yellow Sea, the region is a convergence of ocean and rivers, with major rivers like the Yellow River and the Liao River running through it. The developed water system is conducive to transportation industries such as water transport and shipping, as well as the manufacturing of automobiles and vessels. However, the dense water network also contributes to the accumulation of pollutants, resulting in relatively high concentrations of BFRs detected in this region. The difference between the maximum and minimum concentration values of BFRs is large. The highest pollution concentration point is located in the Taihu Basin, with a concentration value of 2934.2 ng/L [88]. The point with the lowest pollution concentration is located in Dalian city, with a concentration value of 1.08 ng/L [89].

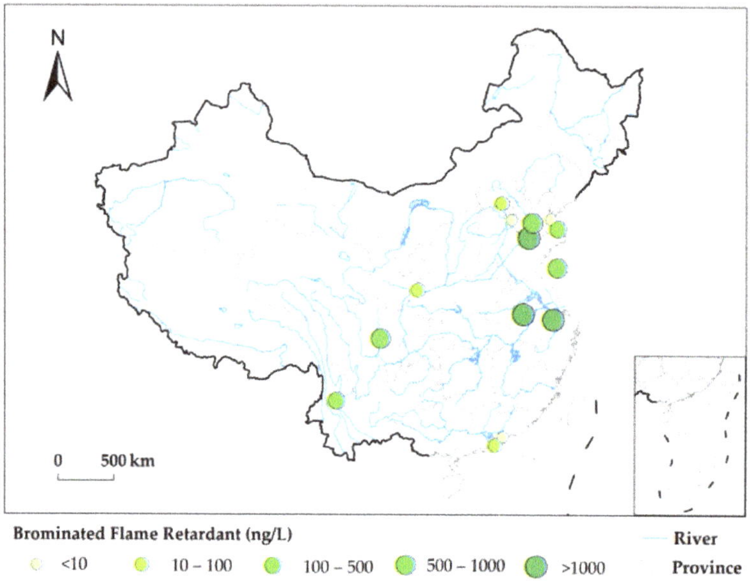

Figure 3. Distribution of Brominated Flame Retardants in China based on Literature Research.

4.3. The Characteristics of PFCs Distribution

The monitoring and research on PFCs are primarily focused on China's eastern, northeastern, southeastern, and southern coastal regions, while quantitative monitoring research on PFCs in the overall western regions is relatively limited. In comparison to the western regions, the eastern regions of China generally have a higher level of industrial development, with numerous industrial parks, important industrial cities, and a developed manufacturing industry. The eastern regions are also economically developed areas, with abundant water resources, larger water flow, and denser water networks. As a result, the distribution density of monitoring sites gradually decreases from the eastern coastal areas to the western inland areas (Figure 4). PFCs are widely distributed, and even in Tibet, PFCs have been found at the level of 0.322 ng/L [90]. In addition, PFCs have also been found in the Ulansuhai Nur [91] and Hohhot City [92] in Inner Mongolia, with concentrations of 263.45 ng/L and 1.8 ng/L, respectively. The pollution levels of PFCs are mostly at the intermediate level, with fewer areas at the maximum level. The areas with dense quantitative monitoring of PFCs are mainly concentrated in the Pearl River Delta, Yangtze River Delta, Dalian Bay, and Bohai Bay in China. These areas are located in the coastal and inland interchange zone, where the industry is developed, and emerging pollutants such as PFCs are easily generated.

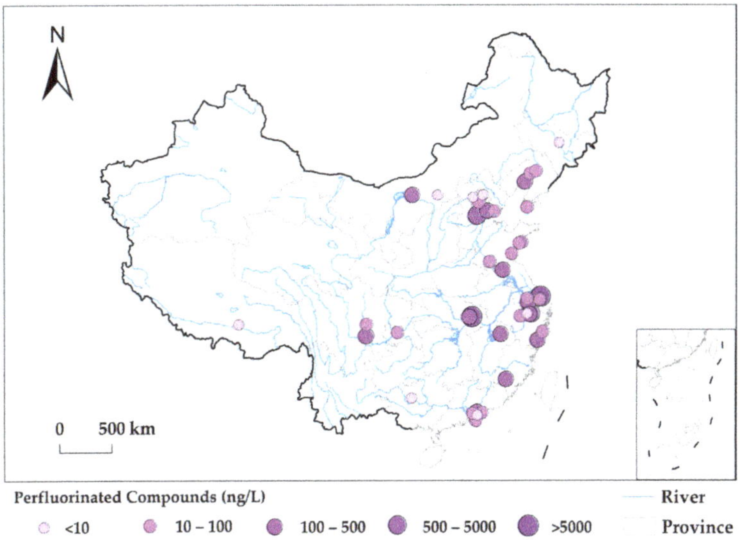

Figure 4. Distribution of Perfluorinated Compounds in China based on Literature Research.

4.4. The Characteristics of MPs Distribution

Of the four emerging pollutants, MPs are the most monitored. MPs are widely distributed in China, and quantitative monitoring studies of MPs have been conducted in remote areas such as Qinghai–Tibet Plateau [93] and Xinjiang [94] regions. The areas with the most intensive quantitative monitoring of MPs are located in the eastern, southern, and central parts of China, which are also the areas with the highest monitoring concentration levels (Figure 5). The eastern, central, and southern regions of China have numerous rivers, lakes, and reservoirs forming a complex water network system. They also have large water bodies such as the East China Sea, South China Sea, and Taihu Lake, which are of great significance for regional economic development and water resource utilization. These regions are economically developed areas with a high level of industrialization. They are major manufacturing clusters, with many industrial parks and important industrial cities involved in various industries including manufacturing, petrochemicals, electronics, and automotive manufacturing. These areas have a large number of manufacturing enterprises and supply chains, playing a vital role in China's economic development. Due to intense industrial activities and urbanization, high concentrations of microplastics (MPs) in water bodies have led to water quality deterioration. This is mainly caused by industrial wastewater, agricultural non-point source pollution, and urban domestic sewage. There is little quantitative monitoring research on MPs in the northeastern region of China. The dense areas of MPs distribution include the Pearl River Delta, Yangtze River Delta, Beijing–Tianjin–Hebei region, and other areas with dense populations and advanced industries. These areas are also located in the downstream areas, where MPs gradually accumulate during river flow and accumulate to the maximum concentration at the estuary. This shows that the concentration of emerging pollutants is closely related to the level of local economic development, natural geographical conditions, and geographical location. The abundance of MPs in the Yellow River in China is the highest, with 930,000 items/m^3 [95], while that in Qinghai Lake is the smallest, with 0.031 items/m^3 [96].

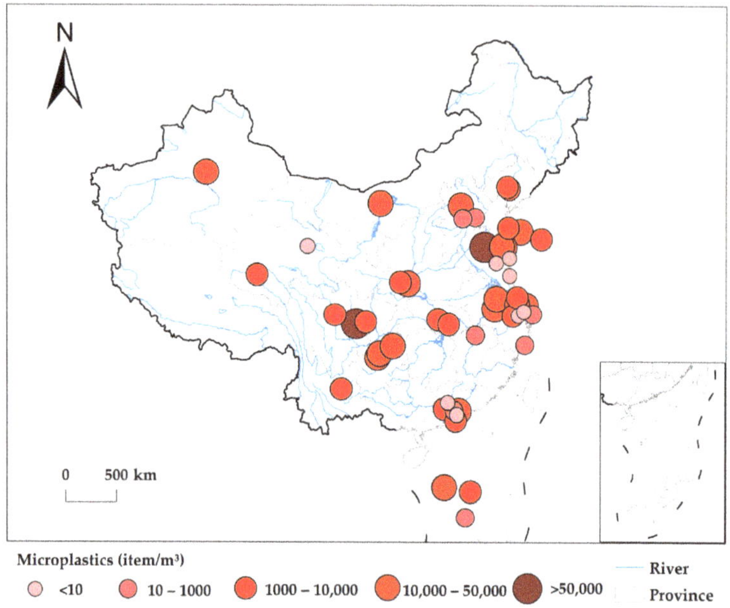

Figure 5. Distribution of Microplastics in China based on Literature Research.

4.5. Overall Characteristics

In this study, 225 points with quantitative monitoring results were extracted by referring to the relevant literature (Supplementary Materials Tables S1–S4) [47,84,86,87,89–295]. The points of the four types of emerging pollutants are mainly concentrated in economically developed areas such as the Yangtze River Delta, the Pearl River Delta, and Beijing–Tianjin–Hebei region. These regions are highly developed in industrialization and urbanization, serving as important manufacturing centers in China, with strong industrial agglomeration effects. They have a large-scale economy and a complete industrial chain. In terms of geographical location, they are all located in river basins or estuarine areas, with abundant water resources. However, with the acceleration of industrial and urbanization processes, these regions are also facing challenges of water resources supply–demand imbalance and water pollution issues. Therefore, these areas have a variety of emerging pollutants with high pollution concentrations. From a national perspective, the quantity and concentration of various emerging pollutants are increasing from west to east, and there are more types of emerging pollutants in coastal areas and their concentration is higher than that in central and western regions. In recent decades, China has vigorously built urban agglomerations and accelerated industrial clusters. In particular, the economic scale of coastal cities continues to expand, and the emission of emerging pollutants is also rising. In addition, emerging pollutants are prone to accumulate in the process of river flow. The eastern region is located at the mouth of the river, so the concentration of emerging pollutants accumulates to the highest value here. The presence of emerging pollutants has also been detected in the relatively economically backward regions of Xinjiang and Tibet. Although the current pollutants in the two regions are single and the pollution concentration is low, it also reflects, to a certain extent, the widespread problem of the distribution of emerging pollutants in China.

5. Conclusions and Discussion

5.1. Emerging Pollutants Are Widely Distributed in China

The distribution of emerging pollutants basically covers most areas of China. Relevant research results show that emerging pollutants exist in all regions of China, except parts of the northwest. Compared to traditional pollutants such as sulfur dioxide and nitrogen oxides, most emerging pollutants are more persistent, accumulative, and migratory. They can be adsorbed on suspended particles in the air or water and migrate long distances to reach remote polar regions. The detection of EDCs in Xinjiang and Tibet proves this point. Therefore, it can be inferred that emerging pollutants are distributed in most areas of China. The lack of research in the northwest is more due to the sparse distribution of rivers, relatively small population, and thus fewer researchers choosing this area as their research area.

5.2. The Distribution of Emerging Pollutants Has a Certain Correlation with the Level of Economic Development

The analyzed results indicate that the key pollution areas of emerging pollutants are mainly concentrated in economically developed regions such as the Yangtze River Delta, Pearl River Delta, and Beijing–Tianjin–Hebei region. Due to the production and use of various chemicals, a large amount of emerging pollutants is generated. The quantity and types of chemicals produced and used in economically developed areas are significantly higher than in other areas. In the absence of a comprehensive chemical management system and classification registration system, the extensive use of chemicals by enterprises and the public has led to relatively high levels of emerging pollutants.

5.3. The Concentration of Emerging Pollutants in China Is Relatively High

The average values of emerging pollutants extracted from the 225 points studied are as follows: EDCs 5044.16 ng/L, BFRs 828.43 ng/L, PFCs 669.69 ng/L, and MPs 44,145.57 item/m^3. These values are significantly higher than the pollution levels in other countries. As researchers tend to choose areas with severe pollution as case studies, these average values are significantly higher than the current average levels of emerging pollutants in China. However, these data are sufficient to prove that the pollution levels in China's key pollution areas are relatively high. Although there is little qualitative monitoring research on emerging pollutants in some areas of China, such as Xinjiang, the monitoring concentration of new pollutants in Xinjiang is also at a high level. In the eastern regions of China, the quantitative monitoring concentrations of emerging pollutants such as MPs, PFCs, and EDCs are all at high levels. In addition, the quantitative monitoring concentration of MPs in the eastern regions of China is at the highest level in the world among existing studies. Therefore, the concentration of emerging pollutants in China is relatively high.

5.4. The Geographical Distribution of Emerging Pollutants Is Quite Distinct

The density and concentration of emerging pollutants in China decrease from the southeast coastal areas to the northwest inland areas of China. Since the level of economic development in China also decreases from the southeast coast to the northwest inland, and the economic level is related to the scale of industrial development, there is a correlation between the distribution of new pollutants and the level of economic development in China. As a result, the higher the level of economic development in the region, the more dense the distribution of new pollutants and the greater the concentration.

5.5. Discussion

First, we inferred the basic spatial distribution pattern of emerging pollutants in China based on the literature data. After understanding the spatial distribution pattern of emerging pollutants in China, government departments can develop region-specific measures. For high-pollution areas, the government can implement measures such as restricting industrial emissions, strengthening law enforcement, and promoting clean production in high-polluting enterprises. For low-pollution areas, efforts can be focused on enhancing environmental monitoring and protection and encouraging the development of green industries. As the spread of emerging pollutants often transcends the boundaries of a single region, cross-regional cooperation is needed. The government can work with neighboring regions, relevant departments, and international organizations to formulate cross-regional measures to deal with pollution, and share data, technology, and experience so as to comprehensively solve the problem of pollution by emerging pollutants. The government should also establish relevant laws, regulations, and policies to support the implementation of targeted measures, while strengthening law enforcement efforts to crack down on illegal activities.

Second, we proved that China exhibits higher levels of emerging pollutant concentrations compared to other parts of the world. As one of the largest manufacturing countries in the world, China has high industrial and energy consumption. Some industrial production processes produce more emissions of emerging pollutants, resulting in high levels of pollutant concentrations. Under such circumstances, the public should ensure their own drinking water and food safety, such as prioritizing treated drinking water and choosing to buy food that meets food safety standards, in order to reduce the risk of ingesting pollutants. The public should also actively participate in environmental protection actions and support the government in implementing targeted measures to reduce the generation of emerging pollutants. Enterprises should carry out technological upgrading and improvement measures, promote the recycling of resources and the economical use of energy, strengthen their internal environmental management and supervision, and set up sound internal auditing and monitoring mechanisms. Enterprises should also actively participate in environmental protection actions, support environmental organizations and projects, promote the development of green industries, and reduce the pollution of water bodies by emerging pollutants.

Third, we demonstrated the pollution caused by emerging pollutants has a wide range of transmission. In the future, water body models and hydrological data should be utilized to analyze the transmission pathways and diffusion characteristics of emerging pollutants, revealing the conduction laws and behaviors of emerging pollutants in different water bodies. It is important to investigate the influences of water flow, water body characteristics, and human activities on their transmission and strengthen the research on the conductivity of emerging pollutants. The core objective of this study is to extract historical concentration values of emerging pollutants in order to obtain the spatial distribution patterns and concentration distributions of emerging pollutants in China. By doing so, it aims to increase the attention of the Chinese government towards emerging pollutants and provide references for the formulation of policies and regulations regarding emerging pollutants in China.

Hence, it is suggested that on one hand, key regions such as the Yangtze River Delta, Pearl River Delta, and Beijing–Tianjin–Hebei region should be selected to carry out pilot surveys of new types of pollutants. The focus should be on understanding the production and usage of products containing typical emerging pollutants in industries, agriculture, and daily life, compiling a list of typical pollutant emissions, and strengthening the prevention and control of emerging pollutants. On the other hand, research on the mechanisms of emerging pollutant generation and migration and the development of monitoring technologies should be strengthened. A comprehensive monitoring system should be gradually established to fully grasp the data on the pollution levels of emerging pollutants.

Supplementary Materials: The following supporting information can be downloaded at: https://www.mdpi.com/article/10.3390/w15213782/s1, Table S1: Distribution Area and Concentration of Microplastics; Table S2: Distribution Area and Concentration of Endocrine Disruptors; Table S3: Distribution Area and Concentration of Perfluorinated Compounds; Table S4: Distribution Area and Concentration of Brominated Flame Retardants. References [97–295] are citied in the Supplementary Materials.

Author Contributions: M.Z. conducted the data collection, data analysis, data plotting, and wrote the paper. Y.S., B.X., and B.L. contributed to the writing of the paper. All authors have read and agreed to the published version of the manuscript.

Funding: This research was funded by [the Program for Key Science and Technology Innovation Team in Shaanxi Province] grant number [2014KCT-27].

Institutional Review Board Statement: Not applicable.

Informed Consent Statement: Not applicable.

Data Availability Statement: Derived data supporting the findings of this study are available from the corresponding author on request.

Acknowledgments: All authors thank the anonymous reviewers and the editor for the constructive comments on the earlier version of the manuscript.

Conflicts of Interest: The authors declare no conflict of interest.

References

1. Vitousek, P.M.; Porder, S.; Houlton, B.Z.; Chadwick, O.A. Terrestrial phosphorus limitation: Mechanisms, implications, and nitrogen—phosphorus interactions. *Ecol. Appl.* **2010**, *20*, 5–15. [CrossRef]
2. Schmid, P.; Kohler, M.; Gujer, E.; Zennegg, M.; Lanfranchi, M. Persistent organic pollutants, brominated flame retardants and synthetic musks in fish from remote alpine lakes in Switzerland. *Chemosphere* **2007**, *67*, S16–S21. [CrossRef] [PubMed]
3. Dixit, R.; Wasiullah, X.; Malaviya, D.; Pandiyan, K.; Singh, U.B.; Sahu, A.; Shukla, R.; Singh, B.P.; Rai, J.P.; Sharma, P.K.; et al. Bioremediation of heavy metals from soil and aquatic environment: An overview of principles and criteria of fundamental processes. *Sustainability* **2015**, *7*, 2189–2212. [CrossRef]
4. Joyce, S. The dead zones: Oxygen-starved coastal waters. *Environ. Health Perspect.* **2000**, *108*, A120–A125. [CrossRef]
5. Sonone, S.S.; Jadhav, S.; Sankhla, M.S.; Kumar, R. Water contamination by heavy metals and their toxic effect on aquaculture and human health through food Chain. *Lett. Appl. NanoBioScience* **2020**, *10*, 2148–2166. [CrossRef]
6. Madhav, S.; Ahamad, A.; Singh, A.K.; Kushawaha, J.; Chauhan, J.S.; Sharma, S.; Singh, P. *Sensors in Water Pollutants Monitoring: Role of Material*; Springer: Singapore, 2020; pp. 43–62. [CrossRef]
7. Khalil, B.; Ouarda, T.B.; St-Hilaire, A.; Chebana, F. A statistical approach for the rationalization of water quality indicators in surface water quality monitoring networks. *J. Hydrol.* **2010**, *386*, 173–185. [CrossRef]
8. Zhong, S.; Li, J.; Zhao, R. Does environmental information disclosure promote sulfur dioxide (SO_2) remove? New evidence from 113 cities in China. *J. Clean. Prod.* **2021**, *299*, 126906. [CrossRef]
9. Shaw, S.; Van Heyst, B. Nitrogen Oxide (NO_x) emissions as an indicator for sustainability. *Environ. Sustain. Indic.* **2022**, *15*, 100188. [CrossRef]
10. Li, G.; Fang, C.; Wang, S.; Sun, S. The effect of economic growth, urbanization, and industrialization on fine particulate matter ($PM_{2.5}$) concentrations in China. *Environ. Sci. Technol.* **2016**, *50*, 11452–11459. [CrossRef]
11. Vigiak, O.; Grizzetti, B.; Udias-Moinelo, A.; Zanni, M.; Dorati, C.; Bouraoui, F.; Pistocchi, A. Predicting biochemical oxygen demand in European freshwater bodies. *Sci. Total Environ.* **2019**, *666*, 1089–1105. [CrossRef]
12. Li, D.; Xu, X.; Li, Z.; Wang, T.; Wang, C. Detection methods of ammonia nitrogen in water: A review. *TrAC-Trend. Anal. Chem.* **2020**, *127*, 115890. [CrossRef]
13. Krieger, J.; Higgins, D.L. Housing and health: Time again for public health action. *Am. J. Public Health* **2022**, *92*, 758–768. [CrossRef]
14. Yao, Y.; Li, M.; Pan, L.; Duan, Y.; Duan, X.; Li, Y.; Sun, H. Exposure to organophosphate ester flame retardants and plasticizers during pregnancy: Thyroid endocrine disruption and mediation role of oxidative stress. *Environ. Int.* **2021**, *146*, 106215. [CrossRef]
15. Hu, W.; Gao, P.; Wang, L.; Hu, J. Endocrine disrupting toxicity of aryl organophosphate esters and mode of action. *Crit. Rev. Environ. Sci. Technol.* **2023**, *53*, 1–18. [CrossRef]
16. Rainieri, S.; Conlledo, N.; Langerholc, T.; Madorran, E.; Sala, M.; Barranco, A. Toxic effects of perfluorinated compounds at human cellular level and on a model vertebrate. *Food Chem. Toxicol.* **2017**, *104*, 14–25. [CrossRef]
17. Suja, F.; Pramanik, B.K.; Zain, S.M. Contamination, bioaccumulation and toxic effects of perfluorinated chemicals (PFCs) in the water environment: A review paper. *Water Sci. Technol.* **2009**, *60*, 1533–1544. [CrossRef]
18. Darnerud, P.O. Toxic effects of brominated flame retardants in man and in wildlife. *Environ. Int.* **2003**, *29*, 841–853. [CrossRef]
19. Pittinger, C.A.; Pecquet, A.M. Review of historical aquatic toxicity and bioconcentration data for the brominated flame retardant tetrabromobisphenol A (TBBPA): Effects to fish, invertebrates, algae, and microbial communities. *Environ. Sci. Pollut. Res.* **2018**, *25*, 14361–14372. [CrossRef]
20. Ziani, K.; Ioniță-Mîndrican, C.B.; Mititelu, M.; Neacsu, S.M.; Negrei, C.; Morosan, E.; Drăgănescu, D.; Preda, O.T. Microplastics: A real global threat for environment and food safety: A state of the art review. *Nutrients* **2023**, *15*, 617. [CrossRef]
21. Urbina, M.A.; Correa, F.; Aburto, F.; Ferrio, J.P. Adsorption of polyethylene microbeads and physiological effects on hydroponic maize. *Sci. Total Environ.* **2020**, *741*, 140216. [CrossRef]
22. Zhang, K.; Wen, Z. Review and challenges of policies of environmental protection and sustainable development in China. *J. Environ. Manag.* **2008**, *88*, 1249–1261. [CrossRef]
23. Rillig, M.C. Microplastic in terrestrial ecosystems and the soil? *Environ. Sci. Technol.* **2012**, *46*, 6453–6454. [CrossRef]

24. Taniyasu, S.; Kannan, K.; Horii, Y.; Hanari, N.; Yamashita, N. A survey of perfluorooctane sulfonate and related perfluorinated organic compounds in water, fish, birds, and humans from Japan. *Environ. Sci. Technol.* **2003**, *37*, 2634–2639. [CrossRef]
25. Jaspers, V.; Covaci, A.; Maervoet, J.; Dauwe, T.; Voorspoels, S.; Schepens, P.; Eens, M. Brominated flame retardants and organochlorine pollutants in eggs of little owls (*Athene noctua*) from Belgium. *Environ. Pollut.* **2005**, *136*, 81–88. [CrossRef]
26. Bowerman, W.W.; Best, D.A.; Grubb, T.G.; Sikarskie, J.G.; Giesy, J.P. Assessment of environmental endocrine disruptors in bald eagles of the Great Lakes. *Chemosphere* **2000**, *41*, 1569–1574. [CrossRef]
27. Lei, M.; Zhang, L.; Lei, J.; Zong, L.; Li, J.; Wu, Z.; Wang, Z. Overview of emerging contaminants and associated human health effects. *BioMed Res. Int.* **2015**, *2015*, 404796. [CrossRef]
28. Elobeid, M.A.; Brock, D.W.; Allison, D.B.; Padilla, M.A.; Ruden, D.M. Endocrine disruptors and obesity: An examination of selected persistent organic pollutants in the NHANES 1999–2002 data. *Int. J. Environ. Res. Public Health* **2010**, *7*, 2988–3005. [CrossRef]
29. Eriksson, P.; Jakobsson, E.; Fredriksson, A. Brominated flame retardants: A novel class of developmental neurotoxicants in our environment? *Environ. Health Perspect.* **2001**, *109*, 903–908. [CrossRef]
30. Georgieva, E.; Antal, L.; Stoyanova, S.; Arnaudova, D.; Velcheva, I.; Iliev, I.; Vasileva, T.; Bivolarski, V.; Mitkovska, V.; Chassovnikarova, T.; et al. Biomarkers for pollution in caged mussels from three reservoirs in Bulgaria: A pilot study. *Heliyon* **2022**, *8*, E09069. [CrossRef]
31. Yancheva, V.; Georgieva, E.; Velcheva, I.; Iliev, I.; Stoyanova, S.; Vasileva, T.; Bivolarski, V.; Todorova-Bambaldokova, D.; Zulkipli, N.; Antal, L.; et al. Assessment of the exposure of two pesticides on common carp (*Cyprinus carpio* Linnaeus, 1758): Are the prolonged biomarker responses adaptive or destructive? *Comp. Biochem. Physiol. C Toxicol. Pharmacol.* **2022**, *261*, 109446. [CrossRef]
32. Dong, M.; He, L.; Jiang, M.; Zhu, Y.; Wang, J.; Gustave, W.; Wang, S.; Deng, Y.; Zhang, X.; Wang, Z. Biochar for the removal of emerging pollutants from aquatic systems: A review. *Int. J. Environ. Res. Public Health* **2023**, *20*, 1679. [CrossRef] [PubMed]
33. Rivas-Sanchez, A.; Cruz-Cruz, A.; Gallareta-Olivares, G.; González-González, R.B.; Parra-Saldívar, R.; Iqbal, H.M.N. Carbon-based nanocomposite materials with multifunctional attributes for environmental remediation of emerging pollutants. *Chemosphere* **2022**, *303*, 135054. [CrossRef] [PubMed]
34. Ahmadi, S.; Rezae, A.; Ghosh, S.; Malloum, A.; Banach, A. A review on bioelectrochemical systems for emerging pollutants remediation: A computational approaches. *J. Environ. Chem. Eng.* **2023**, *11*, 110021. [CrossRef]
35. Jiang, M.; He, L.; Niazi, N.K.; Wang, H.; Gustave, W.; Vithanage, M.; Geng, K.; Shang, H.; Zhang, X.; Wang, Z. Nanobiochar for the remediation of contaminated soil and water: Challenges and opportunities. *Biochar* **2023**, *5*, 2. [CrossRef]
36. Williams, H.H.; MacKenzie, K. Marine parasites as pollution indicators: An update. *Parasitology* **2003**, *126*, S27–S41. [CrossRef]
37. Petrie, B.; Barden, R.; Kasprzyk-Hordern, B. A review on emerging contaminants in wastewaters and the environment: Current knowledge, understudied areas and recommendations for future monitoring. *Water Res.* **2015**, *72*, 3–27. [CrossRef]
38. Gavrilescu, M.; Demnerová, K.; Aamand, J.; Agathos, S.; Fava, F. Emerging pollutants in the environment: Present and future challenges in biomonitoring, ecological risks and bioremediation. *New Biotechnol.* **2015**, *32*, 147–156. [CrossRef]
39. Schriks, M.; Heringa, M.B.; van der Kooi, M.M.; de Voogt, P.; van Wezel, A.P. Toxicological relevance of emerging contaminants for drinking water quality. *Water Res.* **2010**, *44*, 461–476. [CrossRef]
40. Adu-Manu, K.S.; Tapparello, C.; Heinzelman, W.; Katsriku, F.A.; Abdulai, J.D. Water quality monitoring using wireless sensor networks: Current trends and future research directions. *ACM Trans. Sens. Netw.* **2017**, *13*, 1–41. [CrossRef]
41. Gao, G.; Xiao, K.; Chen, M. An intelligent IoT-based control and traceability system to forecast and maintain water quality in freshwater fish farms. *Comput. Electron. Agric.* **2019**, *166*, 105013. [CrossRef]
42. Glasgow, H.B.; Burkholder, J.A.M.; Reed, R.E.; Lewitus, A.J.; Kleinman, J.E. Real-time remote monitoring of water quality: A review of current applications, and advancements in sensor, telemetry, and computing technologies. *J. Exp. Mar. Biol. Ecol.* **2004**, *300*, 409–448. [CrossRef]
43. Richardson, S.D. Disinfection by-products and other emerging contaminants in drinking water. *TrAC Anal. Chem.* **2003**, *22*, 666–684. [CrossRef]
44. Stahl, T.; Mattern, D.; Brunn, H. Toxicology of perfluorinated compounds. *Environ. Sci. Eur.* **2011**, *23*, 38. [CrossRef]
45. Prata, J.C.; da Costa, J.P.; Lopes, I.; Duarte, A.C.; Rocha-Santos, T. Environmental exposure to microplastics: An overview on possible human health effects. *Sci. Total Environ.* **2020**, *702*, 134455. [CrossRef]
46. Fan, D.; Yin, W.; Gu, W.; Liu, M.; Liu, J.; Wang, Z.; Shi, L. Occurrence, spatial distribution and risk assessment of high concern endocrine-disrupting chemicals in Jiangsu Province, China. *Chemosphere* **2021**, *285*, 131396. [CrossRef]
47. Wang, B.; Cao, M.; Zhu, H.; Chen, J.; Wang, L.; Liu, G.; Gu, X.; Lu, X. Distribution of perfluorinated compounds in surface water from Hanjiang River in Wuhan, China. *Chemosphere* **2013**, *93*, 468–473. [CrossRef]
48. Xia, B.; Sui, Q.; Sun, X.; Zhu, L.; Wang, R.; Cai, M.; Chen, B.; Qu, K. Microplastic pollution in surface seawater of Sanggou Bay, China: Occurrence, source and inventory. *Mar. Pollut. Bull.* **2021**, *162*, 111899. [CrossRef]
49. La Farre, M.; Pérez, S.; Kantiani, L.; Barceló, D. Fate and toxicity of emerging pollutants, their metabolites and transformation products in the aquatic environment. *Trends Anal. Chem.* **2008**, *27*, 991–1007. [CrossRef]
50. Houtman, C.J. Emerging contaminants in surface waters and their relevance for the production of drinking water in Europe. *J. Integr. Environ. Sci.* **2010**, *7*, 271–295. [CrossRef]

51. Field, J.A.; Johnson, C.A.; Rose, J.B. *What Is "Emerging"?* American Chemical Society: Washington, DC, USA, 2006; p. 7105. Available online: https://pubs.acs.org/doi/pdf/10.1021/es062982z (accessed on 12 October 2021).
52. Deblonde, T.; Cossu-Leguille, C.; Hartemann, P. Emerging pollutants in wastewater: A review of the literature. *Int. J. Hyg. Environ. Health* **2011**, *214*, 442–448. [CrossRef]
53. Li, Q.; Yu, F.; Cao, G. Progress in the treatment of new pollutants and research on the treatment strategies during the "Fourteenth Five-Year Plan" period and in the long term. *Environ. Prot.* **2021**, *49*, 12–19.
54. Sauvé, S.; Desrosiers, M. A review of what is an emerging contaminant. *Chem. Cent. J.* **2014**, *8*, 15. [CrossRef]
55. Geissen, V.; Mol, H.; Klumpp, E.; Umlauf, G.; Nadal, M.; Ploeg, M.; Zee, S.E.A.T.M.; Ritsema, C.J. Emerging pollutants in the environment: A challenge for water resource management. *Int. Soil Water Conserv. Res.* **2015**, *3*, 57–65. [CrossRef]
56. Bell, K.Y.; Wells, M.J.M.; Traexler, K.A.; Pellegrin, M.L.; Morse, A.; Bandyet, J. Emerging pollutants. *Water Environ. Res.* **2011**, *83*, 1906–1984. [CrossRef]
57. Thomaidis, N.S.; Asimakopoulos, A.G.; Bletsou, A.A. Emerging contaminants: A tutorial mini-review. *Glob. NEST J.* **2012**, *14*, 72–79. Available online: https://www.researchgate.net/publication/234168573 (accessed on 12 October 2021).
58. Hussain, S.M.S.; Adewunmi, A.A.; Mahboob, A.; Murtaza, M.; Zhou, X.; Kamal, M.S. Fluorinated surfactants: A review on recent progress on synthesis and oilfield applications. *Adv. Colloid Interface Sci.* **2022**, *303*, 102634. [CrossRef]
59. Yao, W.; Li, Y.; Huang, X. Fluorinated poly (meth) acrylate: Synthesis and properties. *Polymer* **2014**, *55*, 6197–6211. [CrossRef]
60. Feng, W.; Long, P.; Feng, Y.; Li, Y. Two-dimensional fluorinated graphene: Synthesis, structures, properties and applications. *Adv. Sci.* **2016**, *3*, 1500413. [CrossRef]
61. Wilkie, C.A.; Morgan, A.B. (Eds.) *Fire Retardancy of Polymeric Materials*; CRC Press: Boca Raton, FL, USA, 2009.
62. Hörold, S. Phosphorus-based and intumescent flame retardants. *Polym. Green Flame Retard.* **2014**, *6*, 221–254. [CrossRef]
63. Weil, E.D.; Levchik, S.V. Flame retardants in commercial use or development for textiles. *J. Fire Sci.* **2008**, *26*, 243–281. [CrossRef]
64. Martín-Pozo, L.; Gómez-Regalado, M.C.; Moscoso-Ruiz, I.; Zafra-Gómez, A. Analytical methods for the determination of endocrine disrupting chemicals in cosmetics and personal care products: A review. *Talanta* **2021**, *234*, 122642. [CrossRef] [PubMed]
65. Encarnação, T.; Pais, A.A.; Campos, M.G.; Burrows, H.D. Endocrine disrupting chemicals: Impact on human health, wildlife and the environment. *Sci. Prog.* **2019**, *102*, 3–42. [CrossRef] [PubMed]
66. González-Pleiter, M.; Edo, C.; Velázquez, D.; Casero-Chamorro, M.C.; Leganés, F.; Quesada, A.; Fernández-Piñas, F.; Rosal, R. First detection of microplastics in the freshwater of an Antarctic Specially Protected Area. *Mar. Pollut. Bull.* **2020**, *161*, 111811. [CrossRef] [PubMed]
67. Dumont, M.; Tuzet, F. Light-absorbing Particles in Snow and Climate. In *Chemistry in the Cryosphere (In 2 Parts)*; World Scientific: Singapore, 2022; pp. 795–830. [CrossRef]
68. Yuan, Z.; Pei, C.; Li, H.; Lin, L.; Liu, S.; Hou, R.; Liao, R.; Xu, X. Atmospheric microplastics at a southern China metropolis: Occurrence, deposition flux, exposure risk and washout effect of rainfall. *Sci. Total Environ.* **2023**, *869*, 161839. [CrossRef]
69. Enyoh, C.E.; Verla, A.W.; Verla, E.N.; Ibe, F.C.; Amaobi, C.E. Airborne microplastics: A review study on method for analysis, occurrence, movement and risks. *Environ. Monit. Assess.* **2019**, *191*, 668. [CrossRef]
70. Yang, J.; Zhao, Y.; Li, M.; Du, M.; Li, X.; Li, Y. A review of a class of emerging contaminants: The classification, distribution, intensity of consumption, synthesis routes, environmental effects and expectation of pollution abatement to organophosphate flame retardants (OPFRs). *Int. J. Mol. Sci.* **2019**, *20*, 2874. [CrossRef]
71. Engwa, G.A.; Ferdinand, P.U.; Nwalo, F.N.; Unachukwu, M.N. Mechanism and health effects of heavy metal toxicity in humans. In *Poisoning in the Modern World—New Tricks for an Old Dog?* IntechOpen: London, UK, 2019; Volume 10, pp. 70–90.
72. Pereira, L.C.; de Souza, A.O.; Bernardes, M.F.F.; Pazin, M.; Tasso, M.J.; Pereira, P.H.; Dorta, D.J. A perspective on the potential risks of emerging contaminants to human and environmental health. Environ. *Sci. Pollut. Res.* **2015**, *22*, 13800–13823. [CrossRef]
73. Fenton, S.E.; Ducatman, A.; Boobis, A.; DeWitt, J.C.; Lau, C.; Ng, C.; Smith, J.S.; Roberts, S.M. Per-and polyfluoroalkyl substance toxicity and human health review: Current state of knowledge and strategies for informing future research. *Environ. Toxicol. Chem.* **2021**, *40*, 606–630. [CrossRef]
74. Hallam, T.G.; Clark, C.E.; Lassiter, R.R. Effects of toxicants on populations: A qualitative approach I. Equilibrium environmental exposure. *Ecol. Model.* **1983**, *18*, 291–304. [CrossRef]
75. Conti, M.E.; Cecchetti, G. Biological monitoring: Lichens as bioindicators of air pollution assessment—A review. *Environ. Pollut.* **2001**, *114*, 471–492. [CrossRef]
76. Matamoros, V.; Arias, C.A.; Nguyen, L.X.; Salvadó, V.; Brix, H. Occurrence and behavior of emerging contaminants in surface water and a restored wetland. *Chemosphere* **2012**, *88*, 1083–1089. [CrossRef] [PubMed]
77. Riva, F.; Zuccato, E.; Davoli, E.; Fattore, E.; Castiglioni, S. Risk assessment of a mixture of emerging contaminants in surface water in a highly urbanized area in Italy. *J. Hazard. Mater.* **2019**, *361*, 103–110. [CrossRef] [PubMed]
78. Richardson, S.D.; Kimura, S.Y. Water analysis: Emerging contaminants and current issues. *Anal. Chem.* **2019**, *92*, 473–505. [CrossRef] [PubMed]
79. Rodil, R.; Quintana, J.B.; López-Mahía, P.; Muniategui-Lorenzo, S.; Prada-Rodríguez, D. Multi-residue analytical method for the determination of emerging pollutants in water by solid-phase extraction and liquid chromatography–tandem mass spectrometry. *J. Chromatogr. A* **2009**, *1216*, 2958–2969. [CrossRef] [PubMed]

80. Rodriguez-Mozaz, S.; de Alda, M.J.L.; Barceló, D. Advantages and limitations of on-line solid phase extraction coupled to liquid chromatography–mass spectrometry technologies versus biosensors for monitoring of emerging contaminants in water. *J. Chromatogr. A* **2007**, *1152*, 97–115. [CrossRef]
81. Robles-Molina, J.; Lara-Ortega, F.J.; Gilbert-López, B.; García-Reyes, J.F.; Molina-Díaz, A. Multi-residue method for the determination of over 400 priority and emerging pollutants in water and wastewater by solid-phase extraction and liquid chromatography-time-of-flight mass spectrometry. *J. Chromatogr. A* **2014**, *1350*, 30–43. [CrossRef]
82. Paszkiewicz, M.; Godlewska, K.; Lis, H.; Caban, M.; Białk-Bielińska, A.; Stepnowski, P. Advances in suspect screening and non-target analysis of polar emerging contaminants in the environmental monitoring. *Trends Anal. Chem.* **2022**, *154*, 116671. [CrossRef]
83. Gong, J.; Ran, Y.; Chen, D.; Yang, Y. Typical Endocrine Disruptors in Rivers of the Pearl River Delta. In Proceedings of the 13th Annual Meeting of the Chinese Society of Mineralogy, Petrology and Geochemistry, Guangzhou, China, 7 April 2011; Chinese Society of Mineralogy, Petrology and Geochemistry: Guiyang, China, 2011; p. 1.
84. Pan, G.; Zhou, Q.; Luan, X.; Fu, Q. Distribution of perfluorinated compounds in Lake Taihu (China): Impact to human health and water standards. *Sci. Total Environ.* **2014**, *487*, 778–784. [CrossRef]
85. Guo, C.; Zhang, Y.; Xu, J.; Li, L. Levels and Distribution Characteristics of Perfluorinated Compounds in the Water Environment of Taihu Lake and Dianchi Lake. In Proceedings of the 6th National Environmental Chemistry Conference and Exhibition of Environmental Science Instruments and Analytical Instruments, Shanghai, China, 22–24 September 2011; Chinese Chemical Society (Environmental Chemistry Committee of the Chinese Chemical Society, Environmental Chemistry Branch of the Chinese Society for Environmental Sciences, Analytical Toxicology Committee of the Chinese Society of Toxicology): Nanjing, China, 2011; p. 1.
86. Hou, D. Distribution Characteristics and Ecological Effects of Persistent Toxic Substances in Western Inner Mongolia. Doctoral Dissertation, Inner Mongolia University, Inner Mongolia, China, 2014. [CrossRef]
87. Chen, X. In Vitro Biological Toxicity Effects and Ecological Risk Assessment of Water and Sediment in the Yangtze and Dongjiang River Basins. Master's Thesis, University of Chinese Academy of Sciences, Beijing, China, 2016.
88. Zhu, B.; Gao, Z.; Wang, J.; Hu, G.; Yu, N.; Wei, S. Pollution and Risk Assessment of Polybrominated Diphenyl Ethers and Organic Phosphorus Flame Retardants in the Major Inflow Rivers of Lake Taihu. *Environ. Monit. Forewarning* **2020**, *12*, 8.
89. Wu, X. Study on the Pollution Characteristics and Water-Air Exchange of Polycyclic Aromatic Hydrocarbons and Organic Flame Retardants in the Nearshore Area of Dalian. Master's Thesis, Dalian University of Technology, Liaoning, China, 2018.
90. Sun, D.; Gong, P.; Wang, X.; Wang, C. Characterisation of spatial and temporal distribution of perfluorinated compounds in Lhasa River. *Chin. Environ. Sci.* **2018**, *38*, 4298–4306. [CrossRef]
91. Shi, R.; Mao, R.; Zhang, M.; Lu, Y.; Song, S.; Zhao, J. Distribution, sources and ecological risk of perfluorinated compounds in surface water of the Ulansuhai Lake Basin. *Environ. Sci.* **2021**, *42*, 663–672. [CrossRef]
92. Wang, T.; Khim, J.S.; Chen, C.; Naile, J.E.; Lu, Y.; Kannan, K.; Park, J.; Luo, W.; Jiao, W.; Hu, W.; et al. Perfluorinated compounds in surface waters from Northern China: Comparison to level of industrialization. *Environ. Int.* **2012**, *42*, 37–46. [CrossRef] [PubMed]
93. Zhang, S.; Pan, X.; Lin, L.; Tao, J.; Liu, M. Preliminary investigation on the composition and distribution characteristics of microplastics in the water bodies of the Yangtze River source area. *J. Yangtze River Sci. Res. Inst.* **2021**, *38*, 12–18. [CrossRef]
94. Wang, G.; Lu, J.; Tong, Y.; Liu, Z.; Zhou, H.; Xiayihazi, N. Occurrence and pollution characteristics of microplastics in surface water of the Manas River Basin, China. *Sci. Total Environ.* **2020**, *710*, 136099. [CrossRef]
95. Han, M.; Niu, X.; Tang, M.; Zhang, B.; Wang, G.; Yue, W.; Kong, X.; Zhu, J. Distribution of microplastics in surface water of the lower Yellow River near estuary. *Sci. Total Environ.* **2020**, *707*, 135601. [CrossRef]
96. Xiong, X.; Zhang, K.; Chen, X.; Shi, H.; Luo, Z.; Wu, C. Sources and distribution of microplastics in China's largest inland lake–Qinghai Lake. *Environ. Pollut.* **2018**, *235*, 899–906. [CrossRef]
97. Dai, L.; Wang, Z.; Guo, T.; Hu, L.; Chen, Y.; Chen, C.; Yu, G.; Ma, L.; Chen, J. Pollution characteristics and source analysis of microplastics in the Qiantang River in southeastern China. *Chemosphere* **2022**, *293*, 133576. [CrossRef]
98. Yuan, W.; Christie-Oleza, J.A.; Xu, E.G.; Li, J.; Zhang, H.; Wang, W.; Lin, L.; Zhang, W.; Yang, Y. Environmental fate of microplastics in the world's third-largest river: Basin-wide investigation and microplastic community analysis. *Water Res.* **2022**, *210*, 118002. [CrossRef]
99. Zhang, Y.; Peng, Y.; Xu, S.; Zhang, S.; Zhou, G.; Yang, J.; Li, H.; Zhang, J. Distribution characteristics of microplastics in urban rivers in Chengdu city: The influence of land-use type and population and related suggestions. *Sci. Total Environ.* **2022**, *846*, 157411. [CrossRef]
100. Qu, J.; Wu, P.; Pan, G.; Li, J.; Jin, H. Microplastics in seawater, sediment, and organisms from Hangzhou Bay. *Mar. Pollut. Bull.* **2022**, *181*, 113940. [CrossRef]
101. Yu, X.; Zhao, Y.; Zhang, C.; Yang, C.; Ouyang, Z.; Liu, P.; Guo, X.; Zhu, L. Abundance and characteristics of microplastics in the surface water and sediment of parks in Xi'an city, Northwest China. *Sci. Total Environ.* **2022**, *806*, 150953. [CrossRef] [PubMed]
102. An, L.; Cui, T.; Zhang, Y.; Liu, H. A case study on small-size microplastics in water and snails in an urban river. *Sci. Total Environ.* **2022**, *847*, 157461. [CrossRef] [PubMed]
103. Li, Y.; Zhang, Y.; Chen, G.; Xu, K.; Gong, H.; Huang, K.; Yan, M.; Wang, J. Microplastics in surface waters and sediments from Guangdong Coastal Areas, South China. *Sustainability* **2021**, *13*, 2691. [CrossRef]

104. Huang, D.; Li, X.; Ouyang, Z.; Zhao, X.; Wu, R.; Zhang, C.; Lin, C.; Li, Y.; Guo, X. The occurrence and abundance of microplastics in surface water and sediment of the West River downstream, in the south of China. *Sci. Total Environ.* **2021**, *756*, 143857. [CrossRef]
105. Li, J.; Ouyang, Z.; Liu, P.; Zhao, X.; Wu, R.; Zhang, C.; Lin, C.; Li, Y.; Guo, X. Distribution and characteristics of microplastics in the basin of Chishui River in Renhuai, China. *Sci. Total Environ.* **2021**, *773*, 145591. [CrossRef]
106. Lam, T.W.L.; Fok, L.; Lin, L.; Xie, Q.; Li, H.; Xu, X.; Yeung, L.C. Spatial variation of floatable plastic debris and microplastics in the Pearl River Estuary, South China. *Mar. Pollut. Bull.* **2020**, *158*, 111383. [CrossRef]
107. Jian, M.; Zhang, Y.; Yang, W.; Zhou, L.; Liu, S.; Xu, E. Occurrence and distribution of microplastics in China's largest freshwater lake system. *Chemosphere* **2020**, *261*, 128186. [CrossRef]
108. Mao, Y.; Li, H.; Gu, W.; Yang, G.; Liu, Y.; He, Q. Distribution and characteristics of microplastics in the Yulin River, China: Role of environmental and spatial factors. *Environ. Pollut.* **2020**, *265*, 115033. [CrossRef]
109. Liu, Y.; Zhang, J.; Cai, C.; He, Y.; Chen, L.; Xiong, X.; Huang, H.; Tao, S.; Liu, W. Occurrence and characteristics of microplastics in the Haihe River: An investigation of a seagoing river flowing through a megacity in northern China. *Environ. Pollut.* **2020**, *262*, 114261. [CrossRef]
110. Wu, P.; Tang, Y.; Dang, M.; Wang, S.; Jin, H.; Liu, Y.; Jing, H.; Zheng, C.; Yi, S.; Cai, Z. Spatial-temporal distribution of microplastics in surface water and sediments of Maozhou River within Guangdong-Hong Kong-Macao Greater Bay Area. *Sci. Total Environ.* **2020**, *717*, 135187. [CrossRef]
111. Zhou, G.; Wang, Q.; Zhang, J.; Li, Q.; Wang, Y.; Wang, M.; Huang, X. Distribution and characteristics of microplastics in urban waters of seven cities in the Tuojiang River basin, China. *Environ. Pollut.* **2020**, *189*, 109893. [CrossRef] [PubMed]
112. Wang, T.; Hu, M.; Song, L.; Yu, J.; Liu, R.; Wang, S.; Wang, Z.; Sokolova, I.M.; Huang, W.; Wang, Y. Coastal zone use influences the spatial distribution of microplastics in Hangzhou Bay, China. *Environ. Pollut.* **2020**, *266*, 115137. [CrossRef] [PubMed]
113. Mao, R.; Hu, Y.; Zhang, S.; Wu, R.; Guo, X. Microplastics in the surface water of Wuliangsuhai Lake, northern China. *Sci. Total Environ.* **2020**, *723*, 137820. [CrossRef] [PubMed]
114. Huang, Y.; Yan, M.; Xu, K.; Nie, H.; Gong, H.; Wang, J. Distribution characteristics of microplastics in Zhubi Reef from South China Sea. *Environ. Pollut.* **2019**, *255*, 113133. [CrossRef]
115. Fan, Y.; Zheng, K.; Zhu, Z.; Chen, G.; Peng, X. Distribution, sedimentary record, and persistence of microplastics in the Pearl River catchment, China. *Environ. Pollut.* **2019**, *251*, 862–870. [CrossRef]
116. Zhang, J.; Zhang, C.; Deng, Y.; Wang, R.; Ma, E.; Wang, J.; Bai, J.; Wu, J.; Zhou, Y. Microplastics in the surface water of small-scale estuaries in Shanghai. *Mar. Pollut. Bull.* **2019**, *149*, 110569. [CrossRef]
117. Ding, L.; Mao, R.f.; Guo, X.; Yang, X.; Zhang, Q.; Yang, C. Microplastics in surface waters and sediments of the Wei River, in the northwest of China. *Sci. Total Environ.* **2019**, *667*, 427–434. [CrossRef]
118. Ding, J.; Jiang, F.; Li, J.; Wang, Z.; Sun, C.; Wang, Z.; Fu, L.; Ding, N.X.; He, C. Microplastics in the coral reef systems from Xisha Islands of South China Sea. *Environ. Sci. Technol.* **2019**, *53*, 8036–8046. [CrossRef]
119. Di, M.; Wang, J. Microplastics in surface waters and sediments of the Three Gorges Reservoir, China. *Sci. Total Environ.* **2018**, *616*, 1620–1627. [CrossRef]
120. Sun, X.; Liang, J.; Zhu, M.; Zhao, Y.; Zhang, B. Microplastics in seawater and zooplankton from the Yellow Sea. *Environ. Pollut.* **2018**, *242*, 585–595. [CrossRef]
121. Wang, W.; Yuan, W.; Chen, Y.; Wang, J. Microplastics in surface waters of dongting lake and hong lake, China. *Sci. Total Environ.* **2018**, *633*, 539–545. [CrossRef] [PubMed]
122. Hu, L.; Chernick, M.; Hinton, D.E.; Shi, H. Microplastics in small waterbodies and tadpoles from Yangtze River Delta, China. *Environ. Sci. Technol.* **2018**, *52*, 8885–8893. [CrossRef] [PubMed]
123. Su, L.; Xue, Y.; Li, L.; Yang, D.; Kolandhasamy, P.; Li, D.; Shi, H. Microplastics in taihu lake, China. *Environ. Pollut.* **2016**, *216*, 711–719. [CrossRef] [PubMed]
124. Zhao, S.; Zhu, L.; Wang, T.; Li, D. Suspended microplastics in the surface water of the Yangtze Estuary System, China: First observations on occurrence, distribution. *Mar. Pollut. Bull.* **2014**, *86*, 562–568. [CrossRef]
125. Cheng, J.; Guo, X.; Li, Y. Characteristics of microplastic pollution in Forty Mile Bay in the northern Yellow Sea. *Mar. Environ. Sci.* **2021**, *40*, 1–7. [CrossRef]
126. Zhao, H.; Gui, F.; Zhao, S. Characteristics of microplastic pollution in aquaculture waters and nearshore tidal flats. *Anhui Agric. Sci.* **2019**, *47*, 55–58. [CrossRef]
127. Luo, Y.; Lin, Q.; Jia, F.; Xu, G.; Li, F. Distribution characteristics of microplastics in four seawater bathing beaches in Qingdao. *Environ. Sci.* **2019**, *40*, 2631–2638. [CrossRef]
128. Wang, Z.; Yang, F.; Yang, W.; Li, W.; Yang, J.; Qin, Y.; Li, H. Occurrence characteristics and mass estimation of microplastics in drainage ditches of Hetao Irrigation District in Inner Mongolia. *Environ. Sci.* **2020**, *41*, 4590–4598. [CrossRef]
129. Li, Z.; Gao, C.; Yang, J.; Wu, L.; Zhang, S.; Liu, Y.; Jin, D. Distribution characteristics of microplastics in surface waters and sediments of Haizhou Bay, Lianyungang. *Environ. Sci.* **2020**, *41*, 3212–3221. [CrossRef]
130. Liu, S.; Jian, M.; Zhou, L.; Li, W.; Wu, X.; Rao, D. Characteristics of microplastic pollution in the habitat of migratory birds in Poyang Lake wetland. *Environ. Sci.* **2019**, *40*, 2639–2646. [CrossRef]

131. Yin, S.; Jia, F.; Liu, X.; Li, F. Distribution of microplastics in surface seawater and tidal flats sediments near Qingdao and their influencing factors. *J. Environ. Sci.* **2021**, *41*, 1410–1418. [CrossRef]
132. Zhang, Q.; Diao, X.; Xie, J.; Chen, Y. Distribution characteristics of microplastics in water and sediments from mariculture areas in eastern Hainan. *J. Hainan Univ. (Nat. Sci. Ed.).* **2020**, *38*, 159–165. [CrossRef]
133. Xiong, K.; Zhao, X.; Zhou, Q.; Fu, C.; Tu, C.; Li, L.; Luo, Y. Characteristics of microplastic pollution in water and sediment in Sanggou Bay, Yellow Sea. *Mar. Environ. Sci.* **2019**, *38*, 198–204. [CrossRef]
134. Hu, D. Pollution Characteristics of Microplastics in Dongting Lake and Their Contribution to Total Organic Carbon. Master's Thesis, Hunan University, Changsha, China, 2020. [CrossRef]
135. Xiong, K. Research on the Characteristics of Microplastic Pollution and Its Influencing Factors in Sanggou Bay, Yellow Sea. Master's Thesis, Zhejiang Ocean University, Zhoushan, China, 2019. [CrossRef]
136. Dai, Z. Research on the Distribution of Microplastics and Their Influencing Factors in the Bohai Sea. Master's Thesis, University of Chinese Academy of Sciences, Beijing, China, 2018.
137. Wang, J. Research on the Pollution Characteristics of Microplastics and Their Adsorption Behaviours in Beijiang River Water and Sediments. Master's Thesis, Guangdong University of Technology, Guangzhou, China, 2017. [CrossRef]
138. Bai, L. Current Status of Microplastics in the Nearshore Waters of Tianjin and Their Effects on Adsorption and Photolysis of Potassium Penicillin. Master's Thesis, Tianjin University, Tianjin, China, 2019. [CrossRef]
139. Zheng, L.; Liu, Y.; Li, X.; Cao, Y.; Chen, Y.; Tang, J.; Sui, Y. Characteristics of riverine microplastic distribution in Songjiang area of Shanghai. *Mar. Fish.* **2021**, *43*, 241–246. [CrossRef]
140. Yuan, H.; Hou, L.; Liang, Q.; Li, J.; Ren, J. Correlation between microplastic pollution and eutrophication in nearshore waters of Dianchi Lake. *Environ. Sci.* **2021**, *42*, 3166–3175. [CrossRef]
141. Li, J.; Ling, W.; Shen, X.; He, Y.; Xu, X.; Chen, X.; An, L. Characteristics and distribution of microplastics in surface waters of Shuangtaizi River and Daliao River. *J. Ecotoxicol.* **2021**, *16*, 192–199. [CrossRef]
142. Zhou, X.; Zhao, W.; Li, T.; Wu, W.; Guo, Y.; Yang, C. Distribution and compositional characteristics of microplastics in surface waters of nearshore waters of Zhejiang Province. *J. Zhejiang Univ. Agric. Life Sci.* **2021**, *47*, 371–379. [CrossRef]
143. Pan, X.; Lin, L.; Zhang, S.; Zhuo, W.; Tao, J.; Li, D. Characteristics of microplastic composition and distribution in Danjiangkou Reservoir and its inlet tributaries. *Environ. Sci.* **2021**, *42*, 1372–1379. [CrossRef]
144. Wang, Z.; Yang, J.; Yang, F.; Yang, W.; Li, W.; Li, X. Distribution characteristics of microplastics in the Wuliangsu Sea ice cap and their response relationship with salinity and chlorophyll a. *Environ. Sci.* **2021**, *42*, 673–680. [CrossRef]
145. Zhang, X.; Zhang, M.; Wei, A.; Zheng, J.; Fu, D.; Zhou, C.; Li, H.; Han, Z. Composition and characterisation of microplastics in surface seawater of nearshore waters of Jiangsu Province. *Environ. Monit. Early Warn.* **2021**, *13*, 9–13.
146. Ye, Q. Characterisation of microplastic pollution in typical urban water bodies in Nanjing. *Energy Environ. Prot.* **2020**, *34*, 79–83. [CrossRef]
147. Zhang, S.; Lin, L.; Pan, X.; Dong, L. Characteristics of microplastics in water bodies of Han River (Danjiangkou Damxia-Xinglong section). *Environ. Sci. Res.* **2022**, *35*, 1203–1210. [CrossRef]
148. Zhang, Y.; Cao, J.; Ke, T.; Tao, Y.; Wu, W.; Wang, P.; Zhou, M.; Chen, L. Pollution Characteristics and Risk Prediction of Endocrine Disruptors in Lakes of Wuhan. *Toxics* **2022**, *10*, 93. [CrossRef] [PubMed]
149. Wang, S.; Rao, Z.; Guo, F.; Liu, C.; Zhan, N. Characterisation of endocrine disruptors in Wuxi-Changzhou groundwater and health risk assessment. *Environ. Sci.* **2021**, *42*, 166–174. [CrossRef]
150. Xu, Y.; Sun, S.; Lu, N.; Sun, L.; Zhao, Q.; Jia, R. Distributional characteristics of endocrine disruptors in the tributaries of Xiaoqing River (Jinan section). *J. Ecotoxicol.* **2021**, *16*, 178–186. [CrossRef]
151. Qi, K. Preliminary study on the pollution of nonylphenol polyoxyethylene ether and its degradation products in the Yellow River (Lanzhou section). *Leather Prod. Environ. Prot. Technol.* **2021**, *2*, 98–99.
152. X, X.; Gong, J.; L, C.; Zhou, Y.; Du, Y.; Wu, C. Environmental endocrine disruptors and their risks in drinking water sources of Pearl River Delta rivers. *J. Ecol. Environ.* **2020**, *29*, 996–1004. [CrossRef]
153. Huang, W.; Bao, Y.; Hu, X.; Yin, D. Distribution Characteristics and Ecological Risk Assessment of 31 Endocrine Disruptors in the Upstream Water Source of Huangpu River. *Environ. Chem.* **2020**, *39*, 1488–1495. [CrossRef]
154. Zhao, H.; Zhao, S. Investigation on the Pollution Status of Endocrine Disruptors in the Aquaculture Area of Dongji Island. *Anhui Agric. Sci.* **2019**, *47*, 74–77. [CrossRef]
155. Zhou, Y.; Xiong, X.; Nie, Y.; Wu, Y.; Zhang, L.; Gong, J.; Chen, Y.; Xu, L.; Yong, Y.; Lin, C. Distribution Characteristics and Risk Assessment of Environmental Endocrine Disruptors in Chengdu Rivers. In Proceedings of the 10th National Conference on Environmental Chemistry (NCEC2019), Tianjin, China, 15 August 2019; Chinese Chemical Society, Chinese Society for Environmental Sciences: Guangzhou, China, 2011; p. 93.
156. Lin, C.; Gong, J.; Xiong, X.; Zhou, Y.; Wu, C. Pollution and Risk of Corticosteroids in Drinking Water Sources of the Pearl River Delta. *China Environ. Sci.* **2019**, *39*, 4752–4761. [CrossRef]
157. Gong, J.; Lin, C.; Xiong, X.; Chen, D.; Chen, Y.; Zhou, Y.; Wu, C. Contamination and Risks of Environmental Corticosteroids in Rivers of the Pearl River Delta. In Proceedings of the 10th National Conference on Environmental Chemistry (NCEC2019), Tianjin, China, 15 August 2019; Chinese Chemical Society, Chinese Society for Environmental Sciences: Guangzhou, China, 2011; p. 16.
158. Liu, C.; Zhang, W.; Shan, B. Spatial Distribution and Risk Assessment of Endocrine Disruptors in Typical River Sections of the Pearl River Estuary. *J. Environ. Sci.* **2018**, *38*, 115–124. [CrossRef]

159. Wang, S. Distribution Characteristics, Ecological Risks, and Biological Effects of Endocrine Disruptors in the Typical Environment of the Ba River on Hemiculter Leucisculus. Master's Thesis, Northwest A&F University, Xianyang, China, 2018.
160. Fan, J.; Wang, S.; Tang, J.; Dai, Y.; Wang, L.; Long, S.; He, W.; Liu, S.; Wang, J.; Yang, Y. Spatial and Temporal Distribution Characteristics and Environmental Risk Assessment of Six Endocrine Disruptors in the Liuxi River in Guangzhou City. *Environ. Sci.* **2018**, *39*, 1053–1064. [CrossRef]
161. Shi, B.; Wang, Z.; Liu, J.; Chen, Q.; Sun, Q.; Hu, L. Pollution Characteristics of Three Estrogens in Drinking Water Sources of the Jiangsu Section of the Yangtze River. *J. Environ. Sci.* **2018**, *38*, 875–883. [CrossRef]
162. Han, X. Pollution Characteristics and Environmental Risks of UV Absorbers in Road Dust and Typical Water Bodies in Hefei City. Master's Thesis, North China Electric Power University, Beijing, China, 2018.
163. Chen, M.; Guo, M.; Liu, D.; Li, J.; Zhang, S.; Hu, L. Pollution Characteristics of Typical Endocrine Disruptors in the Water and Sediments of Taihu Lake and Its Tributaries. *China Environ. Sci.* **2017**, *37*, 4323–4332. [CrossRef]
164. Diao, P. Pollution and Risk Assessment of Phenolic Endocrine Disrupt in the Water and Aquatic Animals of the Pearl River Estuary. Master's Thesis, Jinan University, Guangzhou, China, 2017.
165. Yang, X.; Zhou, J.; Yu, J.; Li, K.; Luo, J.; Wang, L.; Xu, J. Investigation on the Pollution Status of Nonylphenol, an Endocrine Disruptor, in a River in Southwest China. *Shanxi Med. J.* **2017**, *46*, 2155–2157.
166. Yan, Y. Pollution Characteristics and Ecological Risk Assessment of Phenolic Endocrine Disruptors in Rivers: A Case Study of Zaohe River and Bahe River in Xi'an City. Master's Thesis, Chang'an University, Xi'an, China, 2017.
167. Wang, H. Pollution Distribution and Migration Simulation of Phthalate Esters in Typical Wetlands of Sanjiang Plain. Doctoral Dissertation, Northeast Forestry University, Harbin, China, 2017.
168. Du, J.; Hu, H.; Yang, J.; Li, C.; Han, W.; Zhu, X. Monitoring and Risk Assessment of Endocrine Disruptors in Groundwater in Xuzhou Area. *Environ. Monit. Manag. Technol.* **2016**, *28*, 38–40. [CrossRef]
169. Wang, B.; Dong, F.; Chen, S.; Bai, Y.; Zhu, J.; Dai, Q.; Tan, J.; Fu, X. Pollution Characteristics and Source Analysis of Phenolic EDCs in Panlong River Water and Sediment. *J. Saf. Environ.* **2016**, *16*, 288–294. [CrossRef]
170. Zhao, Y. Research on Coupling Degradation Technology of Advanced Oxidation and Biological Pretreatment for Trace Organic Pollutants in Raw Water of Taihu Lake. Master's Thesis, Southeast University, Nanjing, China, 2016. [CrossRef]
171. Chen, X.; Zhao, J.; Liu, Y.; Jiang, Y.; Yang, Y.; Ying, G. Pollution Characteristics and Ecological Risk of Endocrine Disrupting Chemicals in the Middle and Lower Reaches of the Yangtze River. *Chin. J. Ecotoxicol.* **2016**, *11*, 191–203. [CrossRef]
172. Tan, R. Distribution Characteristics and Risk Assessment of Typical EDCs in the Hun River System. Master's Thesis, Capital University of Economics and Business, Beijing, China, 2016. [CrossRef]
173. Wang, Y. Concentration Levels and Health Risk Assessment of Volatile Halogenated Hydrocarbons and Benzene Series Compounds in the Qiantang River. Master's Thesis, Zhejiang University, Hangzhou, China, 2015.
174. Liu, Z.; Guan, S.; Lin, Y.; Ding, B.; Cha, X.; Deng, Z. Research on Paraben Preservatives in Rivers of Guangzhou City. J. Ecol. *Environ. Sci.* **2015**, *7*, 1197–1201. [CrossRef]
175. Wang, K. Distribution Characteristics and Fate of Typical PPCPs in Surface Water of Shanghai. Master's Thesis, Donghua University,, Shanghai, China, 2015.
176. Guo, W. Establishment and Optimization of Detection Methods for Glucocorticoids in Surface Water and Their Application in the Qinghe River in Beijing. Master's Thesis, Nanjing Normal University, Nanjing, China, 2015. [CrossRef]
177. Xiong, J.; Qian, S.; Xie, Y.; Zhao, Y.; Luo, B.; Yuan, H. Distribution characteristics of endocrine disruptors in the Tuojiang River system during flood season. *China Environ. Monit.* **2014**, *30*, 53–57. [CrossRef]
178. Tang, N. Monitoring of environmental hormones and related substances in the water and sediment of the Songhua River. *Heilongjiang Environ. Bull.* **2014**, *1*, 62–64. [CrossRef]
179. Ji, F. Current Status of Estrogen Pollution in Water Bodies in Hefei and Its Removal in Biotreatment Processes. Master's Thesis, Hefei University of Technology, Hefei, China, 2014. [CrossRef]
180. Chen, R. Distribution Characteristics and Risk Assessment of Nonylphenol and Octylphenol in the Water and Sediment of the Pearl River Estuary. Master's Thesis, Jinan University, Guangzhou, China, 2014.
181. Jiang, N. Distribution and Removal Characteristics of Benzotriazole Pollutants in the Harbin Section of the Songhua River. Master's Thesis, Harbin Institute of Technology, Harbin, China, 2014. [CrossRef]
182. Wen, X. Distribution Patterns of Estrogen and Progestogen in the Water of the Songhua River and Water Treatment Units. Master's Thesis, Harbin Institute of Technology, Harbin, China, 2014. [CrossRef]
183. Wang, L. Distribution and Sources of Bisphenol A and Estradiol in Three Regions of Zhejiang Province. Master's Thesis, Wenzhou Medical University, Wenzhou, China, 2014. [CrossRef]
184. Yang, J. Distribution and Environmental Behavior of Typical Endocrine Disruptors in the Water Environment of the Pearl River Delta. Doctoral Dissertation, University of Academy of Sciences, Beijing, China, 2013.
185. Yue, H.; Yang, Y.; Liu, M.; Nie, M.; Shi, H. Distribution characteristics of endocrine disrupting compounds in water and surface sediments of the Huangpu River source water area. In Proceedings of the 2013 Annual Conference of the Chinese Geographical Society (East China Region), Fuzhou, China, 1 November 2013; Geographical Society of China: Shanghai, China, 2011. p. 9.
186. Gong, D. Research on Multi-Medium Residues and Environmental Behavior of Pyrethroid Pesticides in the Liangtan River Basin. Master's Thesis, Chongqing University, Chongqing, China, 2013. [CrossRef]

187. Zhao, M. Study on the Exposure of Tetrabromobisphenol A in Environmental Media in Erhai Lake. Master's Thesis, Changsha University of Science and Technology, Changsha, China, 2013. [CrossRef]
188. Wang, Z.; Zhang, Y.; Zhang, Y.; Wang, S.; Wu, P.; Huang, J. Spatial distribution and risk assessment of phenolic endocrine disruptors in the Yili River of the Taihu Lake Basin. *Environ. Sci. Res.* **2012**, *25*, 1351–1358. [CrossRef]
189. Shen, J.; Wang, X.; Zhang, Z.; Hu, X. Distribution characteristics of five environmental endocrine disruptors in the Luoshi River. *Environ. Sci. Res.* **2012**, *25*, 495–500. [CrossRef]
190. Lu, P. Investigation and Analysis of Environmental Endocrine Disruptors in the Main and Tributaries of the River in Xi'an. Master's Thesis, Chang'an University, Xi'an, China, 2012.
191. Liu, C. Distribution Characteristics and Ecological Risk Assessment of Typical Phenolic Endocrine Disruptors in the Mouth of the Daliao River. Master's Thesis, Ocean University of China, Qingdao, China, 2012. [CrossRef]
192. Zhang, Y. Investigation of Bisphenol A Pollution in Major Water Bodies, Drinking Water, and Food Contact Materials in Suzhou City and Research on Its Removal Methods. Master's Thesis, Soochow University, Suzhou, China, 2012. [CrossRef]
193. Zhang, Z.; Feng, Y.; Gao, P.; Sun, Q.; Ren, N. Investigation of endocrine disruptors and estrogen activity in Songhua River water. *J. Harbin Inst. Technol.* **2011**, *43*, 58–62.
194. Zhang, Z. Study on the Environmental Exposure Level of Endocrine Disruptors in the Songhua River Basin. Doctoral Dissertation, Harbin Institute of Technology, Harbin, China, 2011. [CrossRef]
195. Pan, X.; Huang, B.; Liu, J.; Wang, B.; Wan, X.; Fang, K.; Wang, R.; Jin, M. Study on endocrine disruptors in the Dianchi Lake water system. In Proceedings of the 6th National Conference on Environmental Chemistry, Shanghai, China, 21 September 2011; Chinese Chemical Society, Chinese Society for Environmental Sciences: Kunming, China, 2011. p. 346.
196. Wu, S. Ecological Risk Assessment and Source Analysis of Steroid Hormones in Liaodong Bay. Master's Thesis, Peking University, Beijing, China, 2011.
197. Zhang, D. Study on the Pollution Characteristics of Antibiotics and Estrogens in the Jiulong River Basin. Master's Thesis, University of Chinese Academy of Sciences, Beijing, China, 2011.
198. Xue, N.; Xu, X.; Liu, X. Distribution and sources of pesticide endocrineors in Guanting Reservoir, Beijing. *Environ. Sci.* **2006**, *27*, 2081–2086. [CrossRef]
199. Shao, X.; Ma, J. Preliminary investigation on 13 endocrine disrupting chemicals in Songhua River water. *J. Environ. Sci.* **2008**, *28*, 1910–1915. [CrossRef]
200. Ma, X.; Gao, N.; Li, Q.; Xu, B.; Le, L.; Wu, J. Investigation on the status of endocrine disrupting chemicals in the raw water and water treatment process of Huangpu River. *China Water Wastewater* **2006**, *22*, 1–4. [CrossRef]
201. Xue, X.; Wu, F.; Deng, N. Investigation and analysis of endocrine disrupting substances in rivers and lakes in Wuhan area. *J. Luoyang Univ.* **2005**, *20*, 33–36. [CrossRef]
202. Cheng, C.; Chen, Z.; Bi, C.; Xu, S.; Wang, D. Analysis of pollution characteristics and health risk assessment of hexachlorocyclohexanes and dioxin-like compounds in the Huangpu River water source in Shanghai. *J. Agro-Environ. Sci.* **2008**, *27*, 705–710. [CrossRef]
203. Zheng, Y. Distribution Characteristics of Typical Endocrine Disruptors in the Main Stream of Haihe River in Vallisneria Spiralis. Master's Thesis, Tianjin University, Tianjin, China, 2009. [CrossRef]
204. Shen, G.; Zhang, Z.; Yu, G.; Li, F.; Li, X. Pollution status of nonylphenols and octylphenols in the Haihe River and Bohai Bay in summer. *China Environ. Sci.* **2005**, *25*, 733–736. [CrossRef]
205. Xu, Y. Overview of Organic Pollution in the Water Bodies of the Main Urban Area of Chongqing in the Upper Reaches of the Yangtze River and Preliminary Exploration of the Environmental Migration Behavior of Typical Organic Pollutants. Master's Thesis, Southwest University, Chongqing, China, 2010. [CrossRef]
206. Y, L. Study on the Pollution Characteristics of Persistent Organic Pollutants in Zhalong Wetland. Master's Thesis, Harbin Engineering University, Harbin, China, 2008. [CrossRef]
207. Dong, J. Pollution of Chlorophenols and Bisphenol A in the Water of the Pearl River Estuary Region and Their Degradation by Iron Oxides. Doctoral Dissertation, Sun Yat-sen University, Guangzhou, China, 2007.
208. Shao, B.; Hu, J.; Yang, M. Investigation on the pollution status of nonylphenol in the water environment of Jialing River and Yangtze River in Chongqing Basin. *J. Environ. Sci.* **2002**, *22*, 12–16. [CrossRef]
209. Sun, P.; Li, Z.; Wang, X.; Cui, W.; Fu, M. Distribution characteristics of nonylphenol in the estuary of the Yellow River. *Coastal* **2007**, *26*, 1722. [CrossRef]
210. Cheng, C.; Chen, Z.; Bi, C.; Xu, S.; Shi, G. Analysis of residues of hexachlorocyclohexane and dichlorodiphenylichloroethane in multi-media in the Huangpu River water source area. *Urb. Environ. Urban Ecol.* **2007**, *20*, 28–31.
211. Zhou, X. Study on the Levels of Estrogens in the Li Village River-Jiaozhou Bay Water Area in Qingdao. Doctoral Dissertation, Ocean University of China, Qingdao, China, 2009. [CrossRef]
212. Zhang, C. Distribution Characteristics of Heavy Metals Such as Hg and As in Shanghai's Drinking Water Sources and Their Risk Assessment. Master's Thesis, East China Normal University, Shanghai, China, 2008. [CrossRef]
213. Fu, M. Concentration Distribution and Preliminary Ecological Risk Assessment of Alkylphenols in Nearshore Marine Anduarine Environments. Master's Thesis, Ocean University of China, Shandong, China, 2007. [CrossRef]
214. Wan, Y. Monitoring of Environmental Hormones in Surface Water in Beijing and Health Risk Assessment. Master's Thesis, Beijing University of Technology, Beijing, China, 2009.

215. Wang, Y. Distribution Characteristics and Ecological Risk Assessment of Tetrabromobisphenol A and Its Brominated Metabolites in the Tributary Watershed of Mi River. Master's Thesis, Shandong Agricultural University, Shandong, China, 2022. [CrossRef]
216. Yang, K. Preliminary Exploration on the Distribution Characteristics and Migration Behavior of Environmental Corticosteroids and Environmental Estrogens in Rivers and Tidal Flats of the Pearl River Delta. Master's Thesis, Guangzhou University, Guangdong, China, 2022. [CrossRef]
217. Hu, J.; Lan, S.; Kang, G.; Gao, Y.; Wen, Z.; Lu, X.; Zhang, X.; Yu, Y. Pollution characteristics, sources, and ecological risks of typical toxic organic pollutants in the Dongjiang River Basin. *Environ. Sci.* **2022**, *42*, 147–155. [CrossRef]
218. Yang, Y.; Lu, G.; Yang, W.; Pan, J.; Gu, B.Z.; Shanxia, S.; Lin, Q. Distribution characteristics of perfluorinated compounds in water environment and biological samples in Shenyang. *J. Environ. Sci.* **2010**, *30*, 2097–2107. [CrossRef]
219. Xu, H.; Zhu, J.; Lei, C.; Xu, X.; Wang, W.; Lu, Y.; Zhang, D. The investigation of perfluorinated compounds in surface waters of the Xixi wetland, Hangzhou, China. *Bull. Environ. Contam. Toxicol.* **2016**, *97*, 770–775. [CrossRef]
220. Chai, J.; Lei, P.; Xia, X.; Xu, G.; Wang, D.; Sun, R.; Gu, J.; Tang, L. Pollution patterns and characteristics of perfluorinated compounds in surface water adjacent potential industrial emission categories of Shanghai, China. *Ecotoxicol. Environ. Saf.* **2017**, *145*, 659–664. [CrossRef]
221. Wu, J.; Junaid, M.; Wang, Z.; Sun, W.; Xu, N. Spatiotemporal distribution, sources and ecological risks of perfluorinated compounds (PFCs) in the Guanlan River from the rapidly urbanizing areas of Shenzhen, China. *Chemosphere* **2020**, *245*, 125637. [CrossRef]
222. Zhou, Z.; Shi, Y.; Li, W.; Xu, L.; Cai, Y. Perfluorinated compounds in surface water and organisms from Baiyangdian Lake in North China: Source profiles, bioaccumulation and potential risk. *Bull. Environ. Contam. Toxicol.* **2012**, *89*, 519–524. [CrossRef]
223. Cao, Y.; Cao, X.; Wang, H.; Wan, Y.; Wang, S. Assessment on the distribution and partitioning of perfluorinated compounds in the water and sediment of Nansi Lake, China. *Environ. Monit. Assess.* **2015**, *187*, 1–9. [CrossRef]
224. Qi, Y.; Huo, S.; Hu, S.; Xi, B.; Su, J.; Tang, Z. Identification, characterization, and human health risk assessment of perfluorinated compounds in groundwater from a suburb of Tianjin, China. *Environ. Earth Sci.* **2016**, *75*, 1–12. [CrossRef]
225. Chen, C.; Wang, T.;; Khim, J.S. Perfluorinated compounds in water and sediment from coastal regions of the northern Bohai Sea, China. *Chem. Ecol.* **2011**, *27*, 165–176. [CrossRef]
226. Dong, W.; Liu, B.; Song, Y.; Zhang, H.; Li, J.; Cui, X. Occurrence and partition of perfluorinated compounds (PFCs) in water and sediment from the Songhua River, China. *Arch. Environ. Contam. Toxicol.* **2018**, *74*, 492–501. [CrossRef]
227. Shi, Y.; Pan, Y.; Wang, J.; Cai, Y. Distribution of perfluorinated compounds in water, sediment, biota and floating plants in Baiyangdian Lake, China. *J. Environ. Monit.* **2012**, *14*, 636–642. [CrossRef] [PubMed]
228. Wang, T.; Chen, C.; Naile, J.E.; Khim, J.S.; Giesy, J.P.; Lu, Y. Perfluorinated compounds in water, sediment and soil from Guanting Reservoir, China. Bull. Environ. *Contam. Toxicol.* **2011**, *87*, 74–79. [CrossRef]
229. Zhang, G.; Pan, Z.; Wu, Y.; Shang, R.; Zhou, X.; Fan, Y. Distribution of perfluorinated compounds in surface water and soil in partial areas of Shandong province, China. *Soil Sediment Contam.* **2019**, *28*, 502–512. [CrossRef]
230. Pan, Y.; Shi, Y.; Wang, J.; Jin, X.; Cai, Y. Pilot investigation of perfluorinated compounds in river water, sediment, soil and fish in Tianjin, China. *Bull. Environ. Contam. Toxicol.* **2011**, *87*, 152–157. [CrossRef]
231. Wang, T.; Lu, Y.; Chen, C.; Naile, J.E.; Khim, J.S.; Park, J.; Luo, W.; Jiao, W.; Hu, W.; Giesy, J.P. Perfluorinated compounds in estuarine and coastal areas of north Bohai Sea, China. *Mar. Pollut. Bull.* **2011**, *62*, 1905–1914. [CrossRef]
232. Chen, S.; Jiao, X.; Gai, N.; Li, X.; Wang, X.; Lu, G.; Piao, H.; Rao, Z.; Yang, Y. Perfluorinated compounds in soil, surface water, and groundwater from rural areas in eastern China. *Environ. Pollut.* **2016**, *211*, 124–131. [CrossRef]
233. Li, F.; Sun, H.; Hao, Z.; He, N.; Zhao, L.; Zhang, T.; Sun, T. Perfluorinated compounds in Haihe River and Dagu drainage canal in Tianjin, China. *Chemosphere* **2011**, *84*, 265–271. [CrossRef]
234. Yang, L.; Zhu, L.; Liu, Z. Occurrence and partition of perfluorinated compounds in water and sediment from Liao River and Taihu Lake, China. *Chemosphere* **2011**, *83*, 806–814. [CrossRef] [PubMed]
235. So, M.K.; Taniyasu, S.; Yamashita, N.; Giesy, J.P.; Zheng, J.; Fang, Z.; Im, S.H.; Lam, P.K.S. Perfluorinated compounds in coastal waters of Hong Kong, South China, and Korea. *Environ. Sci. Technol.* **2004**, *38*, 4056–4063. [CrossRef] [PubMed]
236. So, M.K.; Miyake, Y.; Yeung, W.Y.; Ho, Y.M.; Taniyasu, S.; Rostkowski, P.; Yamashita, N.; Zhou, B.S.; Shi, X.J.; Wang, J.X.; et al. Perfluorinated compounds in the Pearl river and Yangtze river of China. *Chemosphere* **2007**, *68*, 2085–2095. [CrossRef]
237. Sun, H.; Li, F.; Zhang, T.; Zhang, X.; He, N.; Song, Q.; Zhao, L.; Sun, L.; Sun, T. Perfluorinated compounds in surface waters and WWTPs in Shenyang, China: Mass flows and source analysis. *Water Res.* **2011**, *45*, 4483–4490. [CrossRef] [PubMed]
238. Fang, S.; Li, C.; Bian, Y.; Wang, D.; Hao, Y.; Yin, H.; Sun, J. Pollution characteristics and emission fluxes of perfluorinated compounds in the Minjiang River Basin. Chin. *Environ. Sci.* **2019**, *39*, 2983–2989. [CrossRef]
239. Park, H.; Chen, S.; Jiao, X.; Gai, N.; Yin, X.; Lu, G.; Wang, X.; Tan, K.; Pan, J. Distribution of perfluorinated compounds in water bodies during the abundant water period of the Grand Canal. *Chin. Environ. Sci.* **2016**, *36*, 3040–3047. [CrossRef]
240. Qiao, X.; Zhao, X.; Guo, R.; Wang, X.; Hao, S.; Li, X.; Liu, Y. Distribution characteristics and risk evaluation of perfluorinated compounds in the water environment of typical karst areas. *Environ. Sci. Res.* **2019**, *32*, 2148–2156. [CrossRef]
241. Li, J.; Gao, Y.; Wang, Z.; Wang, B.; Hao, H.; Xu, Y.; Zhu, T.; Xu, N.; Ni, J. Risk assessment of perfluorinated compounds in Han River water and sediments. *J. Peking Univ. (Nat. Sci. Ed.)* **2017**, *53*, 913–920. [CrossRef]
242. Wu, Q.; Wu, Q.; Song, S.; Ren, J.; Yang, S.; Wu, Y. Spatial distribution, sources and risk assessment of perfluorinated compounds in major rivers and soils in Tianjin. *Environ. Sci.* **2021**, *42*, 3682–3694. [CrossRef]

243. Zhang, H.; Wang, S.; Yu, Y. Concentrations of typical perfluorinated compounds and pollution contributions of their precursors in river water of Le'an River. *Environ. Sci.* **2020**, *41*, 3204–3211. [CrossRef]
244. Xia, X.; Wu, M.; Xu, G.; Sun, R.; Tang, L. Environmental behavioural properties of perfluorinated compounds in ambient waters around characteristic point sources in Shanghai. *J. Shanghai Univ. (Nat. Sci. Ed.)* **2019**, *25*, 266–274. [CrossRef]
245. Liu, B.; Zhang, H.; Xie, L.; Liu, G.; Wang, Y.; Wang, X.; Li, J.; Dong, W. Pollution characteristics of perfluorinated compounds in the nearshore waters of Shenzhen. *Environ. Sci.* **2015**, *6*, 2028–2037. [CrossRef]
246. Wang, S.; Cao, X. Characteristics of typical perfluorinated compounds (PFCs) pollution and potential ecological risks in the water environment of Binhai Tourist Resort in Shandong Province. *Environ. Sci.* **2020**, *41*, 5428–5437. [CrossRef]
247. Du, G.; Jiang, X.; Zhuo, L.; Shi, Y.; Ren, M.; Cai, F.; Zheng, J.; Zhuang, X.; Luo, W. Pollution characteristics and risk evaluation of perfluorinated compounds in water bodies of Chongqing section of Yangtze River Basin. *J. Ecotoxicol.* **2019**, *28*, 2266–2272. [CrossRef]
248. Zhou, Z.; Hu, Y.; Shi, Y.; Cai, Y.; Liang, Y. Perfluorinated compounds (PFCs) pollution levels and their distribution characteristics in the water environment of Wuhan. *J. Ecotoxicol.* **2017**, *12*, 425–433.
249. Gao, J.; Li, W.; Li, G.; Huang, J.; Yu, G. A preliminary study on the contamination level of perfluorinated compounds in groundwater in some areas of Beijing. *J. Ecotoxicol.* **2016**, *11*, 355–363.
250. Zhang, D.; Wang, D.; Zhang, L.; Luo, L. Study on the status of perfluorinated compounds in Meiliang Bay of Taihu Lake. *J. Environ. Sci.* **2012**, *32*, 2978–2985. [CrossRef]
251. Zhang, M.; Tang, C.; Yu, Y.; Xu, J.; Li, H.; Wu, M.; Zhang, W.; Pan, J. Residual levels and distribution characteristics of perfluorinated compounds in surface water of Qiantang River (Hangzhou section). *Environ. Sci.* **2015**, *36*, 4471–4478. [CrossRef]
252. Wei, L.; Shao, X.; Zhang, J.; Wei, Y.; Li, Y.; Ding, G. Pollution level analysis of perfluorinated compounds in water bodies of Shuangtaizi estuary. *J. Environ. Sci.* **2016**, *36*, 1723–1729. [CrossRef]
253. Gong, X.; Li, B.; Liu, Y.; Liu, R.; Song, Y. Pollution level and ecological risk evaluation of typical perfluorinated compounds in water bodies and sediments of the Hun River-Daliao River system. *J. Environ. Sci.* **2015**, *35*, 2177–2184. [CrossRef]
254. Xue, H.; Ding, G.; Zhang, N.; Zhang, J.; Chen, G.; Yao, Z.; Wang, Y.; Ge, L. Study on the pollution level and distribution of perfluoroalkyl compounds in surface seawater of Dalian Bay. *Mar. Environ. Sci.* **2018**, *37*, 252–257. [CrossRef]
255. Wang, S.; Sun, J.; Yang, Y.; Zhang, M. Spatial distribution of perfluoroalkyl acid compounds and their precursor transformations in river water and wastewater plant effluent from a typical tourist city. *Environ. Sci.* **2018**, *39*, 5494–5502. [CrossRef]
256. Zhang, M.; Tang, G.; Zhang, W.; Ma, Q.; Wang, J.; Xu, J.; Chen, F. Study on the pollution level of perfluorinated compounds in surface water of Hangzhou Qingshan Reservoir. In Proceedings of the Chinese Society for Environmental Sciences 2020 Science and Technology Annual Conference, Naijing, China, 21 September 2020; Chinese Society for Environmental Sciences: Hangzhou, China, 2020; pp. 2453–2458.
257. Zheng, J.; Yan, S.; Cai, F.; Luo, W.; Zhuang, X. Perfluorinated compounds pollution characteristics and ecological risk assessment in Chongqing section of Yangtze River Basin. In Proceedings of the 2019 Annual Academic Conference of the Chinese Society of Environmental Science, Xi'an, China, 23 August 2019; Chinese Society for Environmental Sciences: Guangzhou, China, 2019; pp. 3673–3678.
258. Han, T. Distribution Characteristics and Source Analysis of Perfluorinated Compounds in Seawater and Organisms in Jiaozhou Bay. Master's Thesis, The First Institute of Oceanography, State Oceanic Administration, Shangdong, China, 2018.
259. Xue, H. Study on the Distribution and Allocation Behaviour of Perfluorinated Compound Pollution in Dalian Bay. Master's Thesis, Dalian Maritime University, Liaoning, China, 2018. [CrossRef]
260. Wan, Y. Characterisation of Spatial and Temporal Changes of Perfluorinated Compounds in Different Water Body Environments in Qingdao. Master's Thesis, Qufu Normal University, Shandong, China, 2017.
261. Liu, S. Exposure Characteristics and Risk Evaluation of Perfluorinated Compounds in Water Sources in Beijing. Master's Thesis, Northwest University, Shaanxi, China, 2017.
262. Wang, S.; Zhang, L.; Zhang, X.; Sun, J.; Li, W.; Guo, F.; Wu, S. Characteristics of spatial distribution of perfluoroalkyl acid compounds and their precursors in the water body of Si River and effluent of sewage plant. *J. Environ. Sci.* **2018**, *38*, 1549–1557. [CrossRef]
263. Cai, Y. Pollution Distribution and Risk Assessment of Perfluorinated Compounds in the Jiulong River Basin-Estuary System. Master's Thesis, Xiamen University, Fujian, China, 2016.
264. Fang, S.; Wang, K.; Yin, H.; Yang, Y.; Ye, Z. Pollution characteristics and health risk evaluation of perfluorinated compounds in the water phase of Minjiang River Basin. In Proceedings of the 2016 National Advanced Symposium on Water Environment Pollution Control and Ecological Remediation Technology Proceedings, Tengchong, China, 14 December 2016; Chinese Society for Environmental Sciences, Chinese Research Academy of Environmental Sciences: Chengdu, China, 2016; pp. 130–135.
265. Wei, L. Study on the Distribution and Risk Assessment of Perfluorinated Compounds (PFCs) Pollution in the Water Body of Shuangtaizi Estuary. Master's Thesis, Dalian Maritime University, Liaoning, China, 2015. [CrossRef]
266. Gong, X. Pollution Status and Adsorption Characteristics of Typical Perfluorinated Compounds in Water Bodies and Sediments of the Hun River-Daliao River System. Master's Thesis, Shandong University of Technology, Shandong, China, 2015. [CrossRef]
267. Pan, C. Pollution Characteristics of Perfluorinated Compounds in Water Environment and Drinking Water of Typical Rivers in China and Their Exposure Risks. Doctoral Dissertation, University of Chinese Academy of Sciences, Beijing, China, 2014.

268. Pan, C.; Ying, G.; Zhang, Q.; Chen, Z.; Liu, Y.; Peng, F. Characterisation of perfluorinated compounds in five typical rivers in the Pearl River Delta. In Proceedings of the Seventh National Conference on Environmental Chemistry, Guiyang, China, 23 September 2013; Chinese Chemical Society, Chinese Society for Environmental Sciences: Guangzhou, China, 2013; pp. 230–231.
269. Zhao, Z. Distribution and Transport of Perfluoro/Polyfluoroalkyl Compounds in Coastal Zone Environments. Doctoral Dissertation, University of Chinese Academy of Sciences, Beijing, China, 2013.
270. Wei, X. Characteristics of Perfluorinated Compounds in Waste Leachate and River Water in the Three Gorges Reservoir Area. Master's Thesis, Chongqing University, Chongqing, China, 2019.
271. Zhou, Z.; Shi, Y.; Cai, Y. *Study on the pollution level of perfluorinated compounds in Wuhan Townsend Lake area. Proceedings of the 28th Annual Meeting of the Chinese Chemical Society, Chengdu, China, 13 April 2012*; Chinese Chemical Society: Beijing, China, 2012; p. 1.
272. Du, X. Presence Level and Distribution of Semi-Volatile Organic Pollutants and Perfluorinated Compounds in the Water Bodies of the Yangtze River Estuary. Master's Thesis, Dalian University of Technology, Liaoning, China, 2013.
273. Wang, X.; Zhang, H.; Wang, Y.; Luo, J. Perfluoroalkyl acid contamination levels and drinking water exposure risk in seven Chinese watersheds. *Environ. Sci.* **2018**, *2*, 703–710. [CrossRef]
274. Wang, Q. Characterisation of the Distribution of Perfluoroalkyl Compounds in Several Specific Regions of China. Master's Thesis, Nankai University, Tianjing, China, 2019.
275. Li, F.; Sun, H.; Yang, J.; Luo, J.; He, N.; Zhang, X. Distribution, source and mass flow analyses of perfluorinated compounds (PFCs) in surface water of Hun River and Shenyang urban area. In Proceedings of the POPs Forum 2010 and the Fifth National Symposium on Persistent Organic Pollutants, Nanjing, China, 1 May 2010; Chinese Chemical Society, Chinese Society for Environmental Sciences: Tianjing, China, 2010; pp. 74–75.
276. Ye, X. Distribution Characteristics and Sources of Polyfluorinated Compounds in the Aojiang, Oujiang, and Wenruitang River Basins of Wenzhou and Fengjiang Street of Taizhou. Master's Thesis, Wenzhou Medical University, Zhejiang, Chian, 2010. [CrossRef]
277. Wang, X. Pollution characteristics of PBDEs/PFASs in groundwater in a typical activity area in the north. *J. Environ. Sci. Technol.* **2017**, *40*, 79–85.
278. Zhou, Y. Research on Pollution Characteristics and Management Countermeasures of Emerging Pollutants in Yongding River-Yanghe River Basin. Master's Thesis, University of Chinese Academy of Sciences, Beijing, China, 2017.
279. Liu, J. Study on the Distribution Characteristics and Bioconcentration Effect of PFASs in Multi-Media in Baiyangdian. Master's Thesis, Hebei Normal University, Hebei, China, 2019.
280. Ju, X.; Jin, Y.; SITO, K. *Investigation on the Current Status of PFOS and PFOA Pollution in Coastal Surface Water of Dalian. Persistent Organic Pollutants Forum 2007 and Proceedings of the Second National Symposium on Persistent Organic Pollutants, Dalian, China, 1 May 2007*; Chinese Chemical Society: Shenyang, China, 2007; pp. 51–52.
281. Sin, D.; Liu, L.; Fan, H.; Liu, S. Characterisation of persistent organic matter (POP) pollution in water bodies of the Liaohe River Basin. *J. Hefei Coll. (Gen. Ed.)* **2018**, *35*, 31–37. [CrossRef]
282. Chen, D.; Zhang, Z.; Zhao, W.; Li, J.; Jiao, X. Distribution of typical perfluorinated compounds in groundwater and ecological risks in Beijing reclaimed water irrigation area. *Rock and Mineral Testing* **2022**, *41*, 499–510. [CrossRef]
283. Huang, J.; Wu, W.; Huang, T.; Chen, S.; Xiang, S.; Pang, Y. Characteristics, sources and health risk assessment of perfluorinated compounds in surface water and sediments of Lake Luoma. *Environ. Sci.* **2022**, *43*, 3562–3574. [CrossRef]
284. Song, J.; Wang, Y.; Sun, J.; Fang, S. Pollution characterisation and source analysis of typical and emerging perfluorinated/polyfluorinated compounds in the Tuojiang River Basin. *Environ. Sci.* **2022**, *43*, 4522–4531. [CrossRef]
285. Feng, H.; Cheng, Y.; Ruan, Y.; Tsui, M.M.P.; Wang, Q.; Jin, J.; Wu, R.; Zhang, H.; Lam, P.K.S. Occurrence and spatial distribution of legacy and novel brominated flame retardants in seawater and sediment of the South China sea. *Environ. Pollut.* **2021**, *271*, 116324. [CrossRef]
286. Hou, L.; Jiang, J.; Gan, Z.; Dai, Y.; Yang, P.; Yan, Y.; Ding, S.; Su, S.; Bao, X. Spatial distribution of organophosphorus and brominated flame retardants in surface water, sediment, groundwater, and wild fish in Chengdu, China. *Arch. Environ. Contam. Toxicol.* **2019**, *77*, 279–290. [CrossRef] [PubMed]
287. Zhang, J. Research on Pollution Characteristics and Ecological Risk Evaluation of New Brominated Flame Retardants in the Water Environment of Longgang District, Shenzhen. Master's Thesis, Chang'an University, Shaanxi, China, 2019.
288. Yuan, X. Study on the Distribution Characteristics and Adsorption Behaviour of Typical Brominated Flame Retardants in the Shaanxi Section of Weihe River. Master's Thesis, Chang'an University, Shaanxi, China, 2020.
289. Liu, L. Characterisation of Spatial and Temporal Distribution of Currently Used Pesticides and Halogenated Flame Retardants in the Bohai Sea Region. Master's Thesis, University of Chinese Academy of Sciences (Yantai Institute of Coastal Zone Research, Chinese Academy of Sciences), Yantai, China, 2018.
290. Zhuang, Y. Distribution and Source Analysis of Organophosphate Flame Retardants in Taihu Lake and Surrounding River Waters. Master's Thesis, Nanjing University, Nanjing, China, 2019.
291. Wang, R. Preliminary Study on Organophosphate Flame Retardants in Major Coastal Rivers of the Bohai Sea. Master's Thesis, Yantai Institute of Coastal Zone Research, Chinese Academy of Sciences, Beijing, China, 2015.
292. Dai, L.; Zhao, W.; Wu, Y.; Sun, Y.; Zhu, L. Distribution and bioconcentration of tetrabromobisphenol A in the Bohai and Yellow Seas during spring and autumn. *J. Ocean Univ. China (Nat. Sci. Ed.)* **2021**, *51*, 66–72. [CrossRef]

293. Zhang, P.; Li, Y.; Li, J.; Li, Y. Distribution of tetrabromobisphenol A concentration and spatial and temporal distribution characteristics in sediments and water bodies of Chaohu Lake. *Inner Mongol. Sci. Econ.* **2011**, *5*, 51–55. [CrossRef]
294. Xie, Q. Determination of Organohalogen Pollutants in Typical Surface Water in Beijing Urban Area and Study of Their Species Distribution. Master;s Thesis, Chengdu University of Technology, Sichuan, China, 2015.
295. Chao, T. Characterisation and Risk Assessment of Organophosphate Pollution in Natural Water Bodies and Culture Water Bodies of Surrounding Farms in the Adjacent Waters of the Yellow River Estuary. Master's Thesis, Xihua University, Sichuan, China, 2023. [CrossRef]

water

MDPI

Article

Possible Pollution of Surface Water Bodies with Tequila Vinasses

Allan Tejeda [1], Arturo Montoya [1,2], Belkis Sulbarán-Rangel [2] and Florentina Zurita [1,*]

1 Environmental Quality Research Center, Centro Universitario de la Cienega, University of Guadalajara, Ocotlan 47820, Jalisco, Mexico; allan.tejeda@academicos.udg.mx (A.T.); arturo.montoya@academicos.udg.mx (A.M.)

2 Department of Water and Energy, Centro Universitario de Tonala, University of Guadalajara, Tonala 45425, Jalisco, Mexico; belkis.sulbaran@academicos.udg.mx

* Correspondence: florentina.zurita@academicos.udg.mx

Abstract: The aim of this study was to evaluate the water quality in two streams of the Valles region of Jalisco, Mexico and fully determine if they are being used as tequila vinasse disposal sites. Three sampling campaigns were carried out at eight different points of the two streams that run near tequila factories (TFs). Different physicochemical parameters of water quality were measured: chemical oxygen demand (COD); biochemical oxygen demand (BOD_5); total suspended solids (TSSs); total phosphates; fats, oils, and grease (FOG); Kjeldal nitrogen; nitrite; nitrate; pH; conductivity; temperature; dissolved oxygen (DO); and turbidity. Also, the analysis of samples of tequila vinasses (TVs) diluted with tap water were carried out to have a reference for the level of pollution in the streams. Furthermore, due to the fact that COD could be considered the main indicator of pollution with TVs, a linear regression was performed between COD concentrations and the percentage of dilution of TVs (with tap water). A positive correlation was found between these two variables, and based on this analysis, the vinasse content was estimated at each sampling point of the streams. It was found that on average, a volume of 8.5 ± 6.3% and 11.5 ± 4.9% of TVs were present in each sampling point of the Atizcoa and Jarritos Streams, respectively. Additionally, it was found that, in general, the concentration of pollutants increased as the streams passed through the TFs, particularly the Atizcoa Stream. According to the Water National Commission criteria, most of the points would be classified as highly polluted, since they reach concentrations of COD and BOD_5 up to 6590 mg/L and 3775 mg/L, respectively, temperature values up to 37 °C, and DO values of 0.5 mg/L. Therefore, it was confirmed that the streams are being used as tequila vinasse disposal sites. Due to the above, there is an urgent need for tequila companies to implement treatment systems for the vinasse generated, since under current conditions, the monitored streams are practically devoid of aquatic life.

Keywords: turbidity; surface water pollution; tequila industry; Atizcoa stream; Jarritos Stream

Citation: Tejeda, A.; Montoya, A.; Sulbarán-Rangel, B.; Zurita, F. Possible Pollution of Surface Water Bodies with Tequila Vinasses. *Water* **2023**, *15*, 3773. https://doi.org/10.3390/w15213773

Academic Editor: Bommanna Krishnappan

Received: 26 September 2023
Revised: 19 October 2023
Accepted: 26 October 2023
Published: 28 October 2023

1. Introduction

Tequila production is an iconic activity of great social and economic importance for the producing regions of Mexico, particularly for the state of Jalisco [1,2]. However, throughout the tequila production process, a liquid residue is generated at the bottom of the stills during the distillation process of the fermented must known as tequila vinasse [3,4]. In general, the number of tequila factories (TFs) that treat their vinasses prior to discharge is low, since it is mainly large companies that do so [5]. The rest of the companies do not treat their effluents, mainly justifying this decision with the economic limitations (especially micro- and small companies) of paying for the construction, operation, maintenance, and personnel of a conventional wastewater treatment plant [4]. Therefore, the untreated vinasse ends up being discharged into the soils or surface water bodies (rivers, streams, and lakes) located in the tequila production regions, without complying with the requirements established by current Mexican regulations [2,6,7].

Tequila vinasses are characterized by their typical dark reddish-brown color and an alcohol-caramel smell, with high temperatures of up to 90 °C, an acidic pH of approximately 3.4 to 4.5, and high electrical conductivity (2.4 to 5.8 mS/cm) [3,8]. In addition, vinasses contain fats, oils, and grease (FOG) in concentrations of 10 to 100 mg/L, total chemical oxygen demand (COD) from 60,000 to 100,000 mg/L, total biochemical oxygen demand (BOD_5) from 35,000 to 60,000 mg/L, total solids from 25,000 to 50,000 mg/L, phosphates from 100 to 700 mg/L, and total nitrogen from 20 to 50 mg/L, among others [9–11].

These characteristics of tequila vinasses make them a highly dangerous waste for surface aquatic ecosystems if they are discharged without any treatment or with inadequate treatment. Different studies have shown evidence for the negative consequences for the environment of vinasses from other types of industries but with similar characteristics to tequila vinasses [12]. Among them, the evaluation of the toxicity of sugarcane vinasse from the production of refined sugar and ethanol stands out [4]. It was found that vinasse has toxic and cytotoxic potential in fish liver and that this depends on the concentration of vinasse in bodies of water [13,14]. Gunkel et al. [15] evaluated the Ipojuca river in northeastern Brazil, which receives runoff from vinasse irrigation from sugarcane crops, and revealed that vinasse was the main source of contamination in the river, because it causes an increase in the temperature and acidification of the water and increases the turbidity and depletion of dissolved oxygen [13,15].

On the other hand, the Valles region of the State of Jalisco, located in the region with Denomination of Origin to produce Tequila in Mexico [16], stands out as the region with the highest number of tequila factories (44), with the municipality of Tequila having 22 TFs in 2021 [2]. In this way, the main economic activity in both the region and the municipality is the production of tequila [17]. This municipality is part of the Tequila Route in the Agave Landscape that attracts hundreds of both national and international tourists per year [18]. Furthermore, due to the increase in tequila consumption around the world in recent years [16], the number of tequila factories is increasing, which implies a greater generation of tequila vinasses. Due to the physicochemical characteristics of the vinasses and their possibly inadequate disposal, the Valles region is considered an environmental risk area since there are no effective vinasse disposal plans [5].

In addition, in general, given the inappropriate management of wastewater in Mexico, the surface water exhibits different degrees of deterioration. An example of this is the Santiago River, which is considered one of the most polluted and deteriorated rivers in Mexico since it receives constant discharges of domestic and industrial wastewater, with contributions of up to 4.22 ton/day of COD, 1.87 ton/day of BOD_5, and 4.44 tons/day of total suspended solids (TSSs) [19,20]. The Santiago River has a length of 562 km, originates in Lake Chapala, and flows into the Pacific Ocean in the state of Nayarit [21]; part of its route takes place in the Valles region of the state of Jalisco. Tequila vinasses are among the industrial wastewaters that the Santiago River receives, since these are discharged without treatment or with incomplete treatment in different streams that end up flowing into this river [5]. In Mexico, despite the increasing production of tequila [16] and the consequent generation of vinasse, as well as the common perception that it is discharged without treatment, there are no studies that fully demonstrate such discharges and their impacts.

Therefore, the aim of this research was to evaluate and analyze the quality of water in two streams in the municipality of Tequila in the Valles region of the state of Jalisco, Mexico and to comprehensively demonstrate whether or not they are being used as sites of final disposal of tequila vinasses. In this way, we expect to make visible the negative impacts of managing tequila vinasse in a very important tequila-producing region. This study was carried out by measuring water quality parameters at different points in the streams during the dry season and in different samples of tequila vinasse diluted with tap water (as a frame of reference) to analyze the level of contamination in the streams.

2. Materials and Methods

2.1. Study Area

The surface waters monitored were the Atizcoa and Jarritos Streams located in the municipality of Tequila, belonging to the Valles region of the state of Jalisco in Mexico. The Atizcoa Stream originates in the Tequila volcano and runs approximately 16.5 km until it empties into the Santiago River [22] as part of the Lerma Santiago basin (Figure 1). These streams are the most important in the town of Tequila, Jalisco. The Jarritos Stream is smaller than the Atizcoa Stream and there is no information available on where it originates or its source; apparently, it originates within the same town of Tequila, near some tequila factories where the first monitoring point was located.

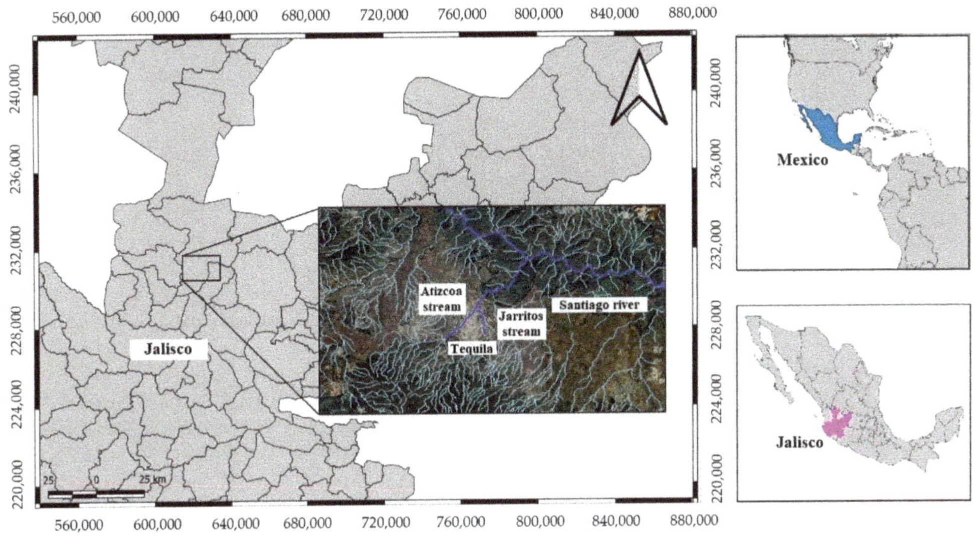

Figure 1. Geographical location of the study area in the municipality of Tequila.

2.2. Monitoring of Atizcoa and Jarritos Stream

The monitoring was carried out in the dry season to avoid the dilution of probable contaminants. Three sampling campaigns were carried out every 15 days (2 February, 16 February and 1 March 2021) at 8 different points in the water bodies identified as the Atizcoa and Jarritos Streams. Figure 2 shows the location of the streams, the sampling points and evidence of the location of tequila factories near the streams. The 8 sampling points were located along the two streams; 4 of them in the Atizcoa Stream, 3 points in the Jarritos and 1 point after the union of both streams. The points were chosen for their proximity to tequila companies and the ease and accessibility of taking water samples. These points were numbered in ascending order in each of the streams as they moved downstream. In the Atizcoa Stream, the 4 points were identified as A1 (located before the stream passed through the tequila factories), A2, A3 (located very close to several tequila factories), and A4 (located after the stream passed through the tequila factories). In the Jarritos Stream, the 3 points were identified as J1, J2 (located near some tequila factories), and J3 (after the stream passed through the factories). Finally, the point after the union of both streams was named AJ.

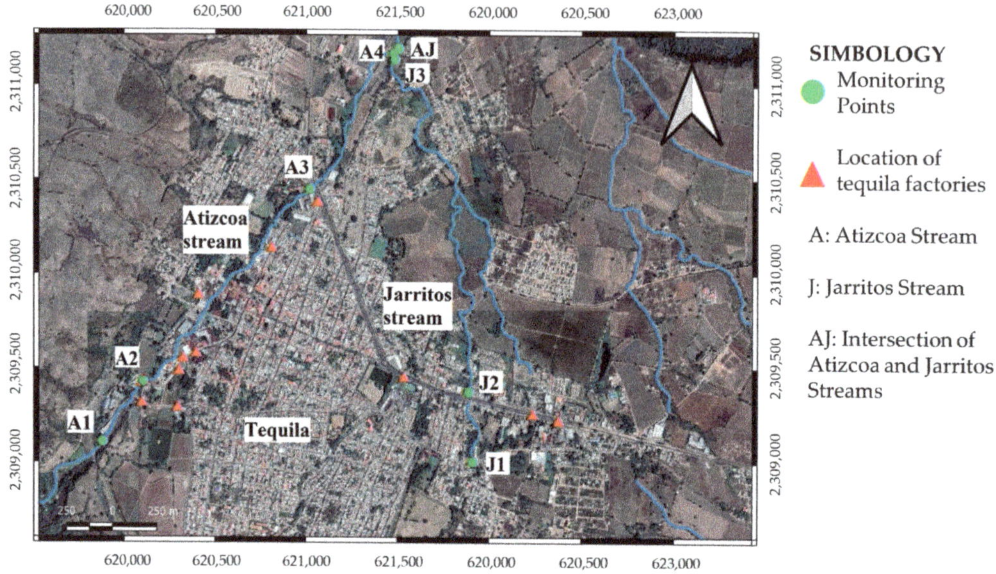

Figure 2. Sampling points in the Atizcoa and Jarritos Streams and evidence of the presence of tequila factories around them in the town of Tequila.

With regard to the monitored parameters, some of them were measured directly in the monitored streams, that is, the measurements were made on-site. These parameters were pH, conductivity, dissolved oxygen, and temperature. For this, a HACH model HQ40d portable meter was used with specific INTELLICAL probes for each of the parameters. The other parameters, that is, turbidity, FOG, total suspended solids (TSSs), total phosphates, nitrite, nitrate, ammonia, organic nitrogen, COD, and BOD_5 were determined at the Environmental Quality Research Center of the University of Guadalajara Campus Ciénega, located in the municipality of Ocotlán, Jalisco. The samples were preserved at 4 °C until processing and the techniques used were based on the Mexican standards, which in turn are based on the standard methods for the analysis of water and wastewater [23]. It is worth mentioning that the parameters were selected from those contaminants usually found in tequila vinasses. Furthermore, to have a reference to compare the concentration of each of the contaminants in the monitored streams, most of these parameters were evaluated in vinasse samples diluted with tap water (TW). Furthermore, the flow rate along the streams was estimated approximately using the well-known velocity/area method, which consists of measuring the mean velocity of the flow and the cross-sectional area of the stream (Flow rate (Q) = A (area of the cross-section transverse) × V (velocity of the water at the surface)) [24].

2.3. Statistical Analysis

Because COD is the parameter that best reflects contamination by tequila vinasse, a linear regression was carried out from control samples (tequila vinasse diluted with tap water) to determine the relationship between vinasse concentration (independent variable) and COD, and then, according to the model, estimate the presence of vinasse in the water courses. Additionally, a randomized block experimental design was used to analyze changes in water quality parameters at each stream sampling point. For the analyses, the response variables were the on-site measurements as well as the laboratory measured quality parameters mentioned in the last section. The treatments were all sampling points along each stream, specifically, A1, A2, A3, A4, and AJ for Atizcoa que Stream and J1, J2, J3, and AJ for Jarritos Stream. The date of the sampling campaign was used as a blocking factor. The linear regression and analysis of variance (ANOVA) were performed using STATGRAPHICS CENTURION XV.II with a significant level of $p = 0.05$. Specifically, when the ANOVA revealed significant differences, multiple comparisons were made using the least significant difference (LSD) test in order to determine the difference between means.

3. Results and Discussion

3.1. Measurements of Pollutants in Diluted Vinasse Samples as a Frame of Reference

Table 1 shows the concentrations of the different contaminants measured in diluted vinasse samples (with tap water). It is evident that contaminant concentrations were very low or absent in tap water and increased dramatically when the percentage of TV increased. In the case of pH, the value in tap water was 8.58 and was reduced to a value close to 4 when the samples contained between 25% and 100% vinasse. Due to the fact that, to the best of our knowledge, this is the first study in which TV discharges to surface waters are evaluated, there is no reference study to compare our results, so we consider these results with diluted vinasses to be suitable as a reference.

Table 1. Pollutant concentrations in samples of diluted tequila vinasses with tap water a.

	Dilutions of Tequila Vinasses with Tap Water				
Parameter	**0% (TW)**	**25%**	**50%**	**75%**	**100%(TV)**
Chemical oxygen demand (mg/L)	13.5 ± 2.1	8850 ± 1131	15,225 ± 176.8	25,250 ± 1484.9	35,150 ± 1202.1
Total Kjeldahl nitrogen (mg/L)	0.22 ± 0.02	125.1 ± 4.8	236.5 ± 8.9	284.0 ± 21.7	426.1 ± 2.8
Nitrate (mg/L)	0.65 ± 0.07	450 ±70.7	655 ± 49.5	880.0 ± 28.3	220 ± 169.7
Nitrite (mg/L)	0.004 ± 0.00	2.1 ± 0.00	2.9 ± 0.40	4.55 ± 0.21	6.05 ± 0.49
Total phosphates (mg/L)	1.65 ± 0.78	100 ± 14.1	190 ± 70.7	335 ± 7.1	385 ± 49.5
Total suspended solids (mg/L)	0.0 ± 0.0	3833 ± 235.7	8917 ± 117.9	12,417 ± 1532	17,833 ± 2121
Turbidity (NTU)	0.0 ± 0.0	1300 ± 118	2363 ± 40.3	4031 ± 298.4	6497 ± 1565.3
pH	8.58± 0.00	3.96 ± 0.00	3.8 ±0.00	3.76 ±0.01	3.74 ± 0.01
Electrical conductivity (μs/cm)	72.03 ± 0.10	905.2 ± 0.57	1403.5 ± 0.71	1928.5 ± 2.12	2372 ± 1.41

On the other hand, COD could be considered the main indicator of the presence of vinasse in water bodies, since its value can increase significantly in water bodies that receive vinasse discharges, as a result of its high values in the raw tequila vinasses, from 60,000 to 100,000 mg/L [2]. Additionally, linear regression models are widely used in environmental study cases in order to determine relationships between specific variables and specific industrial/anthropogenic activities or pollutants [25,26]. Figure 3 shows a positive correlation between vinasse concentration (%) and COD with a r^2 of 0.9955. In addition, the variance analysis shows a p value < 0.05, which means that the variables have a significant statistical relationship. The adjusted regression model resulted in the following equation:

$$COD = -439.6 + 346.692\ (\%\ Vinasse)$$

Based on this equation, the vinasse content was estimated in each of the sampling points of the streams, which were also graphed in Figure 3. In this way, the calculated values show the presence of vinasses in the two streams. For Atizcoa, it was estimated that there is an average vinasse content of 8.46 ± 6.3% in each of the sampling points, while in the Jarritos Stream the vinasse content is slightly higher, that is, 11.86 ± 4.9%.

3.2. Parameters Measured on Site

Figure 3 shows the average results of the parameters measured on site. These were temperature, conductivity, pH, and dissolved oxygen. The temperature (Figure 4a) results showed, in the Atizcoa Stream, a considerable increase at point A2 (37 °C) with respect to A1, with a significant difference ($p < 0.05$); this value was reduced as the stream advanced in its course, but even so, it remained above the value found in A1. In the Jarritos Stream, at point J3 there was also a considerable increase in the temperature value compared to J1 and J2 (from 22 to 28 °C), also with significant differences ($p < 0.05$). These increases in temperature demonstrate the discharges of vinasse, since vinasse is generated at 90 °C [10]. As is known, high temperatures have an impact on the physical and chemical properties of water, especially density, viscosity, solubility of dissolved oxygen, and the speed of chemical and biochemical reactions that could occur in the body of water [27].

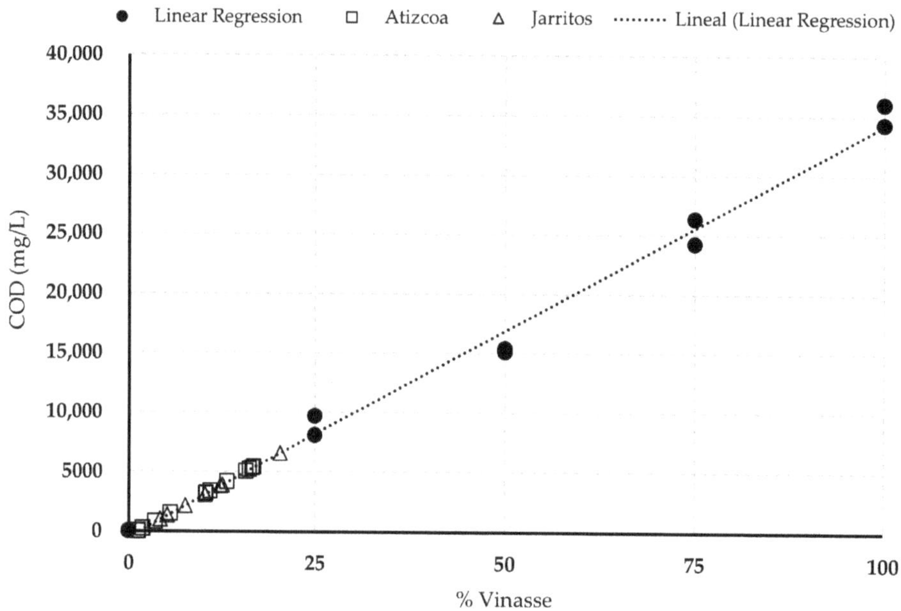

Figure 3. Scatter plot with regression line that shows the relationship between COD and vinasse percentage.

Regarding the pH (Figure 4b), in the Atizcoa Stream, at point A1, the values were in the optimal range for aquatic life (7–7.5) [28]. However, as the stream ran through the other points near the TFs, these values decreased considerably until reaching acidic values (4–5) statistically different in comparison to A1 ($p < 0.05$). Meanwhile, in the Jarritos Stream, slightly acidic pH values (5–6.5) were found at all points, because from the first sampling point there were several TFs that, it seems, discharge their vinasses into the stream. No significant difference ($p > 0.05$) was found between the different sampling points. Consequently, point AJ, which is the union of the two streams, presented acidic pH values (4–5.5), which reflects the presence of vinasse discharges since such values are close to those measured in samples with 25 to 100% vinasse (Table 1). These pH values definitely preclude the existence of aquatic fauna in the two streams, since in general, acute or chronic exposure to acidic values negatively affects their physiological functions; most aquatic animals, including fish, live in a narrow pH range close to neutrality [28].

With respect to the conductivity results (Figure 4c), low values were found in the Atizcoa Stream at point A1, but as the stream ran through the area where the TFs are located (points A2, A3, and A4), these values increased significantly from one point to another ($p < 0.05$), reaching a final average value of 1384.77 ± 63.5 μS/cm at A4; this value is very similar to that found in the dilution of 50% of vinasse with tap water. Regarding the Jarritos Stream, high conductivity values were found at all points (between 750 and 1400 μS/cm) without significant differences ($p < 0.05$). As a result, at the AJ junction point, high values were also found (between 800 and 1400 μS/cm). Such values can only be due to discharges of tequila vinasse, since increases of such magnitude would not be reached if the discharges were domestic wastewater [27]. As the results in Table 1 suggest, tequila vinasses with high electrical conductivity values have the potential to modify the electrical conductivity of uncontaminated waters, such as streams.

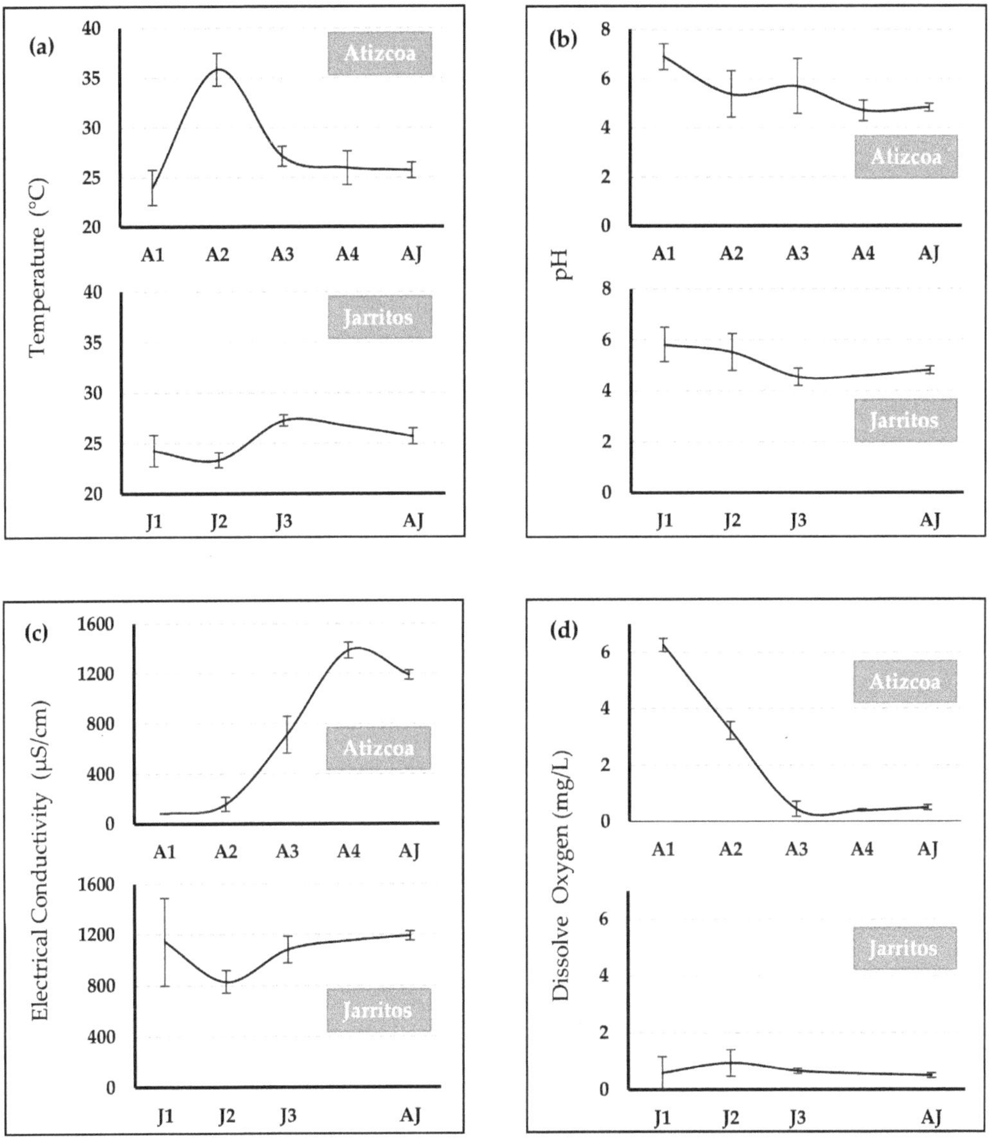

Figure 4. Parameters measured on-site in the Atizcoa and Jarritos Streams (mean ± confidence interval, $p < 0.05$, $n = 3$). (**a**) Temperature; (**b**) pH; (**c**) electrical conductivity; (**d**) dissolved oxygen.

Regarding DO concentrations (Figure 4d), in the Atizcoa Stream, the values were optimal at point A1, but as the stream flowed through the TFs, these values decreased significantly ($p < 0.05$) from 6 to 0.5 mg/L. In contrast, in the Jarritos Stream, at all points the DO concentrations were low, between 1.27 to 0.5 mg/L, without significant difference between them ($p < 0.05$) and, as expected, at point AJ, where both streams join, the DO concentration values were very low (0.5 mg/L). This was to be expected when physically seeing the state of the streams. The noticeable turbidity of the water prevents the sun's rays from penetrating properly, making it impossible for photosynthetic organisms that produce oxygen to be present. The decrease in DO is also due to excess organic matter and high temperatures. The high concentrations of organic matter cause the aerobic microorganisms responsible for its degradation to demand more DO than usual, which leads to its reduction [2,13,20]. In this way, streams develop a toxic environment for aerobic aquatic life, enabling the life of only a small group of anaerobic microorganisms [27].

3.3. Parameters Measured in the Laboratory

The parameters measured in the laboratory are reported in Figure 5 for Atizcoa Stream and Figure 6 for Jarritos Stream.

Regarding the results for COD and BOD_5, which are indicators of the content of organic matter in the bodies of water, it was found that at point A1 of the Atizcoa Stream, the average values were low, that is, 9.7 ± 1.5 mg/L and 7.7 ± 1.9 mg/L, respectively. According to the water quality indices of the Water National Commission (CONAGUA) [1], such values allow the stream to be classified as "acceptable quality" for BOD_5 and "good quality" for COD. However, as the stream flowed through the TFs (points A2, A3, and A4), these values increased significantly ($p < 0.05$) until reaching average values of 4778.5 ± 565.25 mg/L and 2792.5 ± 314.35 for COD and f BOD_5, respectively, at A4. In contrast, in the Jarritos Stream, the concentrations of these pollutants were high in the three sampling points along the sampling campaigns (from 1005 to 6590 mg/L for COD and from 930 to 3346 mg/L for BOD_5) without significant difference between the sampling points ($p < 0.05$). As a result, when the two streams joined, at point AJ, the concentrations were also high (between 3162.5 to 5395 mg/L for COD and between 1919.2 to 3775 mg/L for BOD_5). Such values are unusually high, much higher than the values even for municipal wastewater considered to be of high concentration (400 mg/L for BOD_5 and 1000 mg/L for COD) [29]. According to the estimation using the linear regression model between COD and % vinasse, the two streams had vinasse in different percentages. In general, the polluting potential of vinasses could be up to 100 times higher than that of domestic sewage due mainly due to the low pH, high corrosivity, and BOD_5 concentrations [14]. Evidently, according to the CONAGUA quality indices, the two streams are classified as heavily polluted [1].

With respect to the concentration of FOG, in Atizcoa Stream the average concentrations did not show a significant difference between the sampling points ($p > 0.05$), although the average concentrations were 5.0 ± 3.6 and 56.2 ± 61.0 at A1 and A4, respectively. It is likely that the ANOVA was affected by the small number of samples. In the Jarritos Stream, the three points presented average high concentrations (28.5 ± 12.6 mg/L to 105.8 ± 91.7 mg/L), without significant differences ($p > 0.05$). Point AJ also presented a high concentration of FOG that was in the range of 12.62 to 125.94 mg/L. These values show again that the streams are receiving wastewater discharges, presumably tequila vinasse. According to [2], an average concentration of 119 ± 109 mg/L of FOG was found in the vinasse of 24 tequila factories. It is important to highlight that the presence of FOG in tequila vinasse is due to its content in the agave plant. In agave bagasse (the solid residue after extraction of the cooked juice during the production of tequila) the content of extractives, which includes fats, phenolics, resin acids, waxes, and inorganics, varies between 19 and 57% [30]. FOG are considered a basic contaminant, according to the Official Mexican Standard NOM-001-SEMARNAT 2021, which must be removed or stabilized by conventional processes; in this case, the values found in the two streams were mostly higher than the maximum limits allowed for discharges, which is 15 mg/L [31].

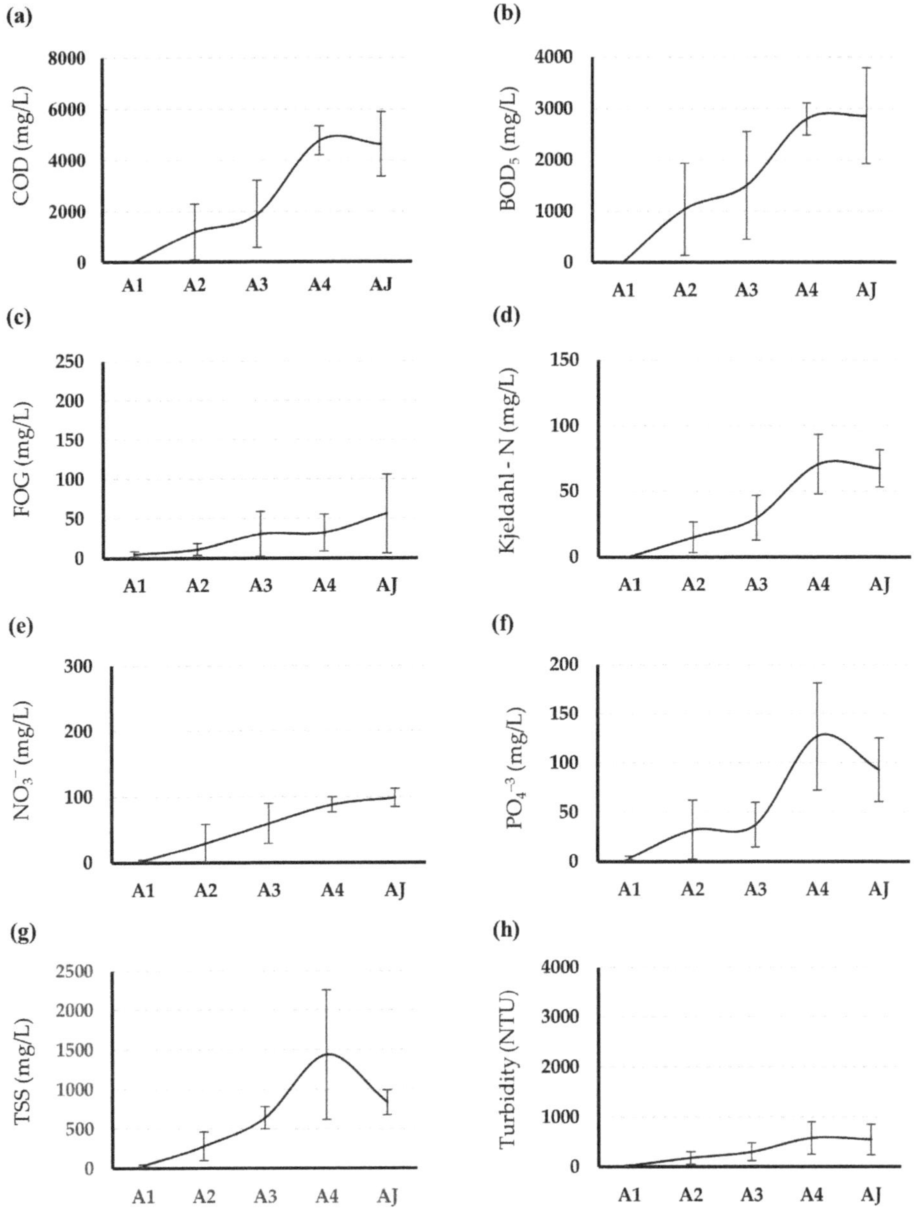

Figure 5. Parameters measured in the laboratory from sampling points of the Atizcoa Stream (mean ± confidence interval, $p < 0.05$, $n = 3$). (**a**) Chemical oxygen demand; (**b**) biochemical oxygen demand; (**c**) fats, oils, and grease; (**d**) Kjeldahl nitrogen; (**e**) nitrate; (**f**) phosphates; (**g**) total suspended solids; (**h**) turbidity.

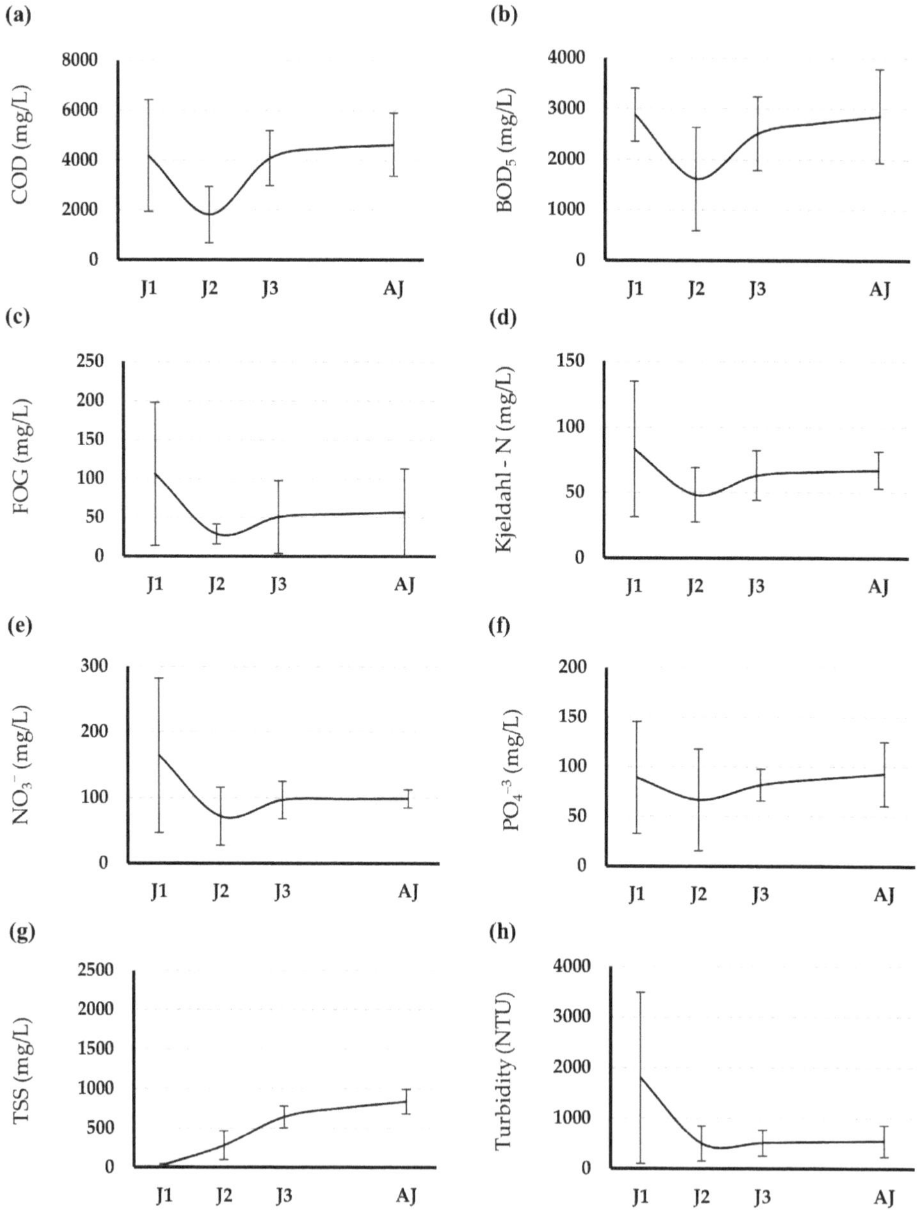

Figure 6. Parameters measured in the laboratory from sampling points of the Jarritos Stream (mean ± confidence interval, $p < 0.05$, $n = 3$). (**a**) Chemical oxygen demand; (**b**) biochemical oxygen demand; (**c**) fats, oils, and grease; (**d**) Kjeldahl nitrogen; (**e**) nitrate; (**f**) phosphates; (**g**) total suspended solids; (**h**) turbidity.

With regard to the results of nitrogen compounds, very high concentrations were found in some monitored points of the streams. Specifically, for Kjeldahl nitrogen, which is the sum of organic nitrogen and ammonia, the concentrations increased significantly from A1 to A4 ($p < 0.5$), reaching an average value of 70.6 mg/L after the stream flowed through the TFs. In the Jarritos Stream, at all points as well as at the AJ junction point, high concentrations were observed (between 28.9 and 131.4 mg/L) in the different sampling campaigns with no significant difference between the sampling points ($p > 0.05$). In this case, some values were similar to those in Table 1 for samples with 25% vinasse. In the case of nitrate, it can be seen in Figure 5 that at point A1 of the Atizcoa Stream, the average concentration was low (2.7 mg/L), but it increased significantly ($p < 0.05$)) as the stream flowed through the TFs. In contrast, in the Jarritos Stream, nitrate concentrations were high at all sampling points without difference ($p > 0.05$). Regarding nitrite, it can be highlighted that, in the eight points, the concentrations were low (between 0.01 and 0.6 mg/L); this was expected, since nitrite is an intermediate compound in the nitrogen transformation reactions [32]. In general, nitrogen is one of the main pollutants that cause eutrophication of surface waters when untreated or poorly treated wastewater is discharged into them [33].

Regarding phosphate concentrations, again, low concentrations were found in A1 of the Atizcoa Stream and increasing concentrations along the course of the stream ($p < 0.05$), as well as permanently high concentrations along the Jarritos Stream ($p > 0.05$). High concentrations of phosphates and nitrate in surface water bodies can cause eutrophication, resulting in the presence of algae and aquatic weeds in water bodies [34]. That being said, in the case of these streams, the high turbidity makes it impossible for sunlight to penetrate, so the presence of algae is not possible. However, when the streams finally flow into the Santiago River, there they can contribute to the eutrophication that is observed in some parts of the river, where it is covered with *Eichornia crassipes* [35].

Turbidity and TSS had the same trend as the aforementioned parameters. In the Atizcoa Stream, it can be seen in Figure 5 that the values of these parameters increased as the stream ran through the TFs. It has been reported that TSS reaches up to 24.7 g/L in tequila vinasses [4], so vinasse discharges significantly impact the concentration of these two parameters. The turbidity in the Atizcoa Stream ranged from 16 to 785 NTU and the TSS ranged from 36 to 829 mg/L. The variation between the sampling points was significant for both parameters ($p < 0.05$). In the Jarritos Stream, high turbidity and TSS values were also observed, mainly on February 16, where point J1 stood out with values of 4490 NTU of turbidity and 6185 mg/L of TSS; at point AJ, turbidity and SST also presented high values, between 227 and 844 NTU for turbidity and between 657.1 and 941.7 mg/L for TSS. In general, the values of both parameters were exceptionally high, indicating that the streams transported mainly vinasses. Some specific values were similar to those of samples with 75% vinasse (Table 1). In this case, there was no significant difference between the sampling points of the stream ($p > 0.05$). By comparison, [27] found mean values of 55.66 ± 4.18 NTU and 31.71 ± 1.41 mg/L for turbidity and TSS, respectively, at a domestic wastewater discharge point in a stream in Mali. Moreover, [36] reported a concentration 49.46 ± 21.59 mg/L of TSS and 76.23 ± 51.27 NTU of turbidity in the Cau River in Vietnam during the dry season. On the other hand, based on the concentration of TSS, the two streams are classified as highly polluted bodies of water (TSS concentration > 400 mg/L), according to the water quality indices of the Water National Commission (CONAGUA) [1].

Regarding the flow rate, it was found that in the Atizcoa Stream, it varied from 23 L/s at the point prior to its passage through the tequila plants (A1) to 53 L/s at point A2 and finally to 107 L/s before joining the Jarritos Stream (A4). Meanwhile, in the Jarritos Stream, the flow rate was lower, varying from 18 L/s to 84 L/s at the last sampling point. This increase in stream flow probably also reflects the incorporation of tequila vinasses.

Finally, the results of the physicochemical analyses of the water samples from the two streams show that their contamination is evident. Therefore, it is suggested that tequila companies in the Valles region implement best practices to mitigate this impact in the short term. It is recommended that tequila vinasses be treated using conventional wastewater treatment processes such as pretreatment, primary treatment, secondary treatment (physicochemical and/or biological), and advanced treatment processes. In cases of companies that cannot afford a conventional wastewater treatment plant, such as micro- and small factories, low-cost alternative technologies are recommended such as constructed wetlands [4] or advanced oxidation processes [37]. In addition, vinasses could be used and valorized for other applications such as the generation of biogas [38], the production of microbiological culture media in bioprocesses and bioremediation [39], or as raw material for the production of animal foods [40] due to the high content of organic matter, water, and nutrients such as potassium, calcium, magnesium, sulfur, and nitrogen.

4. Conclusions

By taking as reference the concentrations of contaminants in samples with different percentages of tequila vinasse as well as the estimation of the vinasse content in each of the monitored points (through a linear regression model developed between COD and % vinasse), we confirmed that the two monitored streams are being used as final disposal sites for tequila vinasse.

The state of total deterioration in which they are found is physically evident, as it is impossible for aquatic life to exist in such conditions. The increase in the concentration and values of parameters that reflect contamination from the discharge of vinasse (temperature, conductivity, turbidity, FOG, TSS, COD, BOD, total phosphate, nitrate, and Kjeldahl nitrogen), as well as the decrease in the concentration and values of important parameters for the development of aquatic life such as pH and DO, were found to be overwhelming once the streams flowed through the tequila factories. Due to the information above, there is an urgent need for tequila companies to implement treatment systems for the vinasse generated. It is recommended that alternative treatment systems be evaluated alongside the conventional and economically accessible ones so that through their implementation the contamination of these bodies of water can be reduced for the benefit of the locality itself.

Author Contributions: Conceptualization, F.Z.; methodology, A.T., A.M. and B.S.-R.; software, A.T.; validation, F.Z. and B.S.-R.; formal analysis, A.T., F.Z. and B.S.-R.; investigation, A.M.; resources, F.Z.; writing—original draft preparation, A.M. and B.S.-R.; writing—review and editing, A.T. and F.Z.; supervision, F.Z.; project administration, F.Z.; funding acquisition, F.Z. All authors have read and agreed to the published version of the manuscript.

Funding: This research was funded by the National Water Commission and National Council of Science and Technology of Mexico through the "FONDO SECTORIAL DE INVESTIGACIÓN Y DESARROLLO SOBRE EL AGUA", CALL 2018 as part of the project "TECNOLOGÍAS SUSTENTABLES PARA TRATAMIENTO DE VINAZAS Y SU REUSO EN ACTIVIDADES PRODUCTIVAS"; grant number A3-S-66470.

Data Availability Statement: Not applicable.

Acknowledgments: We thank the University of Guadalajara through the Centro Universitario de la Ciénega for its support during the surface water monitoring process.

Conflicts of Interest: The authors declare no conflict of interest.

References

1. Water National Comission (Comision Nacional del Agua). *Estadisticas del Agua en México*; Water National Comission: Mexico City, Mexico, 2023.
2. Zurita, F.; Tejeda, A.; Montoya, A.; Carrillo, I.; Sulbarán-Rangel, B.; Carreón-Álvarez, A. Generation of Tequila Vinasses, Characterization, Current Disposal Practices and Study Cases of Disposal Methods. *Water* **2022**, *14*, 1395. [CrossRef]
3. Montoya, A.; Tejeda, A.; Sulbarán-Rangel, B.; Zurita, F. Treatment of tequila vinasse mixed with domestic wastewater in two types of constructed wetlands. *Water Sci. Technol.* **2023**, *87*, 3072–3082. [CrossRef]
4. Zurita, F.; Vymazal, J. Oportunities and challenges of using constructed wetlands for the treatment of high-strength distillery effluents: A review. *Ecol. Eng.* **2023**, *196*, 107097. [CrossRef]
5. Díaz-Vázquez, D.; Carrillo-Nieves, D.; Orozco-Nunnelly, D.A.; Senés-Guerrero, C.; Gradilla-Hernández, M.S. An Integrated Approach for the Assessment of Environmental Sustainability in Agro-Industrial Waste Management Practices: The Case of the Tequila Industry. *Front. Environ. Sci.* **2021**, *9*, 682093. [CrossRef]
6. Tejeda, A.; Valencia-Botín, A.J.; Zurita, F. Resistance evaluation of Canna indica, Cyperus papyrus, Iris sibirica, and Typha latifolia to phytotoxic characteristics of diluted tequila vinasses in wetland microcosms. *Int. J. Phytoremediation* **2023**, *25*, 1259–1268. [CrossRef]

7. Toledo-Cervantes, A.; Villafán-Carranza, F.; Arreola-Vargas, J.; Razo-Flores, E.; Méndez-Acosta, H.O. Comparative evaluation of the mesophilic and thermophilic biohydrogen production at optimized conditions using tequila vinasses as substrate. *Int. J. Hydrogen Energy* **2020**, *45*, 11000–11010. [CrossRef]
8. Kharayat, Y. Distillery wastewater: Bioremediation approaches. *J. Integr. Environ. Sci.* **2012**, *9*, 69–91. [CrossRef]
9. García-Depraect, O.; León-Becerril, E. Fermentative biohydrogen production from tequila vinasse via the lactate-acetate pathway: Operational performance, kinetic analysis and microbial ecology. *Fuel* **2018**, *234*, 151–160. [CrossRef]
10. López-López, A.; Contreras-Ramos, S.M. Tratamiento de Efluentes y aprovechamiento de residuos, capítulo 8. In *Ciencia y Tecnología del Tequila: Avances y Perspectivas*, 2nd ed.; Centro de Investigación y Asistencia en Tecnología y Diseño del Estado de Jalisco (CIATEJ): Guadalajara, Mexico, 2015.
11. Ramírez-Ramírez, A.A.; Sulbarán-Rangel, B.C.; Jáuregui-Rincón, J.; Lozano-Álvarez, J.A.; la Torre, J.A.F.-d.; Zurita-Martínez, F. Preliminary Evaluation of Three Species of Ligninolytic Fungi for Their Possible Incorporation in Vertical Flow Treatment Wetlands for the Treatment of Tequila Vinasse. *Water Air Soil Pollut.* **2021**, *232*, 456. [CrossRef]
12. Ratna, S.; Rastogi, S.; Kumar, R. Current trends for distillery wastewater management and its emerging applications for sustainable environment. *J. Environ. Manag.* **2021**, *290*, 112544. [CrossRef]
13. Correia, J.E.; Christofoletti, C.A.; Marcato, A.C.C.; Marinho, J.F.U.; Fontanetti, C.S. Histopathological analysis of tilapia gills (*Oreochromis niloticus* Linnaeus, 1758) exposed to sugarcane vinasse. *Ecotoxicol. Environ. Saf.* **2017**, *135*, 319–326. [CrossRef]
14. Marinho, J.F.U.; Correia, J.E.; Marcato, A.C.d.C.; Pedro-Escher, J.; Fontanetti, C.S. Sugar cane vinasse in water bodies: Impact assessed by liver histopathology in tilapia. *Ecotoxicol. Environ. Saf.* **2014**, *110*, 239–245. [CrossRef]
15. Gunkel, G.; Kosmol, J.; Sobral, M.; Rohn, H.; Montenegro, S.; Aureliano, J. Sugar Cane Industry as a Source of Water Pollution—Case Study on the Situation in Ipojuca River, Pernambuco, Brazil. *Water Air Soil Pollut.* **2007**, *180*, 261–269. [CrossRef]
16. Tequila Regulatory Council (Consejo Regulador del Tequila). Producción Total de Tequila y Tequila 100% en 2023. Available online: https://www.crt.org.mx/EstadisticasCRTweb/ (accessed on 1 March 2023).
17. Instituto Nacional de Estadística y Geografía; Consejo Agroalimentario de Jalisco. Colección de estudios sectoriales y regionales. In *Conociendo la Industria del Tequila y el Mezcal*; Instituto Nacional de Estadística y Geografía: Aguascalientes, Mexico, 2019.
18. Gómez-Cuevas, K.; Delgado-Cruz, A.; Palmas-Castrejón, Y.D. Originalidad del tequila como símbolo de identidad mexicana. Percepción del turista-consumidor a partir de su exportación. *RIVAR (Santiago)* **2020**, *7*, 59–80. [CrossRef]
19. Almazán-Torres, M.G.; Ordóñez-Regil, E.; Ruiz-Fernández, A.C. Uranium and plutonium in anoxic marine sediments of the Santiago River mouth (Eastern Pacific, Mexico). *J. Environ. Radioact.* **2016**, *164*, 395–399. [CrossRef] [PubMed]
20. Torres, G.; Lozano, F.; Solis, J.; Garcia Velasco, J.; Medina, M.G.; Peregrina-Lucano, A. Air Quality and its Effects on Child Health of a Metropolitan City in Jalisco, Mexico. 2016. Available online: https://www.researchgate.net/publication/340314393_Air_Quality_and_its_Effects_on_Child_Health_of_a_Metropolitan_City_in_Jalisco_Mexico (accessed on 1 March 2023).
21. Rizo-Decelis, L.D.; Andreo, B. Water quality assessment of the Santiago River and attenuation capacity of pollutants downstream Guadalajara City, Mexico. *River Res. Appl.* **2015**, *32*, 1505–1516. [CrossRef]
22. Pérez de la Vega Baños, A. *Efecto de las Descargas de las Industrias Tequileras Sobre las Poblaciones de Macroinvertebrados Bentónicos en el Arroyo Atizcoa*; Municipio de Tequila, Jalisco; Mexico (diciembre, 1986-mayo, 1987) Universidad de Guadalajara: Guadalajara, Mexico, 1988.
23. APHA; AWWA; WEF. *Standard Methods for the Examination of Water and Wastewater*, 21st ed.; American Public Health Association: Washington, DC, USA, 2005.
24. González, A.; Ramírez, J.D. *Manual Piragüero-Medición del Caudal*; LibroArte: Medellin, Colombia, 2014.
25. Fernandes, A.P.; Fonseca, A.R.; Pacheco FA, L.; Fernandes, L.S. Water quality predictions through linear regression—A brute force algorithm approach. *MethodsX* **2023**, *10*, 102153. [CrossRef]
26. Veeramsetty, V.; Shadamaki, N.; Pinninti, R.; Guduri, N.; Ashish, G. Water quality index estimation using linear regression model. *AIP Conf. Proc.* **2022**, *2418*, 040033. [CrossRef]
27. Alves, R.N.; Mariz, C.F., Jr.; de Melo Alves, M.K.; Cavalcanti, M.G.N.; de Melo, T.J.B.; de Arruda-Santos, R.H.; Zanardi-Lamardo, E.; Carvalho, P.S.M. Contamination and Toxicity of Surface Waters Along Rural and Urban Regions of the Capibaribe River in Tropical Northeastern Brazil. *Environ. Toxicol. Chem.* **2021**, *40*, 3063–3077. [CrossRef]
28. Kwong, R.W.; Kumai, Y.; Perry, S.F. The physiology of fish at low pH: The zebrafish as a model system. *J. Exp. Biol.* **2014**, *217*, 651–662. [CrossRef]
29. Tchobanoglous, G.; Burton, F.L.; Stensel, H.D.; Metcalf, E. *Wastewater Engineering: Treatment and Reuse*; McGraw-Hill: New York, NY, USA, 2003.
30. Octavio, G.-D.; Daryl Rafael, O.-L.; Elizabeth, L.-B. A Comprehensive Overview of the Potential of Tequila Industry By-Products for Biohydrogen and Biomethane Production: Current Status and Future Perspectives. In *New Advances on Fermentation Processes*; Rosa María, M.-E., Ed.; IntechOpen: Rijeka, Croatia, 2019; Chapter 6.
31. Secretariat of Environment and Natural Resources (Secretaria del Medio Ambiente y Recursos Naturales). Norma Oficial Mexicana, NOM-001-SEMARNAT-2021 Que Establece los Límites Máximos Permisibles de Contaminantes en las Descargas de Aguas Residuales en Aguas y Bienes Nacionales. 2021. Available online: https://www.dof.gob.mx/nota_detalle.php?codigo=5645374&fecha=11/03/2022#gsc.tab=0 (accessed on 1 March 2023).

32. Del Toro Farías, A.; Zurita Martínez, F. Changes in the nitrification-denitrification capacity of pilot-scale partially saturated vertical flow wetlands (with corncob in the free-drainage zone) after two years of operation. *Int. J. Phytoremediation* **2021**, *23*, 829–836. [CrossRef]
33. Nakase, C.; Zurita, F.; Nani, G.; Reyes, G.; Fernández-Lambert, G.; Cabrera-Hernández, A.; Sandoval, L. Nitrogen Removal from Domestic Wastewater and the Development of Tropical Ornamental Plants in Partially Saturated Mesocosm-Scale Constructed Wetlands. *Int. J. Environ. Res. Public Health* **2019**, *16*, 4800. [CrossRef] [PubMed]
34. Mora, A.; García-Gamboa, M.; Sánchez-Luna, M.S.; Gloria-García, L.; Cervantes-Avilés, P.; Mahlknecht, J. A review of the current environmental status and human health implications of one of the most polluted rivers of Mexico: The Atoyac River, Puebla. *Sci. Total Environ.* **2021**, *782*, 146788. [CrossRef] [PubMed]
35. Hansen, A.M.; González-Márquez, L.C. Scenarios of metal concentrations in the Arcediano Dam (State of Jalisco, Mexico). *J. Environ. Sci. Health Part A* **2010**, *45*, 99–106. [CrossRef]
36. Son, C.T.; Giang, N.T.H.; Thao, T.P.; Nui, N.H.; Lam, N.T.; Cong, V.H. Assessment of Cau River water quality assessment using a combination of water quality and pollution indices. *J. Water Supply Res. Technol. Aqua* **2020**, *69*, 160–172. [CrossRef]
37. Castillo-Monroy, J.; Godínez, L.A.; Robles, I.; Estrada-Vargas, A. Study of a coupled adsorption/electro-oxidation process as a tertiary treatment for tequila industry wastewater. *Environ. Sci. Pollut. Res.* **2021**, *28*, 23699–23706. [CrossRef]
38. Toledo-Cervantes, A.; Méndez-Acosta, H.O.; Arreola-Vargas, J.; Gabriel-Barajas, J.E.; Aguilar-Mota, M.N.; Snell-Castro, R. New insights into microbial interactions and putative competitive mechanisms during the hydrogen production from tequila vinasses. *Appl. Microbiol. Biotechnol.* **2022**, *106*, 6861–6876. [CrossRef]
39. Díaz-Vázquez, D.; Orozco-Nunnelly, D.A.; Yebra-Montes, C.; Senés-Guerrero, C.; Gradilla-Hernández, M.S. Using yeast cultures to valorize tequila vinasse waste: An example of a circular bioeconomy approach in the agro-industrial sector. *Biomass Bioenergy* **2022**, *161*, 106471. [CrossRef]
40. Alemán-Nava, G.S.; Gatti, I.A.; Parra-Saldivar, R.; Dallemand, J.-F.; Rittmann, B.E.; Iqbal, H.M.N. Biotechnological revalorization of Tequila waste and by-product streams for cleaner production—A review from bio-refinery perspective. *J. Clean. Prod.* **2018**, *172*, 3713–3720. [CrossRef]

water

MDPI

Article

Assessment of Sewage Molecular Markers: Linear Alkylbenzenes in Sediments of an Industrialized Region in Peninsular Malaysia

Sadeq A. A. Alkhadher [1,*], Hussein E. Al-Hazmi [2,*], Suhaimi Suratman [3], Mohamad P. Zakaria [4], Najat Masood [5], Bartosz Szeląg [6], Sami M. Magam [7], Ebrahim H. H. Al-Qadami [8], Joanna Majtacz [3], Przemysław Kowal [3,*], Tonni A. Kurniawan [9], Sameer A. M. Abdulrahman [10] and Saeed S. Albaseer [11]

1 Institute of Energy Infrastructure (IEI), Universiti Tenaga Nasional (UNITEN), Putrajaya Campus, Jalan IKRAM-UNITEN, Kajang 43000, Selangor, Malaysia
2 Faculty of Civil and Environmental Engineering, Gdansk University of Technology, Narutowicza 11/12, 80-233 Gdansk, Poland
3 Institute of Oceanography and Environment, Universiti Malaysia Terengganu, Kuala Nerus 21030, Terengganu, Malaysia
4 Institute of Ocean and Earth Sciences (IOES), University of Malaya, Kuala Lumpur 50603, Malaysia
5 Chemistry Department, Faculty of Science, University of Ha'il, Ha'il 1560, Saudi Arabia
6 Faculty of Environmental, Geomatics and Renewable Energy, Kielce University of Technology, 25-314 Kielce, Poland
7 Basic Science Department, Preparatory Year, University of Ha'il, Ha'il 1560, Saudi Arabia
8 Eco Hytech Research Centre, Faculty of Civil Engineering and Built Environment, Universiti Tun Hussein Onn Malaysia, Parit Raja 86400, Johor, Malaysia
9 College of the Environment and Ecology, Xiamen University, Xiamen 361102, China; tonni@xmu.edu.cn
10 Department of Chemistry, Faculty of Education and Sciences-Rada'a, Albaydha University, Albaydha 14517, Yemen
11 Department of Evolutionary Ecology & Environmental Toxicology, Biologicum, Goethe University Frankfurt, 60438 Frankfurt am Main, Germany
* Correspondence: sadeq.abdo@umt.edu.my (S.A.A.A.); hussein.hazmi1@pg.edu.pl (H.E.A.-H.); przkowal@pg.edu.pl (P.K.)

Citation: Alkhadher, S.A.A.; Al-Hazmi, H.E.; Suratman, S.; Zakaria, M.P.; Masood, N.; Szeląg, B.; Magam, S.M.; Al-Qadami, E.H.H.; Majtacz, J.; Kowal, P.; et al. Assessment of Sewage Molecular Markers: Linear Alkylbenzenes in Sediments of an Industrialized Region in Peninsular Malaysia. *Water* **2023**, *15*, 3301. https://doi.org/10.3390/w15183301

Academic Editor: Imokhai Theophilus Tenebe

Received: 19 July 2023
Revised: 9 September 2023
Accepted: 12 September 2023
Published: 19 September 2023

Abstract: In this study, the use of linear alkylbenzenes (LABs) was employed to pinpoint the sources of human activity that cause detrimental impacts on the coastal environment and river ecosystems. LABs were detected using GC–MS in sediment samples assembled from Kim Kim River (KKR) and the Port Dickson coast (PDC). To assess the significance of variations in the distribution and concentrations of LABs across the sampling sites, this study utilized several statistical techniques such as post hoc tests, LSD techniques, analysis of variance (ANOVA), and the Pearson correlation coefficient using a significance level of $p < 0.05$. The degradation levels of LABs and wastewater treatment were assessed in the study using internal congeners (I/E), homologs of C^{13} and C^{12}, and long-to-short-chain (L/S) ratios. The results revealed that the LAB concentrations varied between 88.3 and 112 ng/g dw in KKR and 119 to 256 ng/g dw in the PDC. Most of the surveyed areas exhibited a substantial count of C^{13}–LABs homologs that displayed a significant difference ($p < 0.05$). The I/E ratios ranged from 1.7 to 2.0 in KKR and from 2.0 to 4.1 in the PDC, suggesting that the effluents originated from sources associated with the physical phase and biological phase in wastewater treatment systems (WWTSs). The results revealed that the degradation of LABs varied between 34% and 38% in KKR and between 40% and 64% in the PDC. This study underscores the importance of ongoing improvements to WWTSs and emphasizes the potential of LABs as indicators for monitoring wastewater contamination.

Keywords: linear alkylbenzenes; sediment; internally over exterior; pollutants contained in wastewater; degradation; molecular marker

1. Introduction

Anthropogenic pollution stemming from metropolitan areas is a significant contributor to the degradation of riverine and coastal ecosystems [1–4]. The release of substantial amounts of industrial and household wastewater in these regions significantly affects the biodiversity of the ecosystems [5–8]. Therefore, regular monitoring of aquatic ecosystems and water sources is crucial for maintaining the health and integrity of these environments [9–14]. It provides valuable data for decision making, pollution control, and the preservation of biodiversity, ensuring sustainable management and protection of our precious water resources [15–18]. Obtaining such data is crucial for achieving efficient environmental security and economic management. With this information, strategies and implemented measures can be developed that mitigate the harmful impacts of anthropogenic pollution on these delicate ecosystems, helping to preserve them for future generations [8,19–22].

Linear alkylbenzenes (LABs) can serve as indicators of industrial and domestic pollution. Linear alkylbenzene sulfonates (LASs), which constitute the primary ingredients of detergents, are composed of alkyl chains ranging from C10 to C14 and are referred to as LABs [23]. In the 1960s, the utilization of branched alkylbenzene constituents was replaced by LABs, which were preferred for their cost-effectiveness and biodegradability. However, ref. [24] noted that improper sulfonation has led to the extensive and prevalent discharge of LABs into the aquatic ecosystem through untreated wastewater effluent. In contrast, ref. [25] indicated that LABs exhibit a significant attraction toward industrial effluents, which renders them valuable as indicators of pollution from human-made sources, including both residential and industrial, in aquatic settings.

LAB isomers, which have varied phenyl substitutions and internal/external isomers alongside the straight alkyl chain, serve a purpose in gauging degradation in sediment and suspended particulate matter, as described by [26]. Typically, wastewater treatment systems (WWTSs) contain a series of physical (primary treatment) and biological (secondary treatment) processes designed to remove contaminants from wastewater before it is discharged back into the environment or reused for other purposes [27]. The specific components of a wastewater treatment system can vary depending on the type of contaminants present in the wastewater and the desired level of treatment. According to [28], the identification of industrial and domestic wastewater types discharged into marine ecosystems, including physical (primary) and biological (secondary) systems, as well as the estimation of pollutant residence time, can be accomplished through the utilization of LAB isomers.

Due to the high population in coastal areas of Malaysia, there has been a significant increase in industrial and domestic wastewater contamination. Studies in Asian countries demonstrated a positive correlation between sewage contamination and illnesses [29]. Therefore, it is necessary to investigate the impact of marine sediments and their fate on the possible pollution caused by wastewater to improve water quality and reduce the risk of illness [30,31]. Based on three local media that were recently published, there is water contamination in Simpang Renggam in Sungai Benut, which mostly affects 75,000 residents in Johor state [32]. Thus, the aquatic environment around West and South Malaysia should be frequently monitored to identify the sources of contamination, whether industrial or home wastewater, in order to protect environmental and human health.

Regular monitoring of wastewater sources in West and South Malaysia is crucial for protecting the environment and human health. Although the anthropogenic impacts in KKR and the PDC were investigated, it is essential to extend the research to other regions. This study employed LABs as anthropogenic pollution to evaluate the effects of wastewater sources in the selected areas. This study measured the concentration, distribution, and degradation levels of LABs in the study regions, and investigated the distribution of LAB isomers to enhance the effectiveness of existing wastewater treatment systems (WWTSs).

2. Methods and Materials

2.1. Experimental Design

The research sites that were studied were situated in the southwest region of Malaysia, specifically in the states of Johor and Sembilan, as shown in Figure S1 in the Supplementary Information (SI). Seven sediment samples were taken from sites chosen for their proximity to agricultural and industrial activity, such as Kim Kim River in Johor Bahru State, as well as urban and tourism activity, such as Port Dickson Beach in Sembilan State. Details and a visual representation of the sampling locations used in the study are provided in Table S1 in the SI file. To examine the presence of LABs in these areas and to assess the level of anthropogenic contamination, surface sediments were collected from the top 4 cm of KKR and the PDC using an Ekman dredge sampler. The collected sediment was then put into a stainless steel container that had been cleaned, and the top 4 cm of the sediment cake was sliced, immediately kept in a double-clean Ziploc bag, labeled, and then put on ice at 4 °C in a cooler box for transport to the lab, where it was stored at −20 °C upon arrival. The samples were then put through a freeze-drying procedure to remove any remaining moisture and get them ready for more analysis.

2.2. Chemical Analysis

Two columns were used to extract LABs. The initial one was used for purification to remove polar compounds, and the second was used for fractionation to obtain the desired organic extracts, following the established validation protocols described elsewhere by [33]. To recover dried sediments, 10 g of sediments were mixed with 250 mL dichloromethane (DCM) in a cellulose thimble. This mixture was left for 8–10 h, as per the protocol in [34]. The surrogate standards (SSs), which are similar to the target analytical compounds used in the analytical method for the purpose of identification and could be recovered from the sample matrix with reasonable efficiency, were preloaded in each sample to ensure the precise recovery of the target LABs. The SSs contain 50 µL of 1-C*n* LABs, where 1- denotes each LAB homolog's initial isomer and *n* denotes its carbon number (8–14). Activated copper was added to remove any sulfur that could interfere with the GC chromatograms. The extract was concentrated using a rotary evaporator and then placed on a chromatography column of 5% H_2O-deactivated silica gel (60–200 mesh size, Sigma Chemical, St. Louis, MO, USA), with a diameter of 0.9 cm and height of 9 cm. Hydrocarbons were eluted from the column with a pure hexane/DCM mixture (3:1, v/v) in 20 mL aliquots and then reduced to 1–2 mL. LAB fractions were collected from a fully activated silica gel column with a diameter of 0.47 cm and height of 18 cm using 4 mL of hexane in the second stage. The reduced LAB fractions were transferred to a 2 mL amber vial and dried using a moderate stream of nitrogen. Before performing the GC–MS analysis on the LABs fraction, an internal standard (IS) of biphenyl–d10 with a mass-to-charge ratio (m/z) of 164 was injected.

To detect compounds of LABs, a gas chromatograph from the Agilent Technologies (Santa Clara, CA, USA) 7890A series was utilized, along with a C5975 MSD split/splitless injector. The experiment utilized a 30-meter-long fused silica capillary column with an inner diameter of 0.25 millimeters and a DB–5MS capillary column coated with a 0.25 µm film thickness. Helium was chosen as the carrier gas and was kept at a steady pressure of 60 kg cm^{-2} throughout the experiment.

Further, to obtain the mass spectrum data, the LABs chromatographs were monitored at m/z = 91, 92, and 105 using a selective ion monitoring (SIM) mode. The method utilized for this data operated with specified parameters, such the signal-to-noise ratio (S/N), peak detection limits, and a 5-point calibration curve. LAB peaks in the samples were obtained by comparing the retention times of each target compound to the retention times of the LAB standards run on the same day, which were then confirmed using the m/z value (91, 92, and 105). An external source was employed to carry out the ionization process at a temperature of 200 °C. The GC–MS operating conditions were configured with an electron energy of 70 eV and an electron multiplier voltage of approximately 1250 eV. After injecting the sample using the splitless mode, the injection port temperature was maintained at

300 °C, and a one-minute purge ensued. The temperature T of the column was maintained at 70 °C for two minutes. It was then raised to 150 °C at a rate of 30 °C per minute and subsequently increased at a rate of 4 °C per minute until it reached 310 °C. This final temperature was maintained for 50 min.

2.3. Quality Control and Assurance

LABs are used to determine the degree of degradation due to the properties of external and internal isomers. External isomers degrade faster than internal isomers due to the substitution position of the phenyl group on the straight alkyl chain. Furthermore, LABs are persistent in aquatic environments for extended periods. The I/E ratio can be used to identify the types of wastewater treatment. For instance, raw sewage has a lower I/E ratio, ranging from 0.7 to 0.8. Primary treatment I/E ratios vary between 0.5 and 0.9, while secondary treatment has an I/E ratio greater than 2 [35]. The LABs surrogates (1–C*n*) were successfully recovered in the 60–120% range with minimal loss of the target chemicals during the analytical procedures. The recovery rates for all sediment samples tested were between 87 and 98%.

To ensure that there was no contamination, a blank sample was included in every batch of four samples, which contained all the substances and standards found in typical samples. Each day, the sediment samples were spiked with freshly made surrogate, internal, and native standards at predetermined amounts. The target LAB congeners were located using GC–MS in selected ion monitoring mode at $m/z = 91$, 92, and 105. The quantification of the LAB chemicals was done using a 5-point calibration curve with a LABs standard mixture, with concentrations ranging from 0.25 to 5.0 ppm spiked into the target samples. The limits of quantitation (LOQ) and detection (LOD) were determined based on the lowest concentration levels of each calibration curve divided by the mean sample weight, which were found to be 0.02–0.1 ng/g and 0.1–2 ng/g, respectively, following the method described by [36].

2.4. Statistical Analysis

The statistical analysis process was undertaken using the IBM®–SPSS 25 program. Various techniques, including analysis of variance (ANOVA), post hoc tests, and LSD, were utilized to showcase the significance of fluctuations in the LABs concentrations and distributions across sample sites at a significance level of $p < 0.05$. To evaluate the relationship between sampling stations, the Pearson correlation coefficient was utilized. Additionally, principal component analysis (PCA) was conducted to decrease the number of initial variables, such as the concentration of LABs, and identify a smaller set of underlying factors. The extraction process retained only those elements that had eigenvalues greater than one, which were then subjected to varimax rotation. Standardized regression coefficients were utilized to assess the potential contribution of various sources of LABs.

2.5. Evaluation of Total Organic Carbon (TOC)

The TOC in sediment samples was detected using LECO CR–412 analysis. The samples were dried at 60 °C overnight and then pulverized using a mortar and pestle. In order to remove any inorganic carbon, sediment samples weighing 1–2 g were fully moistened with 2 mL of 1 M HCl, which eliminates carbonates. The sediment samples were then dried once again at 100 °C for 10 h to remove the HCl. Finally, the TOC % was determined using a LECO CR–412 Carbon Analyzer, which had a furnace temperature of 1350 °C and an O_2 boost period of 1 min. The procedure described in [36] was followed.

3. Results and Discussion

3.1. Differences in the Distribution of LABs Concentrations

The LABs formula, depicted in Figure S2, is denoted as "n–Cm", with "n" indicating the position of the benzene ring on the linear alkyl chain and "m" indicating the number of alkyl carbons. The sediment samples that were analyzed consisted of LABs with carbon chain lengths between C10 and C14. Concentrations of LABs in KKR and the PDC were found to be between 88.3 and 112 and between 119 and 256 ng/g dw, respectively. The huge variation, especially among the samples from the PDC, was attributed to anthropogenic activities near the sampling stations, such as urbanization growth, small industries, tourism, and shipping.

According to the study's findings, the significant Pearson correlation between the LAB concentrations in the investigated areas ($r = 0.88$, $p < 0.05$) was most likely caused by the fact that LAB concentrations were higher in areas with a high population density, industrial, agricultural, and tourism activities, as these areas are more likely to have a high volume of wastewater and detergent discharge. Overall, the strong correlation between LAB concentrations and pollution indicates that LABs could serve as a useful measure of both residential and industrial contamination. Table S2 in the SI file shows the details of this correlation. The range of LAB levels was between 88.3 and 256 ng/g dw, with significantly higher concentrations observed at the PDC line, as illustrated in Figure 1. Additionally, the statistical analysis at $p < 0.05$ revealed a significant difference in the LABs concentrations between the various study locations (refer to Table 1).

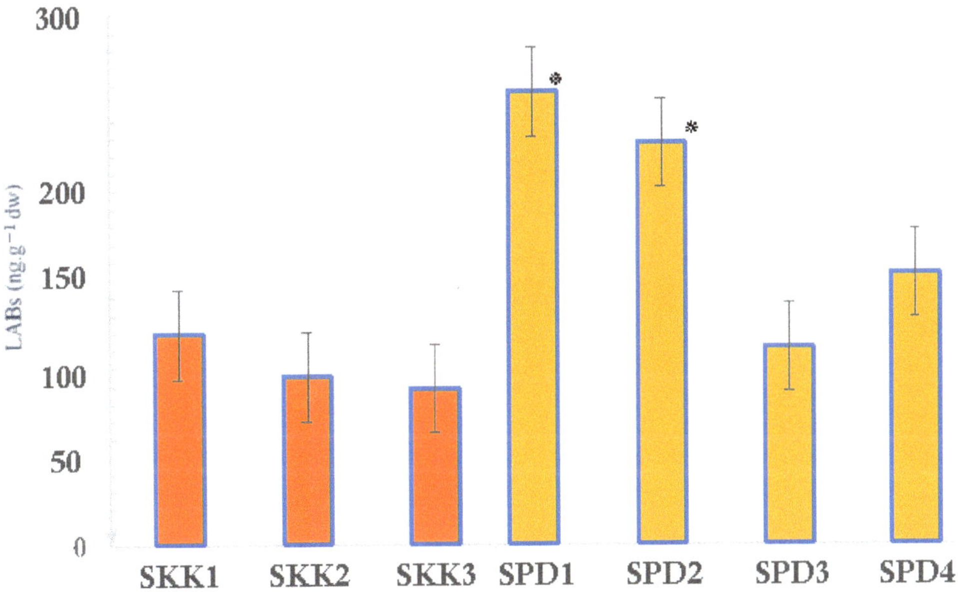

Figure 1. Concentrations of LABs in the Kim Kim River and Port Dickson samples. Standard error bars are shown. (*) indicates the highly significant differences ($p < 0.05$).

The distribution of LABs in riverine environments may be linked to treated effluent runoff and the dilution of organic matter, as indicated by their content. This can affect the distribution of riverine ecosystems, as observed by [37,38]. On the other hand, according to studies by [39,40], the occurrence of LABs in coastal areas is thought to result from high levels of industrialization and urbanization, as well as the lack of piped sewers. In contrast with previous studies by [41,42], which found lower concentrations of LABs at their research sites compared with Port Klang, the Penang Estuary, Malacca, and the Anzali Wetland, refs. [43,44] reported similar lower levels at the respective locations. Nonetheless, the present study identified relatively high quantities of LABs, comparable with those found in Southern Brazil and the Pearl River. The higher levels of LABs detected in this study compared with some other regions worldwide and Malaysia could be attributed to a combination of specific local factors and differences in pollution sources, such as significant industrial, tourism, and urban activities that release LABs into the environment. Port

Dickson is known to be one of Malaysia's famous recreational beaches, with numerous hotels and resorts that offer a wide range of accommodations, as well as the growth in urbanization, small industries, tourism, and shipping [45]. The distribution of LAB isomers and homologs among this trend suggested that their presence was uniformly distributed at the KKR and PDC sites, which implies that the geographic location of the sampling locations may have influenced the distribution of LABs. Additionally, the LABs detected in this study were found to be moderate to low compared with other regions worldwide and Malaysia, suggesting that they may serve as an indicator of wastewater pollution.

Table 1. LABs concentrations (ng/g dw) and relative compound ratios in Kim Kim River and the Port Dickson coast.

Compound	SKK1 [b]	SKK2	SKK3	SPD1	SPD2	SPD3	SPD4
C_{10}-LABs ($ng{\cdot}g^{-1}$ dw) [a]	9.4	7.7	6.8	17.2	15.0	4.2	6.2
C_{11}-LABs ($ng{\cdot}g^{-1}$ dw)	16.3	13.0	12.0	66.7	57.0	19.0	30.7
C_{12}-LABs ($ng{\cdot}g^{-1}$ dw)	25.6	20.7	18.6	68.6	61.0	24.0	38.0
C_{13}-LABs ($ng{\cdot}g^{-1}$ dw)	41.0	32.6	29.5	86.0	78.4	51.2	63.6
C_{14}-LABs ($ng{\cdot}g^{-1}$ dw)	26.8	21.1	21.3	17.4	15.6	13.2	14.8
LABs ($ng{\cdot}g^{-1}$ dw)	119	95.2	88.3	256	227	112	153
LC-LABs ($ng{\cdot}g^{-1}$ dw) [c]	67.8	53.8	51.0	103	94.0	64.5	78.4
SC-LABs ($ng{\cdot}g^{-1}$ dw) [d]	25.8	20.7	19.0	84.0	71.8	23.2	37.0
I/E [e]	2.0	1.8	1.7	4.1	3.7	2.0	2.6
L/S [g]	2.4	2.5	2.6	1.6	1.6	1.7	1.6
C_{13}/C_{12} [h]	5.1	5.2	5.2	9.7	9.5	13.7	14.3
LAB degradation (%) [f]	38	35	34	64	61	40	48
TOC (%) [i]	1.2	1.0	8.1	2.0	2.3	1.5	1.9
TOC (mg/g)	11.5	9.0	81	19.7	22.7	14.7	18.9

Notes: [a] C10-LAB: sum of the 26LAB congeners ranging from 5-C10 to 2-C10. [b] SKK1–3 and SPD1–4: the first letter indicates the station; the second and third letters represent the first letters of location name; and the numbers 1, 2, 3, and 4 indicate the first, second, third, and fourth station for each respective location. [c] LC-LABs: sum of LABs ranging from 6-C^{13} to 2-C^{14}. [d] SC-LABs: sum of LABs ranging from 5-C^{10} to 2-C11. [e] I/E (C^{12}–LABs): ratio of (6-C^{12}LAB + 5-C^{12}LAB) relative to (4-C^{12}LAB + 3-C^{12}LAB + 2-C^{12}LAB). [g] L/S: ratio of (5-C^{13}LAB + 5-C^{12}LAB) relative to (5-C^{11}LAB + 5-C^{10}LAB). [h] C^{13}/C^{12}: ratio of (6-, 5-, 4-, 3-, and 2-C^{13}/(6-, 5-, 4-, 3-, and 2-C^{12}LAB). [f] LAB degradation (%): LAB deg = 81 × log (I/E ratio) +15 (r^2 = 0.96). [i] TOC (%): total organic carbon.

Table 2 indicates that there is a notable disparity in the allocation of LAB homologs among the different sampling locations, with a statistical significance level of $p < 0.05$. The geographical features of different sampling locations, such as proximity to coastlines, rivers, or urban centers, as well as industrial discharges, urban runoff, or agricultural activities and inadequate treatment, could influence the input sources of LAB homologs and their subsequent distribution [46]. The LABs in KKR contained a higher proportion of homologs and isomers with C^{13}, followed by C^{12} and C^{14} (Figure 2), indicating a substantial difference in composition compared with other locations. Conversely, the LABs containing C^{10} and C^{11} were present at low levels, suggesting selective degradation of these compounds during industrial effluent discharge. LC–LABs (long-chain linear alkylbenzenes), such as C^{13} and C^{14}, were found to be the primary component of LABs concentration vs. SC–LABs (short-chain linear alkylbenzenes), like C^{10} and C^{11}, in the river. The chemical test indicated that the concentrations of isomers, such as 6-, 5-C^{13}, and 6-C^{12} were higher than those of other isomers, suggesting that LC–LABs had been transported over a long distance. At the SKK1 station, refs. [47,48] observed that a high concentration of C^{13} homologs was found, suggesting anaerobic degradation of these compounds. Surprisingly, this study revealed a higher concentration of LABs than reported by [49], with numerous LABs found in the mixture of molecules that are isomeric and homologous, particularly LC–LABs, such as C^{13}–LABs.

Table 2. Total LABs concentration in sediments among different locations of Kim Kim River and Port Dickson, Southern Malaysia: (**a**) ANOVA and (**b**) post hoc tests.

(a) ANOVA					
	Sum of Squares	df	Mean Square	F	Sig. *
Between groups	8410	4	210	7.3	0.000
Within groups	8680	30	289		
Total	17,100	34			

(b) Post Hoc Tests

Dependent Variable: LAB Concentration

LSD

(I) Location	(J) Location	Mean Difference (I-J)	Std. Error	Sig.	95% Confidence Interval Lower Bound	95% Confidence Interval Upper Bound
1.00	2.00	−21.2 *	9.1	0.03	−39.7	−2.6
	3.00	−27.2 *	9.1	0.01	−45.7	−8.6
	4.00	−45.1 *	9.1	0.00	−63.7	−26.6
	5.00	−9.1	9.1	0.32	−27.7	9.5
2.00	1.00	21.2 *	9.1	0.03	2.6	39.7
	3.00	−6.0	9.1	0.52	−24.6	12.6
	4.00	−24.0 *	9.1	0.01	−42.5	−5.3
	5.00	12.1	9.1	0.20	−6.5	30.6
3.00	1.00	27.2 *	9.1	0.01	8.6	45.7
	2.00	6.0	9.1	0.52	−12.6	24.6
	4.00	−18.0	9.1	0.06	−36.5	0.6
	5.00	18.0	9.1	0.06	−0.53	36.6
4.00	1.00	45.1 *	9.1	0.00	26.6	63.7
	2.00	24.0 *	9.1	0.01	5.4	42.5
	3.00	18.0	9.1	0.06	−0.6	36.5
	5.00	36.0 *	9.1	0.00	17.4	54.6
5.00	1.00	9.1	9.1	0.32	−9.5	27.7
	2.00	−12.1	9.1	0.20	−30.6	6.5
	3.00	−18.0	9.1	0.06	−36.6	0.53
	4.00	−36.0 *	9.1	0.00	−54.6	−17.4

Notes: * The mean difference was significant at the 0.05 level.

The PDC sediments contained the highest concentration of C^{13}–LABs homologs while having the lowest levels of C^{10} homologs when compared with all other sites in the region. The distribution of LABs was dominated by LC–LABs, like C^{13} and C^{14}, with higher concentrations than SC–LABs, like C^{10} and C^{11}. The order of homologs by concentration was C^{13}, C^{12}, C^{14}, C^{10}, and C^{11}. According to [50], the variation in composition within this range is probably caused by the growing hydrophobic nature of LABs with longer chains. In contrast with the detergents analyzed by [41], the sediments in this region had a reduced amount of C10 homologs in their distribution. Figure 3 illustrates the contrast between the LC–LABs and SC–LABs in terms of the areas studied, allowing for a comparison between them. Therefore, LC homologs were discovered to be more plentiful than SC homologs in the first and second sites. This contrast may be attributed to the greater decline in SC homologs, particularly C^{10} and C^{11}, in the sediments, as documented by [51,52]. LC–LABs, being longer-chain compounds, are generally less volatile and more hydrophobic compared with SC–LABs, and this can result in differences in their environmental fate and transport. LC–LABs may have a greater tendency to adsorb onto particles and sediments due to their higher molecular weight and hydrophobicity. This enhances their sorption and partitioning into sediment matrices, making them more likely to be retained in the sediments over time. LC–LABs also might undergo slower degradation or transformation processes due to their larger size and more complex structure [53,54].

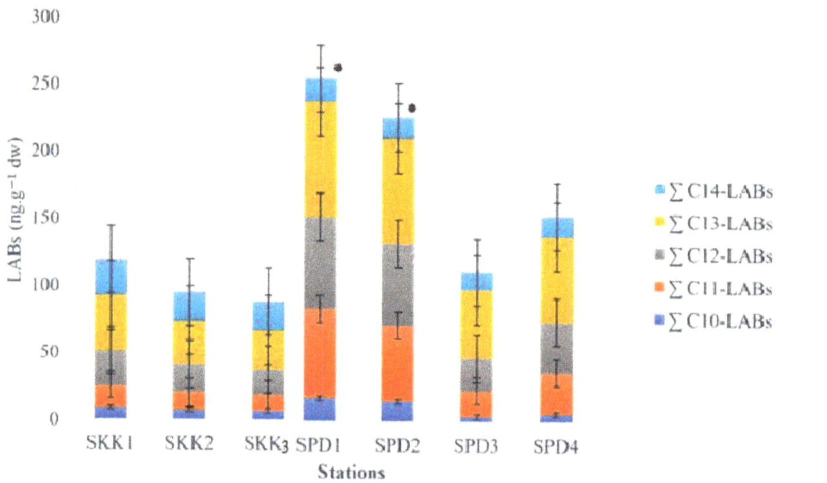

Figure 2. Compositional profiles of LABs in the Kim Kim River and Port Dickson coast sediments. (*) indicates the highly significant differences ($p < 0.05$).

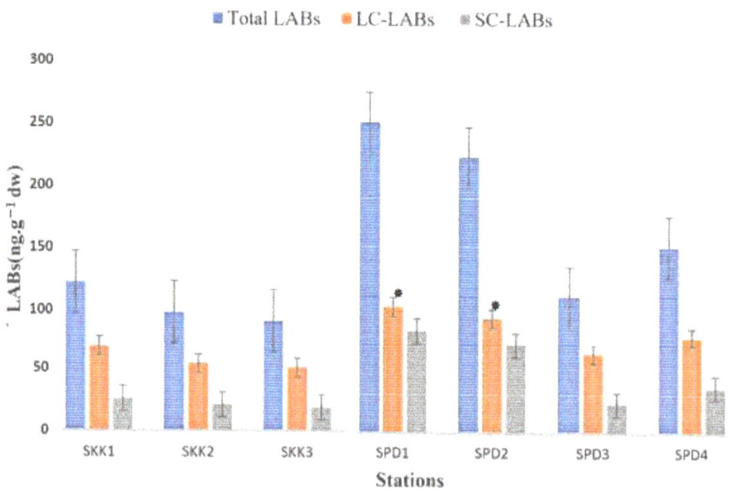

Figure 3. Concentrations of short-chain (SC-LABs), long-chain (LC-LABs), and total linear alkylbenzenes (LABs) in sediments from the Kim Kim River and Port Dickson samples. (*) indicates the highly significant differences ($p < 0.05$).

3.2. Assessing the Efficiency of LABs Removal from Wastewaters

According to [24], the LABs in KKR mainly originated from physical and biological treatment effluents due to the high levels of WWTSs present in the study areas. Figure 4 in their study provides evidence to support this claim. The I/E ratios in KKR varied between SKK2 and SKK1, with an average of 1.8, indicating that primarily treated effluents were released into the river water. This finding is consistent with the study conducted by [55]. On average, the L/S ratio fell between 2.4 and 2.6, with detergents showing a ratio higher than 1.8, indicating that LABs were effectively biodegraded [56]. Furthermore, KKR and the PDC had an average C13/C12 ratio ranging from 5.1 to 5.2, with an average of 5.1, surpassing the ratio of 1.7 recorded in coastal sediments.

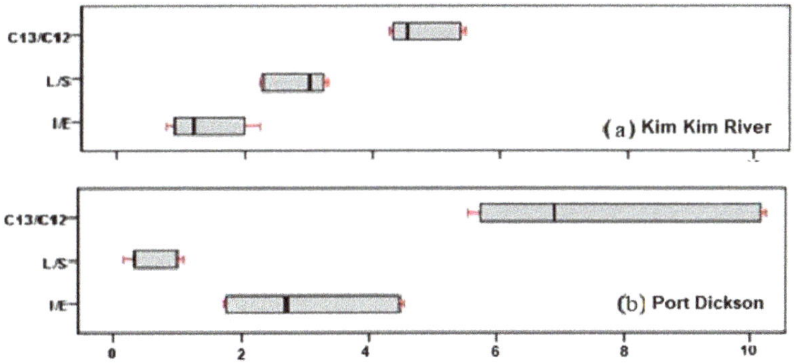

Figure 4. Values of the I/E, L/S, and C^{13}/C^{12} ratios in the samples of Kim Kim River and Port Dickson.

On the other hand, the PDC had an I/E ratio ranging from 2.0 to 4.1, which indicated the discharge of biological treatment effluents in the region. These ratios were considerably higher than those reported by [34] in Selangor's river (0.2–1.0), indicating a substantial discharge of pre-treated effluents in the PDC. Additionally, the degradation of LABs was observed to occur under aerobic conditions. In order to assess the biodegradation of LABs in the studied areas, L/S and C^{13}/C^{12} ratios were used. The presence of LABs supports the idea that untreated effluents were discharged into the PDC, as previously mentioned. According to [57,58], over 80 sewage pipelines were identified as sources of untreated wastewater and effluent discharge from residential areas and hotels, resulting in a deterioration of coastal quality and protection.

The degradation rates of LABs showed variation across different locations, with lower rates found in KKR (SKK3, 34%) and higher rates in the PDC (SPD1, 64%). The results suggest that LABs degrade faster along the PDC shore than in KKR. Over the past few years, waste discharge from ferries and boats, which are used for recreational and fishing purposes, directly into KKR has led to an increase in LABs in the river sediments due to the disposal of detergent waste and washing of boats. These activities have also significantly impacted molecular indices. Untreated wastewater discharge from physical and biological treatments in WWTSs is a significant cause of LAB substances in Malaysia. The results of studies by [59,60] indicate that the KKR region has higher concentrations of industrial effluents, and it discharges LABs from both domestic and industrial wastewater sources. These findings suggest that there could be increased transportation of LABs from their sources to subsequent stages.

3.3. Investigation of TOC

Due to LABs being hydrophobic compounds, they tend to strongly adhere to organic materials in aquatic environments. Consequently, the concentration of LABs found in sediments is closely correlated with the quantity of the TOC available [61]. Thus, a weak correlation ($R_2 = 0.42$) was discovered between the LABs and TOC in sediment samples from

KKR. However, linear regression analysis showed that the TOC did not have a significant impact on the distribution of LABs. The primary factor determining the distribution of LABs in the area was found to be the intensity of their release from various wastewater sources. Previous studies conducted by [34] in Selangor, Peninsular Malaysia, and by [62] in Perak Rivers reported modest correlations between LABs and the TOC, with different R_2 values (0.008 and 0.17, respectively). Nevertheless, the PDC revealed a correlation ($R_2 = 0.64$) between the concentration of LABs and the TOC, indicating that the TOC could be a significant factor in determining the spatial distribution of LABs from urban and industrial sources. This discovery aligns with findings from Dongjiang River ($R_2 = 0.82$), which suggests that industrial and domestic WWTSs are essential sources of organic matter.

3.4. Source Apportionment

During the investigation, significant fluctuations were observed in the I/E ratio, with readings below 1 at many sites, suggesting the discharge of untreated wastewater into the aquatic environment. In order to identify potential sources of LABs in KKR and the PDC shoreline, according to the Kaiser criterion, the data variability at the measurement points can be described by two components. The performed calculations showed that PC1 explains 74.40% and PC2 explains 18.27% of the variability. C10–C14-LABs, LC-LABs, SC-LABs, I/E, L/S, and latitude had the greatest influence on the PC1 component, with the factor loadings ranging from −0.74 to 0.99. However, these variables resulted in a decrease in the PC1 value (Figure 5 and Table S3 in the SI file). On the other hand, the PC2 component was mainly influenced by C13/C12 (−0.815), with a smaller influence from C10-LABs and C14-LABs (0.671–0.711), which led to an increase in the component. The strong correlation between C12–C11-SC-Labs-LC-Labs_Degr was confirmed by the closeness between the variables in Figure 5. Additionally, an increase in longitude and LS led to a decrease in C13/C12. Finally, it was established that there was no correlation between C10 and longitude or latitude.

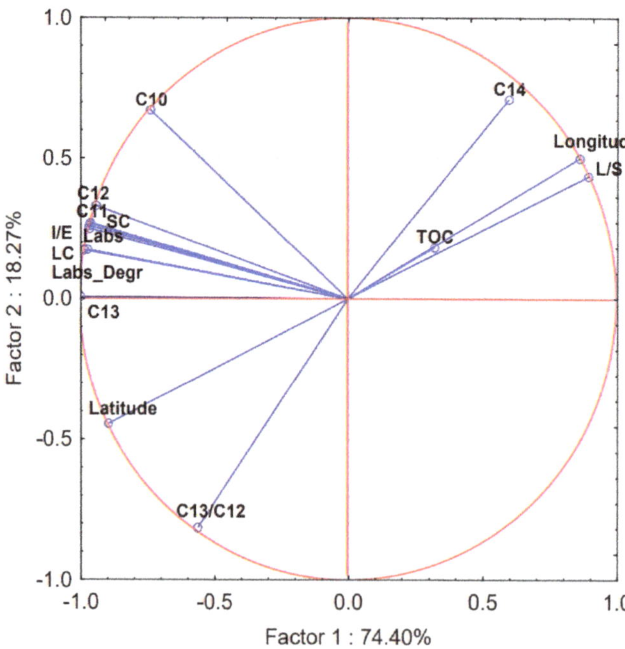

Figure 5. Relationships between factor 1 and factor 2 for loading variables in the sediment samples of Kim Kim River and Port Dickson coast.

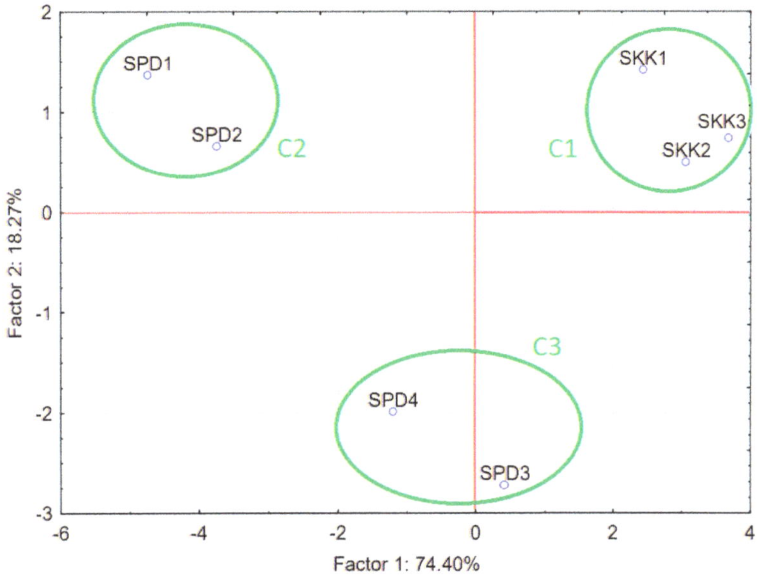

Figure 6. Projection of measurement points on the plane of factors 1 and 2 for loading variables in the sediment samples of Kim Kim River and the Port Dickson coast.

4. Conclusions

The concentration of sedimentary LABs varied greatly among the locations studied, ranging from 88 to 256 ng/g dw. Compared with KKR, the PDC had higher LABs concentrations with more LC homologs than SC homologs. Physical treatment effluents were found to be the essential source of LABs in KKR, while biological treatment effluents dominated the PDC, as indicated by the I/E ratios. The high LABs concentrations in the sediment indicate that the current WWTSs may not be capable of handling the rising population in the area. This research highlights the ongoing problem of untreated wastewater sources, i.e., domestic sewage and industrial influent, which could pose increasing risks as the population expands. According to future projections, the discharge of industrial and urban effluents into rivers and beaches in Peninsular Malaysia is expected to keep rising. This underscores the urgency of enhancing the WWTSs. The findings of this study shed light on the level of pollution in the studied regions, as well as its distribution and sources of organic pollutants (LABs). Public knowledge of the pollution issues in the studied regions can be improved, support for pollution control measures, and environmentally responsible behavior can be encouraged by sharing the findings of environmental monitoring studies. This information can be used to inform decisions made by the authorities about the implementation or modification of pollution control regulations. To decrease sewage contamination in the near future, more effective STPs are required in neighboring areas, together with ongoing monitoring of the marine environment to prevent potential contamination from sewage pollution.

Supplementary Materials: The following supporting information can be downloaded at: https://www.mdpi.com/article/10.3390/w15183301/s1.

Author Contributions: Methodology, formal analysis, investigation, and writing—original draft preparation: S.A.A.A., H.E.A.-H., M.P.Z., N.M., E.H.H.A.-Q., J.M. and B.S. Writing—review and editing and resources: P.K., S.A.M.A., S.S.A. and S.M.M. Supervision: S.S. and T.A.K. All authors have read and agreed to the published version of the manuscript.

Funding: Postdoctoral scheme grants from the University Malaysia Terengganu (UMT) and the Inisiatif Putra Berkumpulan Grant from the University of Perak (UPM) funded this study (9412401).

Data Availability Statement: The metadata used to support the findings of this study were deposited in the University 301 Putra Malaysia repository at http://upm.edu.my/ (accessed on 11 September 2023).

Acknowledgments: The authors thank the University Malaysia Terengganu (UMT) and the Inisiatif Putra Berkumpulan Grant from the University of Perak (UPM) for funding this study.

Conflicts of Interest: The authors declare no conflict of interest.

References

1. Adyasari, D.; Pratama, M.A.; Teguh, N.A.; Sabdaningsih, A.; Kusumaningtyas, M.A.; Dimova, N. Anthropogenic Impact on Indonesian Coastal Water and Ecosystems: Current Status and Future Opportunities. *Mar. Pollut. Bull.* **2021**, *171*, 112689. [CrossRef]
2. Linghu, K.; Wu, Q.; Zhang, J.; Wang, Z.; Zeng, J.; Gao, S. Occurrence, Distribution and Ecological Risk Assessment of Antibiotics in Nanming River: Contribution from Wastewater Treatment Plant and Implications of Urban River Syndrome. *Process Saf. Environ. Prot.* **2023**, *169*, 428–436. [CrossRef]
3. Alkhadher, S.A.A.; Suratman, S.; Al-Hazmi, H.E.; Zakaria, M.P.; Szeląg, B.; Majtacz, J.; Drewnowski, J. Unlocking the Secrets of River Pollution: Analyzing Organic Pollutants in Sediments—Experimental Study. *Water* **2023**, *15*, 2216. [CrossRef]
4. Afify, A.A.; Hassan, G.K.; Al-Hazmi, H.E.; Kamal, R.M.; Mohamed, R.M.; Drewnowski, J.; Majtacz, J.; Mąkinia, J.; El-Gawad, H.A. Electrochemical Production of Sodium Hypochlorite from Salty Wastewater Using a Flow-by Porous Graphite Electrode. *Energies* **2023**, *16*, 4754. [CrossRef]
5. Bogacki, J.P.; Al-Hazmi, H. Automotive Fleet Repair Facility Wastewater Treatment Using Air/ZVI and Air/ZVI/H_2O_2 Processes. *Arch. Environ. Prot.* **2017**, *43*, 24–31. [CrossRef]
6. Al-Hazmi, H.E.; Shokrani, H.; Shokrani, A.; Jabbour, K.; Abida, O.; Mousavi Khadem, S.S.; Habibzadeh, S.; Sonawane, S.H.; Saeb, M.R.; Bonilla-Petriciolet, A.; et al. Recent Advances in Aqueous Virus Removal Technologies. *Chemosphere* **2022**, *305*, 135441. [CrossRef] [PubMed]
7. Al-Hazmi, H.E.; Kot-Wasik, A.; Shokrani, A.; Majtacz, J.; Vatanpour, V.; Munir, M.T.; Habibzadeh, S.; Hejna, A.; Hasanpour, M.; Mohammadi, A.; et al. Diving Boldly into COVID-19 Contaminated Wastewater: Eyes at Nanotechnology-Assisted Solutions. *Clin. Transl. Discov.* **2023**, *3*, e195. [CrossRef]
8. Al-Hazmi, H.; Namiesnik, J.; Tobiszewski, M. Application of TOPSIS for Selection and Assessment of Analytical Procedures for Ibuprofen Determination in Wastewater. *Curr. Anal. Chem.* **2016**, *12*, 261–267. [CrossRef]
9. Pereira, T.D.S.; Al-Hazmi, H.E.; Majtacz, J.; Pires, E.C. Effect of Operating Conditions on N_2O Production in an Anammox Sequencing Batch Reactor Containing Granular Sludge. *Water Sci.* **2019**, *880*, 37–47. [CrossRef]
10. Feng, Y.; Lu, X.; Al-Hazmi, H.; Mąkinia, J. An Overview of the Strategies for the Deammonification Process Start-up and Recovery after Accidental Operational Failures. *Rev. Environ. Sci. Biotechnol.* **2017**, *16*, 541–568. [CrossRef]
11. Al-Hazmi, H.; Lu, X.; Grubba, D.; Majtacz, J.; Kowal, P.; Mąkinia, J. Achieving Efficient and Stable Deammonification at Low Temperatures—Experimental and Modeling Studies. *Energies* **2021**, *14*, 3961. [CrossRef]
12. Lu, X.; Pereira, T.D.S.; Al-Hazmi, H.E.; Majtacz, J.; Zhou, Q.; Xie, L.; Makinia, J. Model-Based Evaluation of N_2O Production Pathways in the Anammox-Enriched Granular Sludge Cultivated in a Sequencing Batch Reactor. *Environ. Sci. Technol.* **2018**, *52*, 2800–2809. [CrossRef] [PubMed]
13. Maktabifard, M.; Al-Hazmi, H.E.; Szulc, P.; Mousavizadegan, M.; Xu, X.; Zaborowska, E.; Li, X.; Mąkinia, J. Net-Zero Carbon Condition in Wastewater Treatment Plants: A Systematic Review of Mitigation Strategies and Challenges. *Renew. Sustain. Energy Rev.* **2023**, *185*, 113638. [CrossRef]
14. Majtacz, J.B.; Kowal, P.; Lu, X.; Al-Hazmi, H.; Makinia, J. Adaptation of the activated sludge to the digestate liquors during the nitrification and denitrification processes. *J. Ecol. Eng.* **2017**, *18*, 104–109. [CrossRef] [PubMed]
15. Al-Hazmi, H.E.; Lu, X.; Grubba, D.; Majtacz, J.; Badawi, M.; Mąkinia, J. Sustainable Nitrogen Removal in Anammox-Mediated Systems: Microbial Metabolic Pathways, Operational Conditions and Mathematical Modelling. *Sci. Total Environ.* **2023**, *868*, 161633. [CrossRef] [PubMed]
16. Derwis, D.; Majtacz, J.; Kowal, P.; Al-Hazmi, H.E.; Zhai, J.; Ciesielski, S.; Piechota, G.; Mąkinia, J. Integration of the Sulfate Reduction and Anammox Processes for Enhancing Sustainable Nitrogen Removal in Granular Sludge Reactors. *Bioresour. Technol.* **2023**, *383*, 129264. [CrossRef]
17. Al-Hazmi, H.E.; Maktabifard, M.; Grubba, D.; Majtacz, J.; Hassan, G.K.; Lu, X.; Piechota, G.; Mannina, G.; Bott, C.B.; Mąkinia, J. An Advanced Synergy of Partial Denitrification-Anammox for Optimizing Nitrogen Removal from Wastewater: A Review. *Bioresour. Technol.* **2023**, *381*, 129168. [CrossRef]
18. Grubba, D.; Yin, Z.; Majtacz, J.; Al-Hazmi, H.E.; Mąkinia, J. Incorporation of the Sulfur Cycle in Sustainable Nitrogen Removal Systems—A Review. *J. Clean. Prod.* **2022**, *372*, 133495. [CrossRef]

19. Al-Hazmi, H.; Grubba, D.; Majtacz, J.; Kowal, P.; Makinia, J. Evaluation of Partial Nitritation/Anammox (PN/A) Process Performance and Microorganisms Community Composition under Different C/N Ratio. *Water* **2019**, *11*, 2270. [CrossRef]
20. Al-Hazmi, H.E.; Yin, Z.; Grubba, D.; Majtacz, J.B.; Mąkinia, J. Comparison of the Efficiency of Deammonification under Different DO Concentrations in a Laboratory-Scale Sequencing Batch Reactor. *Water* **2022**, *14*, 368. [CrossRef]
21. Al-Hazmi, H.E.; Mohammadi, A.; Hejna, A.; Majtacz, J.; Esmaeili, A.; Habibzadeh, S.; Saeb, M.R.; Badawi, M.; Lima, E.C.; Mąkinia, J. Wastewater Reuse in Agriculture: Prospects and Challenges. *Environ. Res.* **2023**, *236*, 116711. [CrossRef]
22. Soleimani-Gorgani, A.; Al-Hazmi, H.E.; Esmaeili, A.; Habibzadeh, S. Screen-Printed Sn-Doped TiO_2 Nanoparticles for Photocatalytic Dye Removal from Wastewater: A Technological Perspective. *Environ. Res.* **2023**, *237*, 117079. [CrossRef] [PubMed]
23. Masood, N.; Alkhadher, S.A.A.; Magam, S.M.; Halimoon, N.; Alsukaibi, A.; Zakaria, M.P.; Vaezzadeh, V.; Keshavarzifard, M.; Maisara, S.; Khaled Bin Break, M. Monitoring of Linear Alkyl Benzenes (LABs) in Riverine and Estuarine Sediments in Malaysia. *Env. Geochem. Health* **2022**, *44*, 3687–3702. [CrossRef] [PubMed]
24. Alkhadher, S.A.A.; Kadir, A.A.; Zakaria, M.P.; Adel, A.-G.; Magam, S.M.S.; Masood, N. Monitoring of Sewage Pollution in the Surface Sediments of Coastal Ecosystems Using Linear Alkylbenzenes (LABs) as Molecular Markers. *J. Soils Sediments* **2020**, *20*, 3230–3242. [CrossRef]
25. Cabral, A.C.; Dauner, A.L.L.; Xavier, F.C.B.; Garcia, M.R.D.; Wilhelm, M.M.; dos Santos, V.C.G.; Netto, S.A.; Martins, C.C. Tracking the Sources of Allochthonous Organic Matter along a Subtropical Fluvial-Estuarine Gradient Using Molecular Proxies in View of Land Uses. *Chemosphere* **2020**, *251*, 126435. [CrossRef] [PubMed]
26. Alkhadher, S.A.A.; Zakaria, M.P.; Yusoff, F.M.; Kannan, N.; Suratman, S.; Keshavarzifard, M.; Magam, S.M.; Masood, N.; Vaezzadeh, V.; Sani, M.S.A. Baseline Distribution and Sources of Linear Alkyl Benzenes (LABs) in Surface Sediments from Brunei Bay, Brunei. *Mar. Pollut. Bull.* **2015**, *101*, 397–403. [CrossRef]
27. Al-Hazmi, H.E.; Grubba, D.; Majtacz, J.; Ziembińska-Buczyńska, A.; Zhai, J.; Mąkinia, J. Combined Partial Denitrification/Anammox Process for Nitrogen Removal in Wastewater Treatment. *J. Environ. Chem. Eng.* **2023**, *11*, 108978. [CrossRef]
28. Alkhadher, S.A.A.; Pauzi Zakaria, M.; Suratman, S.; Alanazi, T.Y.A.; Al-Bagawi, A.H.; Magam, S.M.; Masood, N.; Abdul Kadir, A.; Al-Gheethi, A. Assessment of Sewage Molecular Markers in Port Dickson Coast and Kim Kim River with Sediment Linear Alkylbenzenes. *Polycycl. Aromat. Compd.* **2023**, *43*, 343–355. [CrossRef]
29. Sakai, N.; Shirasaka, J.; Matsui, Y.; Ramli, M.R.; Yoshida, K.; Ali Mohd, M.; Yoneda, M. Occurrence, Fate and Environmental Risk of Linear Alkylbenzene Sulfonate in the Langat and Selangor River Basins, Malaysia. *Chemosphere* **2017**, *172*, 234–241. [CrossRef]
30. Lin, L.; Yang, H.; Xu, X. Effects of Water Pollution on Human Health and Disease Heterogeneity: A Review. *Front. Environ. Sci.* **2022**, *10*, 880246. [CrossRef]
31. Sallan, M.I.B.M.; Al-Hazmi, H.E.; Suratman, S.; Alkhadher, S.A.A.; Szeląg, B.; Majtacz, J.; Kowal, P.; Kurniawan, T.A.; Piechota, G. Anthropogenic Trace Metals in Setiu Wetland: Spatial and Seasonal Distribution and Implications for Environmental Health. *J. Water Process Eng.* **2023**, *55*, 104172. [CrossRef]
32. Al-Gheethi, A.; Rahman, J.A.; Mohamed, R.M.S.R.; Talip, B.; Abdullah, M.E. Bacteria Load Assessment at Sungai Benut in Simpang Renggam, Johor. *IOP Conf. Ser. Earth Environ. Sci.* **2020**, *498*, 012061. [CrossRef]
33. Zakaria, M.P.; Takada, H.; Tsutsumi, S.; Ohno, K.; Yamada, J.; Kouno, E.; Kumata, H. Distribution of Polycyclic Aromatic Hydrocarbons (PAHs) in Rivers and Estuaries in Malaysia: A Widespread Input of Petrogenic PAHs. *Environ. Sci. Technol.* **2002**, *36*, 1907–1918. [CrossRef] [PubMed]
34. Masood, N.; Zakaria, M.P.; Halimoon, N.; Aris, A.Z.; Magam, S.M.; Kannan, N.; Mustafa, S.; Ali, M.M.; Keshavarzifard, M.; Vaezzadeh, V.; et al. Anthropogenic Waste Indicators (AWIs), Particularly PAHs and LABs, in Malaysian Sediments: Application of Aquatic Environment for Identifying Anthropogenic Pollution. *Mar. Pollut. Bull.* **2016**, *102*, 160–175. [CrossRef]
35. Eganhouse, R.P. Molecular Markers and Their Use in Environmental Organic Geochemistry. In *The Geochemical Society Special Publications*; Hill, R.J., Leventhal, J., Aizenshtat, Z., Baedecker, M.J., Claypool, G., Eganhouse, R., Goldhaber, M., Peters, K., Eds.; Geochemical Investigations in Earth and Space Science: A Tribute to Isaac R. Kaplan; Elsevier: Amsterdam, The Netherlands, 2004; Volume 9, pp. 143–158.
36. Tsutsumi, S.; Yamaguchi, Y.; Nishida, I.; Akiyama, K.-I.; Zakaria, M.P.; Takada, H. Alkylbenzenes in Mussels from South and South East Asian Coasts as a Molecular Tool to Assess Sewage Impact. *Mar. Pollut. Bull.* **2002**, *45*, 325–331. [CrossRef]
37. Nelson, D.; Sommers, L. Total carbon, organic carbon and organic matter. In *Methods of Soil Analysis. Part 3: Chemical Methods*; Sparks, D.L., Ed.; Soil Science Society of America: Madison, WI, USA, 1996; pp. 961–1010.
38. Zeng, E.Y.; Vista, C.L. Organic Pollutants in the Coastal Environment off San Diego, California. 1. Source Identification and Assessment by Compositional Indices of Polycyclic Aromatic Hydrocarbons. *Environ. Toxicol. Chem.* **1997**, *16*, 179–188. [CrossRef]
39. Cabral, A.C.; Martins, C.C. Insights about Sources, Distribution, and Degradation of Sewage and Biogenic Molecular Markers in Surficial Sediments and Suspended Particulate Matter from a Human-Impacted Subtropical Estuary. *Environ. Pollut.* **2018**, *241*, 1071–1081. [CrossRef]
40. Shahbazi, A.; Zakaria, M.P.; Yap, C.K.; Tan, S.G.; Surif, S.; Mohamed, C.A.R.; Sakari, M.; Bakhtiari, A.R.; Bahry, P.S.; Chandru, K.; et al. Use of Different Tissues of Perna Viridis as Biomonitors of Polycyclic Aromatic Hydrocarbons (PAHs) in the Coastal Waters of Peninsular Malaysia. *Environ. Forensics* **2010**, *11*, 248–263. [CrossRef]
41. Sakai, N.; Mohd Yusof, R.; Sapar, M.; Yoneda, M.; Ali Mohd, M. Spatial Analysis and Source Profiling of Beta-Agonists and Sulfonamides in Langat River Basin, Malaysia. *Sci. Total Environ.* **2016**, *548–549*, 43–50. [CrossRef]

42. Luo, X.-J.; Chen, S.-J.; Ni, H.-G.; Yu, M.; Mai, B.-X. Tracing Sewage Pollution in the Pearl River Delta and Its Adjacent Coastal Area of South China Sea Using Linear Alkylbenzenes (LABs). *Mar. Pollut. Bull.* **2008**, *56*, 158–162. [CrossRef]
43. Zacchi, F.L.; Flores-Nunes, F.; Mattos, J.J.; Lima, D.; Lüchmann, K.H.; Sasaki, S.T.; Bícego, M.C.; Taniguchi, S.; Montone, R.C.; de Almeida, E.A.; et al. Biochemical and Molecular Responses in Oysters Crassostrea Brasiliana Collected from Estuarine Aquaculture Areas in Southern Brazil. *Mar. Pollut. Bull.* **2018**, *135*, 110–118. [CrossRef]
44. Isobe, K.O.; Zakaria, M.P.; Chiem, N.H.; Minh, L.Y.; Prudente, M.; Boonyatumanond, R.; Saha, M.; Sarkar, S.; Takada, H. Distribution of Linear Alkylbenzenes (LABs) in Riverine and Coastal Environments in South and Southeast Asia. *Water Res.* **2004**, *38*, 2448–2458. [CrossRef] [PubMed]
45. Bakhtiari, A.R.; Javedankherad, I.; Mohammadi, J.; Taghizadeh, R. Distribution of Linear Alkylbenzenes as a Domestic Sewage Molecular Marker in Surface Sediments of International Anzali Wetland in the Southwest of the Caspian Sea, Iran. *Environ.Sci. Pollut. Res.* **2018**, *25*, 20920–20929. [CrossRef] [PubMed]
46. Alkhadher, S.A.A.; Kadir, A.A.; Zakaria, M.P.; Al-Gheethi, A.; Asghar, B.H.M. Determination of Linear Alkylbenzenes (LABs) in Mangrove Ecosystems Using the Oyster Crassostrea Belcheri as a Biosensor. *Mar. Pollut. Bull.* **2020**, *154*, 111115. [CrossRef] [PubMed]
47. Alkhadher, S.A.A.; Suratman, S.; Zakaria, M.P. Lateral Distribution, Environmental Occurrence and Assessment of Organic Pollutants in Surface Sediments of the West and South Peninsular Malaysia. *Water Air Soil Pollut.* **2023**, *234*, 124. [CrossRef]
48. Dauner, A.L.L.; Hernández, E.A.; MacCormack, W.P.; Martins, C.C. Molecular Characterisation of Anthropogenic Sources of Sedimentary Organic Matter from Potter Cove, King George Island, Antarctica. *Sci. Total Environ.* **2015**, *502*, 408–416. [CrossRef]
49. Alkhadher, S.A.A.; Kadir, A.A.; Zakaria, M.P.; Al-Gheethi, A.; Keshavarzifard, M.; Masood, N.; Alenezi, K.M.; Magam, S.M. Linear Alkylbenzenes in Surface Sediments of an Estuarine and Marine Environment in Peninsular Malaysia. *Mar. Pollut. Bull.* **2020**, *153*, 111013. [CrossRef]
50. Sherblom, P.M.; Gschwend, P.M.; Eganhouse, R.P. Aqueous Solubilities, Vapor Pressures, and 1-Octanol-Water Partition Coefficients for C9-C14 Linear Alkylbenzenes. *J. Chem. Eng. Data* **1992**, *37*, 394–399. [CrossRef]
51. Abdo Alkhadher, S.A.; Zakaria, M.P.; Yusoff, F.M.; Kannan, N.; Suratman, S.; Magam, S.M.; Masood, N.; Keshavarzifard, M.; Vaezzadeh, V.; Sani, M.S.A. Distribution and Sources of Linear Alkyl Benzenes (LABs) in Surface Sediments from Johor Bahru Coast and the Kim Kim River, Malaysia. *Environ. Forensics* **2016**, *17*, 36–47. [CrossRef]
52. Magam, S.M.; Zakaria, M.P.; Halimoon, N.; Aris, A.Z.; Kannan, N.; Masood, N.; Mustafa, S.; Alkhadher, S.; Keshavarzifard, M.; Vaezzadeh, V.; et al. Evaluation of Distribution and Sources of Sewage Molecular Marker (LABs) in Selected Rivers and Estuaries of Peninsular Malaysia. *Environ. Sci. Pollut. Res.* **2016**, *23*, 5693–5704. [CrossRef]
53. Alkhadher, S.A.A.; Suratman, S.; Zakaria, M.P. Occurrence and Assessment of Organic Pollutants Residues in the Aquatic Environment of the Coastal Sediments. *Sustainability* **2023**, *15*, 8365. [CrossRef]
54. Alkhadher, S.A.A.; Suratman, S.; Zakaria, M.P. Lateral Distribution, Environmental Occurrence, and Assessment of Organic Pollutants in Surface Sediments of the East Malaysia. *Environ. Monit. Assess.* **2023**, *195*, 720. [CrossRef]
55. Gustafsson, Ö.; Long, C.M.; MacFarlane, J.; Gschwend, P.M. Fate of Linear Alkylbenzenes Released to the Coastal Environment near Boston Harbor. *Environ. Sci. Technol.* **2001**, *35*, 2040–2048. [CrossRef] [PubMed]
56. Ni, H.-G.; Lu, F.-H.; Wang, J.-Z.; Guan, Y.-F.; Luo, X.-L.; Zeng, E.Y. Linear Alkylbenzenes in Riverine Runoff of the Pearl River Delta (China) and Their Application as Anthropogenic Molecular Markers in Coastal Environments. *Environ. Pollut.* **2008**, *154*, 348–355. [CrossRef] [PubMed]
57. Hamzah, A.; Kipli, S.H.; Ismail, S.R.; Una, R.; Sarmani, S. Microbiological Study in Coastal Water of Port Dickson, Malaysia. *Sains Malays.* **2011**, *40*, 93–99.
58. Thomes, M.W.; Vaezzadeh, V.; Zakaria, M.P.; Bong, C.W. Use of Sterols and Linear Alkylbenzenes as Molecular Markers of Sewage Pollution in Southeast Asia. *Environ. Sci. Pollut. Res.* **2019**, *26*, 31555–31580. [CrossRef]
59. Cabral, A.C.; Stark, J.S.; Kolm, H.E.; Martins, C.C. An Integrated Evaluation of Some Faecal Indicator Bacteria (FIB) and Chemical Markers as Potential Tools for Monitoring Sewage Contamination in Subtropical Estuaries. *Environ. Pollut.* **2018**, *235*, 739–749. [CrossRef] [PubMed]
60. Zhang, K.; Wang, J.-Z.; Liang, B.; Shen, R.-L.; Zeng, E.Y. Assessment of Aquatic Wastewater Pollution in a Highly Industrialized Zone with Sediment Linear Alkylbenzenes. *Environ. Toxicol. Chem.* **2012**, *31*, 724–730. [CrossRef]
61. Harwood, J.J. Molecular Markers for Identifying Municipal, Domestic and Agricultural Sources of Organic Matter in Natural Waters. *Chemosphere* **2014**, *95*, 3–8. [CrossRef]
62. Wang, X.-C.; Zhang, Y.-X.; Chen, R.F. Distribution and Partitioning of Polycyclic Aromatic Hydrocarbons (PAHs) in Different Size Fractions in Sediments from Boston Harbor, United States. *Mar. Pollut. Bull.* **2001**, *42*, 1139–1149. [CrossRef]

water

MDPI

Article

An Integrated GIS and Machine-Learning Technique for Groundwater Quality Assessment and Prediction in Southern Saudi Arabia

Mustafa El-Rawy [1,2,*], Okke Batelaan [3], Fahad Alshehri [4,*], Sattam Almadani [4], Mohamed S. Ahmed [4] and Ahmed Elbeltagi [5]

1 Civil Engineering Department, Faculty of Engineering, Minia University, Minia 61111, Egypt
2 Civil Engineering Department, College of Engineering, Shaqra University, Dawadmi 11911, Saudi Arabia
3 National Centre for Groundwater Research and Training, College of Science and Engineering, Flinders University, GPO Box 2100, Adelaide, SA 5001, Australia; okke.batelaan@flinders.edu.au
4 Abdullah Alrushaid Chair for Earth Science Remote Sensing Research, Geology and Geophysics Department, College of Science, King Saud University, Riyadh 11451, Saudi Arabia
5 Agricultural Engineering Department, Faculty of Agriculture, Mansoura University, Mansoura 35516, Egypt; ahmedelbeltagy81@mans.edu.eg
* Correspondence: mustafa.elrawy@mu.edu.eg (M.E.-R.); falshehria@ksu.edu.sa (F.A.)

Abstract: One of the most critical stages for developing groundwater resources for drinking water use is assessing the water quality. The use of a Water Quality Index (WQI) is considered an effective method of evaluating water quality. The objective of this research was to evaluate the performance of six multiple artificial intelligence techniques, i.e., linear regression (stepwise), support vector regression SVM (linear and polynomial kernels), Gaussian process regression (GPR), Fit binary tree, and artificial neural network ANN (Bayesian) to predict the WQI in Jizan, Southern Saudi Arabia. A total of 145 groundwater samples were collected from shallow dug wells and boreholes tapping the phreatic aquifer. The WQI was calculated from 11 physicochemical parameters (pH, TDS, Ca^{2+}, Mg^{2+}, Na^+, K^+, Cl^-, $SO_4{}^{2-}$, $HCO_3{}^-$, $NO_3{}^-$, and TH). The spatial distribution results showed that higher values of Cl^- and $SO_4{}^{2-}$ were recorded in the places close to the coastline, indicating the occurrence of seawater intrusion and salinisation. Seven wells had a WQI of greater than 300, indicating that the water was unfit for consumption. The results showed that the GPR, linear regression (stepwise), and ANN models performed best during the training and testing stages, with a high correlation of 1.00 and low errors. The stepwise fitting model indicated that pH, K^+, and $NO_3{}^-$ were the most significant variables, while $HCO_3{}^-$ was a non-significant variable for the WQI. The GPR, stepwise regression, and ANN models performed best during the training and testing stages, with a high correlation and low errors. In contrast, the SVM and Fit binary tree models performed the worst in the training and testing phases. Based on subset regression analysis, the optimum input combination for WQI model prediction was determined as these eight input combinations with high R^2 (0.975–1.00) and high Adj-R^2 (0.974–1.00). The resultant WQI model significantly contributes to sustainable groundwater resource management in arid areas and generates improved prediction precision with fewer input parameters.

Keywords: water quality index; artificial intelligence; support vector machine; Gaussian process regression; stepwise regression

Citation: El-Rawy, M.; Batelaan, O.; Alshehri, F.; Almadani, S.; Ahmed, M.S.; Elbeltagi, A. An Integrated GIS and Machine-Learning Technique for Groundwater Quality Assessment and Prediction in Southern Saudi Arabia. *Water* **2023**, *15*, 2448. https://doi.org/10.3390/w15132448

Academic Editors: Imokhai Theophilus Tenebe and Giuseppe Pezzinga

Received: 27 April 2023
Revised: 22 June 2023
Accepted: 27 June 2023
Published: 4 July 2023

1. Introduction

Groundwater plays an essential role in the overall use and management of water resources. The demand for groundwater for municipal, agricultural, and industrial use has grown gradually during the past decades, especially in arid regions like Saudi Arabia, where groundwater is the primary source of water. In Saudi Arabia, groundwater contributes to nearly 79% of the total water supply, and around 90% is consumed in agricultural activity. Many cities, towns, and villages rely exclusively on wells and natural springs for their municipal water [1,2].

Groundwater quality is determined by the natural and physical state of the water-rock interactions and by the changes induced by human activities [3]. Groundwater contamination is generally due to urbanisation, industrialisation, and agriculture that has gradually developed over the years without considering environmental consequences [4]. Water quality assessments aim to characterise the chemical, physical, and biological conditions of groundwater and identify the source of any possible contamination that causes water quality degradation [5]. Generally, groundwater quality parameters are compared with permissible levels for a particular use to help indicate contamination sources [6–8]. The assessment of groundwater quality depends mainly on laboratory investigations carried out through the measurement of water quality variables, followed by a comparison of the obtained concentrations with the standards and guidelines [9]. Applied methodologies for water quality assessment often combine all the variables and present a final value as a quality index providing meaningful summaries of water quality data useful to technical and policy individuals and the public interested in water quality [10].

Geographical information systems (GIS) can be a great complementary tool for creating and developing spatial representations of water quality assessments [8,11]. Gunduz and Simsek [12] and Usali and Ismail [13] applied a GIS-integrated technique to assess irrigation water quality in respectively, Turkey and Malaysia. They concluded that water quality parameters could be produced in the form of a map using model-based GIS techniques and considered this product as the most suitable method for groundwater potential prediction zoning.

The application of a Water Quality Index (WQI) is considered an effective method for evaluating water quality [14]. A WQI is a premium method for understanding and summarising large numbers of water quality data by integrating complex information and expressions to represent a combined effect of the variables influencing water quality. Thus, a WQI enables effective monitoring and evaluation of groundwater quality. Over the last few decades, WQIs have been widely used for surface water and groundwater quality assessments worldwide [15,16]. There are many water quality indices, such as the Weighted Arithmetic Water Quality Index (WAWQI), National Sanitation Foundation Water Quality Index (NSFWQI), Canadian Council of Ministers of the Environment Water Quality Index (CCMEWQI), and Oregon Water Quality Index (OWQI). National and international organisations have formulated these indices dependent on a number of water quality parameters relative to the specific requirements of a given area [17,18]. Water quality indices have been shown to demonstrate temporal and spatial differences in water quality, even at small concentrations, in an accurate and timely manner [19].

There is a current rise in the use of artificial intelligence (AI) techniques to estimate WQIs [20–24]. Groundwater quality can be understood and monitored using artificial neural networks (ANNs) and used to predict water quality with great success [25–27]. Also, other computational intelligence techniques, such as genetic algorithms (GA), support vector machine (SVM), Fit binary Tree, and Gaussian process regression (GPR), have attracted growing interest in WQI prediction studies [28,29]. The non-linear structure of computational intelligence techniques and their ability to anticipate complex occurrences, handle massive datasets of varying sizes, and accommodate missing data are all advantages. Additionally, artificial intelligence approaches have been shown to be extremely capable of forecasting water quality [26,27,30–36].

Gazzaz et al. [30] applied a neural network model for calculating a WQI for the Kinta River, Malaysia. The model's WQI predictions were highly correlated with measured WQI values (r = 0.977). El Bilali and Taleb [31] used eight machine learning (ML) models: artificial neural network (ANN), multiple linear regression (MLR), decision tree, Random Forest (RF), support vector machine (SVM), k-nearest neighbour (kNN), stochastic gradient descent (SGD), and adaptive boosting (AdaBoost) to forecast ten irrigation water quality (IWQ) parameters in the Bouregreg watershed, Morocco. The findings of the machine learning models showed that they are cost-effective tools for predicting irrigation water quality. Kulisz et al. [33] developed an ANN model using five parameters (EC, pH, Ca, Mg, and K) to forecast a groundwater WQI in Syczyn, Poland. It was concluded that the ANN tools predicted the WQI at a desirable level of accuracy (r = 0.9992). Kouadri et al. (2021) used eight artificial intelligence algorithms: MLR, RF, M5P model tree, random subspace (RSS), additive regression (AR), ANN, SVR, and locally weighted linear regression (LWLR) to predict a WQI in Illizi region, southeast Algeria. The MLR model had a higher level of accuracy when compared to other models. Gupta et al. [32] employed machine learning algorithms to evaluate a WQI in India's Mid Gangetic Region. They concluded that machine learning models are a suitable alternative for groundwater water quality evaluation and may be applied swiftly utilising a data-driven approach. Setshedi et al. [26] employed an ANN to build the best model for forecasting water quality metrics using data from three district municipalities in the Eastern Cape Province, South Africa. The findings revealed that the ANN model is a valuable and reliable tool for optimising the observational network by identifying key monitoring sites and accurately forecasting the quality of river water variables. Mokhtar et al. [27] applied three artificial intelligence (AI) and four multiple regression models to forecast six irrigation water quality criteria. The findings indicated that these models could be used to make quick decisions about irrigation water quality.

To the best of our knowledge, no research has been published that evaluates the performance of artificial intelligence approaches to predict WQIs in the area of Jazan and Tihama plains in the southwestern part of the Red Sea coast of Saudi Arabia. The choice of the study area takes into consideration its importance to national development in the Kingdom. The study area and its surroundings are considered one of the most promising areas in agricultural and industrial development. Thus, the evaluation of the Water Quality Index in the study area could be useful to help planners and decision-makers to protect groundwater resources from deterioration.

Hence, the goal of this research is to (i) test a number of advanced artificial intelligence techniques in their capacity to estimate a WQI using 11 physicochemical parameters (pH, TDS, Ca^{2+}, Mg^{2+}, Na^{+}, K^{+}, Cl^{-}, $SO_4{}^{2-}$, $HCO_3{}^{-}$, $NO_3{}^{-}$, and TH) collected from 145 groundwater wells in Jizan, Saudi Arabia, and (ii) to select the statistically optimal artificial intelligence model in predicting a WQI. The following steps were taken to achieve this goal. Firstly, the statistical analysis and correlation coefficients for the physicochemical parameters were determined. Secondly, ArcGIS was used to create maps of the spatial distribution of groundwater quality metrics. Thirdly, the Weighted Arithmetic Water Quality Index (proposed by Horton [37]) was used to calculate the WQI. Fourthly, to predict the WQI, multiple artificial intelligence techniques were used (linear regression (stepwise), SVM (linear and polynomial kernels), Gaussian process regression (GPR), Fit binary tree, and ANN (Bayesian)). Finally, the best subset regression analysis was performed to determine the best input combinations for the WQI model. This study presents two essential findings, which are as follows: (1) Creating a single-line linear equation that can be easily applied by water users and decision makers when all parameters are available (11 inputs); (2) when data are limited, we used the best subset regression model to extract the optimal input parameters to the ML model for WQ prediction. As a result of this research, two future plans/strategies for water quality simulations will be developed.

2. Materials and Methods

2.1. Study Area Description

The Jizan study area is located in the southwest corner of Saudi Arabia, directly north of the border with Yemen. It is located between longitude 41°56′18″ E and 43°15′58″ E and latitude 16°23′8″ N and 17°53′56″ N (Figure 1). The study area covers about 10,753 km^2. Jazan City is located on the Red Sea coast and serves a large agricultural heartland with a population of 319,119 as of 2021. Based on climate data for Jazan from 1985 to 2010 (Figure 2), Jazan has a hot desert climate with an average annual temperature of more than 30 °C. The weather is extremely hot all year, with daily lows averaging over 25 °C and highs averaging over 35 °C even in the coldest month. The average evaporation rate is 2000 mm/year (Source: Jeddah Regional Climate Center [38]. The southwestern region of Saudi Arabia is rich in rainfall compared to other areas of the Kingdom of Saudi Arabia, with average annual precipitation in the range of 400–700 mm/year [39]. The watersheds collect these precipitations that exclusively occur during the winter season from the adjoining hills and channel the collected runoff toward the Red Sea as surface runoff and/or infiltration into the near-surface aquifers [40,41]. The importance of the study area for national and economic development was a driving goal for this research.

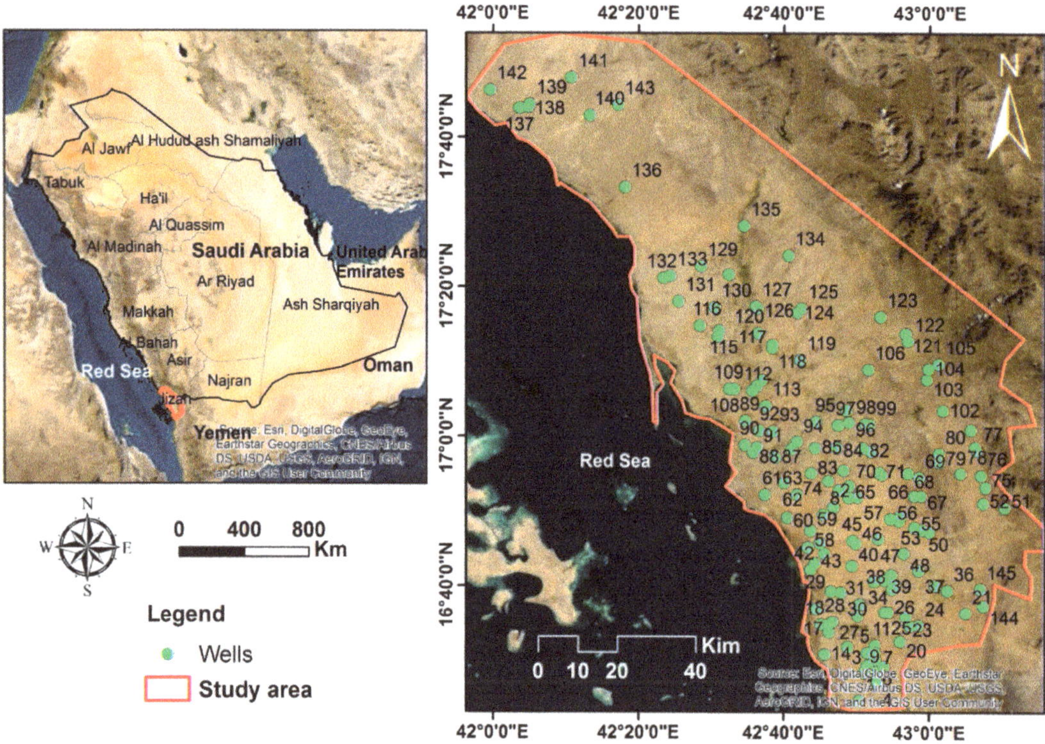

Figure 1. Location of the study area and wells.

The study area spans the western margin of the Proterozoic Arabian Shield and the eastern margin of the Cenozoic Red Sea basin. The Cenozoic rocks are represented by the clastic sedimentary succession underlying the black basaltic sheet of lava flows. The Quaternary deposits cover about half of the Jazan area in the wadi beds and the coastal plain. They consist of interbedded clay and sands, silts, cobbles, and gravels of wadi beds with variable thicknesses from one place to another. The thickness of the alluvial sediments ranges from 10 m towards the foothills to more than 100 m distant from the highlands in the southwest parts [2,42].

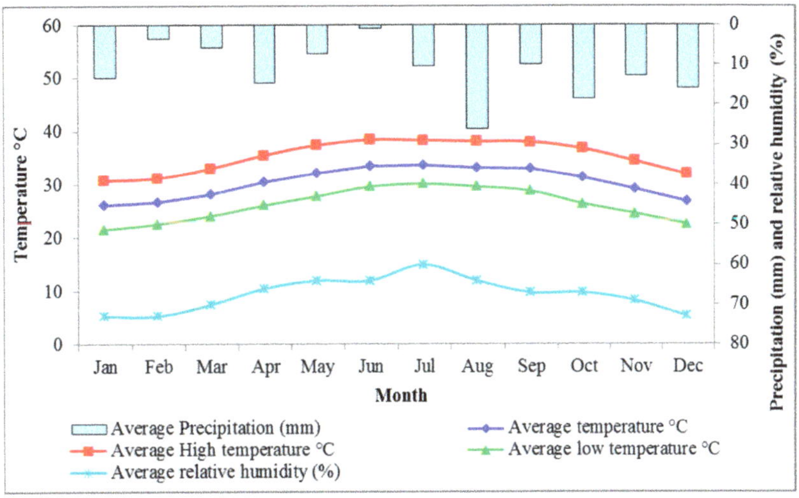

Figure 2. Average temperatures, precipitation, and relative humidity in the study area (Station Jizan, 2021).

2.2. Dataset Collection

A total of 145 groundwater samples were collected and chemically analysed from both shallow dug wells and boreholes tapping the phreatic aquifer (Figure 1). Collected water samples were analysed for major cations (Ca^{2+}, Mg^{2+}, Na^{+}, K^{+}) and anions (HCO_3^{-}, SO_4^{2}, Cl^{-}, CO_3^{2-}) by following standard methods suggested by APHA American Public Health Association (APHA) [43]. Table 1 displays descriptive statistics for physicochemical variables of the groundwater samples and the maximum permissible limits for various parameters, according to the WHO [44]. The inverse distance weighted (IDW) interpolation in GIS was used to map the spatial distribution of the chemical water parameters.

Table 1. Descriptive statistics for physicochemical variables and WHO standards for drinking water.

Element	Min.	Max.	Average	Standard Deviation	WHO Guidelines [44]
pH	6.3	8.7	7.7	0.3	7
TDS (mg/L)	128.0	8320.0	1709.6	1293.0	1000
TH (mg/L)	90.6	3676.6	640.8	526.7	500
Ca^{2+} (mg/L)	23.5	831.7	157.7	131.6	200
Mg^{2+} (mg/L)	4.4	388.8	60.0	55.9	30
Na^{+} (mg/L)	1.6	1291.4	307.6	276.3	200
K^{+} (mg/L)	1.2	188.5	12.4	27.6	12
Cl^{-} (mg/L)	12.8	3669.1	571.6	602.3	250
HCO_3^{-} (mg/L)	9.2	518.1	217.5	89.4	350
SO_4^{2-} (mg/L)	7.2	1098.5	319.8	221.9	350
NO_3^{-} (mg/L)	0.00	34.1	1.8	4.4	50

2.3. Water Quality Index (WQI)

This study uses eleven water quality parameters to calculate a WQI based on the World Health Organization's recommended drinking water quality standards (WHO) [44]. The WQI was analysed using the physicochemical parameters of pH, TDS, Ca^{2+}, Mg^{2+}, Na^{+}, K^{+}, Cl^{-}, SO_4^{2-}, HCO_3^{-}, NO_3^{-}, and TH. The weighted arithmetic WQI (WAWQI), proposed by Horton [37], was used to assess the water quality.

$$WAWQI = \sum_{i=1}^{n} W_i Q_i \quad (1)$$

where W_i is the relative weight of each parameter (Equation (2)), and Q_i (Equation (3)) is the quality rate scale assigned to each parameter by dividing the parameter concentration of the water sample by its respective standard as per the WHO guidelines [44] (Table 1).

$$W_i = \frac{w_i}{\sum_{i=1}^{n} w_i} \quad (2)$$

$$Q_i = \frac{C_i}{S_i} \quad (3)$$

where w_i is the weight of each parameter on a scale of one to five, indicating their relative relevance for drinking water quality, n is the number of parameters, C_i and S_i are respectively the concentration of parameter i, and the standard value of parameter i. Table 2 shows the weights for the various water parameters [33]. Table 3 shows the classification of water quality according to the WAWQI type and range.

Table 2. Physicochemical parameters' weights and relative weights [33].

Parameters	Weight (w_i)	Relative Weight (W_i)
pH	4	0.100
TDS	5	0.125
TH	3	0.075
Ca^{2+}	3	0.075
Mg^{2+}	3	0.075
Na^{+}	4	0.100
K^{+}	2	0.050
Cl^{-}	5	0.125
HCO_3^{-}	1	0.025
SO_4^{2-}	5	0.125
NO_3^{-}	5	0.125

Table 3. Classification of water quality according to the WQI type and range [45].

WAWQI	Water Type
<50	Excellent
50–100	Good
100.1–200	Poor
200.1–300	Very poor
>300.1	Unsuitable

2.4. *Machine Learning Methods*

2.4.1. Multiple Regression

The input parameters of the ANN model were determined using the multiple linear regression model [46]. The purpose of multiple linear regression analysis is to use known independent variables to predict the value of a single dependent variable. The weights of each predictor value indicate how big of an impact it has on the total projection. The independent variables are water quality measures (X_1, X_2,..., X_n) for the dependent WAWQI in this study (Y).

$$Y = a_0 + a_1X_1 + a_2X_2 + a_3X_3 + \cdots + a_iX_i + \cdots + a_nX_n \quad (4)$$

where X_i: is the dependent variable i; n is the number of the dependent variables, a_i is the ith coefficient of the dependent variable X_i, a_0 is the constant term of the model.

2.4.2. Artificial Neural Network (ANN)

ANNs were used to predict the WAWQI using MATLAB's Neural Network library (MathWorks, Natick, MA, USA). The ANN model's input, hidden, and output layers are all separate layers, and each layer contains different types of neurons. The input parameters are entered into the network and stored in input neurons, while the calculated outcomes are attributed in the output layer by output neurons. The hidden layer acts as a mediator to connect the input and output layers [47]. There are many different types of ANNs; one of the most common is the Bayesian regularisation back propagation (BRBP), which is the type applied in this research. The BRBP is a network training function that uses Levenberg Marquardt optimisation to update weight and bias variables. It finds the best mix of squared errors and weights to construct a network that generalises well [48,49].

The input parameters of the ANN model in this work were 11 input neurons, which included physicochemical parameters such as pH, TDS, Ca^{2+}, Mg^{2+}, Na^{+}, K^{+}, Cl^{-}, $SO_4{}^{2-}$, $HCO_3{}^{-}$, $NO_3{}^{-}$, and TH, while the output neurons were the WAWQI (Figure 3). In the hidden layer, nine neurons were used. Moreover, 75% of the dataset was allocated for training the models, and the remaining were considered for testing and validating the models.

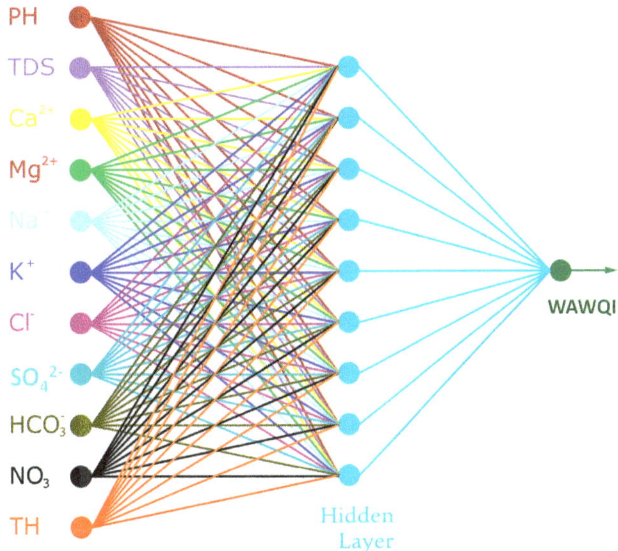

Figure 3. The description of the ANN model for modelling the WAWQI.

2.4.3. Support Vector Machines (SVM)

SVM analysis is a common machine-learning tool for regression analysis and classification [50]. Because it uses kernel functions, SVM regression is classified as a non-parametric approach. The SVM model is used to improve accuracy on low to medium-dimensional data sets. SVM regression is used to find the linear function for training data x of N observations with observed response values y.

$$f(x) = y = \begin{bmatrix} w \\ b \end{bmatrix}^T \begin{bmatrix} x \\ 1 \end{bmatrix} = w^T x + b \ \ x, b \in \mathbb{R}^{N+1} \tag{5}$$

where the parameters w and b are the gradient and the intercept, respectively, and ε represents the tolerance margin, as shown in Figure 4.

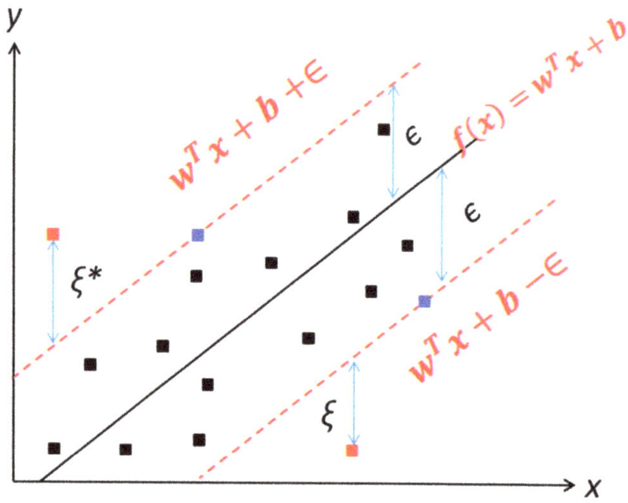

Figure 4. The typical architecture of SVM (one-dimensional linear). ε is the tolerance margin, ξ and ξ* are the control or slack variables of the error from the regression function, points on margins are called support vectors (figure adapted after Cantillo-Luna et al. [51]).

The kernel function determines the non-linear transformation applied to the data before the SVM is trained. In this paper, linear and polynomial kernel functions are used. The memory usage for cubic SVM is higher than linear SVM [52]. The Machine Learning Toolbox in MATLAB provides the following linear and polynomial kernel functions:

$$\text{Linear}: \quad G(x_i, x_j) = x_i' x_j \tag{6}$$

$$\text{Polynomial}: \quad G(x_i, x_j) = \left(1 + x_i' x_j\right)^q, \textit{where } q \textit{ is in the set } \{2, 3, 4, \ldots\} \tag{7}$$

The Gram matrix is an n-by-n matrix with entries $g_{i,j} = G(x_i, x_j)$. Each element $g_{i,j}$ represents the inner product of the predictors as transformed by φ. However, no need to know φ; the Gram matrix can be directly constructed using the kernel function. Non-linear SVM uses this method to determine the best function *f(x)* in the altered predictor space. In this paper, *x* represents the input vector (11 physicochemical parameters), *f(x)* refers to WAWQI, and in the polynomial kernel function, *q* was set as 3 (cubic).

2.4.4. Fit Binary Tree

A binary search tree (BST), also known as an ordered or sorted binary tree, is a rooted binary tree data structure in which each internal node stores a value that is higher than all keys in the node's left subtree but less than those in the node's right subtree. The temporal complexity of operations on the binary search tree is related to the tree's height. Binary search trees provide a binary search for quick data lookup, addition, and removal and may be used to construct dynamic sets and lookup tables. Because the nodes in a BST are arranged so that each comparison skips around half of the remaining tree, the lookup performance is proportional to that of the binary logarithm. In BST, the left sub-tree has elements less than the nodes element, and the right sub-tree has elements greater than the nodes element. A data structure called a BST makes it easy to keep track of a sorted list of numbers. Because each tree node can only have two children, it is known as a binary tree. Because it may be used to search for the presence of a number in O(log(n)) time, it is known as a search tree. BST is a node-based binary tree data structure that has the following properties: The left sub-tree of a node contains only nodes with keys lesser than the node's key; The right sub-tree of a node contains only nodes with keys greater than the node's key; The left and right sub-tree each must also be a binary search tree; there must be no duplicate nodes. Data representation is carried out in the ordered format, and BST does not allow duplicate values. The performance of a binary search tree is determined by the sequence in which the nodes are inserted into the tree; various binary search tree versions may be made with assured worst-case performance. The fundamental operations are search, traverse, insert, and delete. BSTs with assured worst-case complexity outperform an unsorted array, which would need a linear search time. The following pseudocode recursively implements the BST search method (Algorithm 1).

Algorithm 1. Pseudocode recursively for the BST search method.

```
Tree-Search(x, key)
   if x = NIL or key = x.key then
      return x
   if key < x.key then
      return Tree-Search(x.left, key)
   else
      return Tree-Search(x.right, key)
   end if
```

The recursive procedure continues until a NIL is reached or the observed and simulated values are in good agreement.

2.4.5. Gaussian Process Regression (GPR)

The GPR mathematical model is a non-parametric kernel-based probabilistic model [53]. It is important in the field of machine learning programming. The essential understanding of GPR is that the learning sample tracks the past probabilities of the Gaussian process regression. It is based on calculating the consistent subsequent probability and is built using the Bayesian linear regression model. GPR uses the kernel to define the covariance of a prior distribution across the target functions, and the observed training information is used to explain a likelihood function. Based on the Bayes theorem, a (Gaussian) posterior circulation across goal functions is explained, and its mean is used for data prediction. GPR was originally proposed as a 'principled, practically, and probabilistically based approach to kernel-making' [53]. The benefit of GPR over many other methods is that it smoothly integrates hyper-parameter estimates, model training, and risk evaluations; the results are less subjectively impacted and more understandable as a consequence. Gaussian processes (GP) are based on the assumption that the combined dispersion of model output probability is Gaussian [54].

Polynomial kernel (poly kernel) is a kernel feature that is commonly employed with the GPR in the initial variables of a function space to develop non-linear models. The polynomial kernel emerges automatically at the defined characteristics of the input samples to acquire their likeness, as well as combinations of them. In the context of regression analysis, such groups are referred to as interactive features. The enclosed polynomial kernel feature space is similar to a polynomial regression, but it is an educated sum of parameters that do not have a combinative blow-up. When the features' input data is binary (boolean), the features are linked to logical input function conjunctions [55].

The polynomial kernel is well-defined as follows:

$$K(x,Y) = \left(x^T, y + C\right)^d \tag{8}$$

where x and y are vectors in the input space, i.e., vectors of features estimated from workout or trial samples, and $C \geq 0$ is an unlimited parameter balancing the approach of higher-order vs. lower-order polynomial formulations. When C equals zero, the kernel is said to be homogenous.

2.5. Evaluation Indicators

Five statistical indicators were used to assess the performance of the linear regression, ANN, and SVM models: mean error (*ME*), mean absolute error (*MAE*), root mean square error (*RMSE*), mean absolute percentage error (*MAPE*), coefficient of correlation (*R*), and R-squared. The following equations were used to determine these indicators:

$$ME = \frac{1}{N}\sum_{i=1}^{N}(Y_i - Y_i^*) \tag{9}$$

$$MAE = \frac{1}{N}\sum_{i=1}^{N}|Y_i - Y_i^*| \tag{10}$$

$$RMSE = \sqrt{\frac{1}{N}\sum_{i=1}^{N}(Y_i - Y_i^*)^2} \tag{11}$$

$$MAPE = \frac{1}{N}\sum_{i=1}^{N}\frac{|Y_i - Y_i^*|}{Y_i^*} \times 100 \tag{12}$$

$$R = \frac{\sum_{I=i}^{N}(Y_i - \overline{Y})(Y_i^* - \overline{Y^*})}{\sqrt{\sum_{i=1}^{N}(Y_i - \overline{Y})^2\sum_{i=1}^{N}(Y_i^* - \overline{Y^*})^2}} \tag{13}$$

where N is the number of measurements, Y_i is the measured value for each parameter, Y^* is the estimated value for each parameter, $\overline{Y}$ is the mean of the measured values of the Y variables, and $\overline{Y^*}$ is the mean of the estimated values of Y^* variables.

3. Results and Discussion

3.1. Hydrogeological Aspects

The hydrogeological conditions have been studied in the Jazan area, where the groundwater is stored in both the alluvial deposits of the wadi systems and the clastic coarse members of the Cretaceous–Tertiary sedimentary succession [56]. The alluvial aquifer is composed of the Quaternary wadi deposits that enhance seawater intrusion in the coastal aquifer [2,57]. The aquifer's transmissivity ranges from 540 to 5400 m^2/day, with an average of 2190 m^2/day, and specific yield ranges between 0.001 and 0.006, increasing towards west directions, indicating good productivity. The storativity coefficient ranges between 0.01 and 0.25, with an average of 0.13 increasing toward the west direction [58]. Uncontrolled pumping in many locations has caused a cone of depression with the inland movement of the seawater fronts. The main recharge components of the aquifer are local rainfall infiltration that exclusively occurs during floods in the winter season. The shallow unconfined aquifer is subject to over-exploitation from many scattered wells in the area. Discharge of the aquifer includes groundwater pumping from wells to provide an adequate water supply for agricultural and residential areas and evapotranspiration losses in places where the water table is close to the ground surface [2].

Figure 5 shows the hydrogeological conditions ascertained from the fieldwork, including groundwater level compared with the mean sea level (m.asl) and depth of groundwater in the study area. Groundwater occurs at shallow depths, where groundwater levels vary from 10 to 33 m below the ground surface (Figure 5b). The piezometric gradient is inclined towards the west and southwest direction; it varies from 0.005 in the upper parts of the wadi to 0.001 at the beginning of the coastal plain [2]. Generally, the groundwater flow is from the east and northeast to the west and southwest toward the sea (Figure 5a); this might be due to the positive hydraulic gradient set up by the balance between recharge inland and discharge toward the sea. However, excessive fresh groundwater pumping in many areas causes a modification of the natural flow systems (reversing the hydraulic gradients) and, thus, induces seawater intrusion. However, few areas showed characteristic cones of depression.

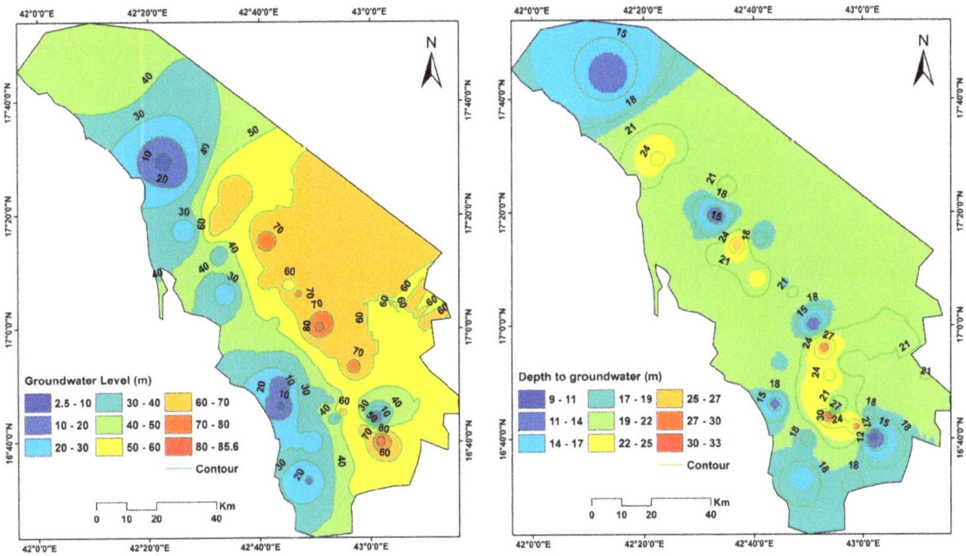

(a) (b)

Figure 5. (**a**) Groundwater level (m.asl), (**b**) depth to groundwater in the study area.

3.2. Statistical Analysis

Table 4 presents the descriptive statistics for the 145 water quality samples. Moreover, the correlation matrix is useful since it independently illustrates each parameter's importance and its effect on the hydrochemical relationships. If (*r*) values in Pearson's correlation matrix (Table 5) are +1 or −1, they represent a complete correlation between two variables, i.e., functional dependence. If the values are near zero, there is no significant interaction between the two variables at the $p < 0.05$ level.

Table 4. Descriptive statistics for all input and output variables.

Variable	Mean	SE Mean	StDev	Minimum	Q1	Median	Q3	Maximum
pH	7.66	0.03	0.31	6.33	7.48	7.67	7.82	8.68
TDS	1710	108	1298	128	803	1408	2128	8320
TH	640.8	43.9	528.5	90.6	303.4	473.2	864.0	3676.6
Ca^{2+}	157.7	11.0	132.0	23.4	73.5	116.6	193.0	831.7
Mg^{2+}	60.02	4.66	56.08	4.37	27.03	39.00	83.11	388.80
Na^{+}	307.6	23.0	277.3	1.6	116.4	242.1	365.1	1291.3
K^{+}	12.44	2.30	27.68	1.17	3.52	5.47	7.62	188.46
Cl^{-}	571.6	50.2	604.4	12.8	173.4	427.9	691.6	3669.1
HCO_3^{-}	217.45	7.45	89.69	9.15	151.94	206.25	266.05	518.06
SO_4^{2-}	319.8	18.5	222.7	7.2	141.2	287.2	427.5	1098.4
NO_3^{-}	1.837	0.363	4.374	0.000	0.330	0.800	1.465	34.140
WAWQI	125.41	7.26	87.40	20.30	64.61	99.46	161.33	592.31

SE Mean: standard error of mean, St Dev: standard deviation, Q1: first quartile, Median: middle number, Q3: third quartile.

Table 5. The Pearson correlation coefficient of the variables, values of R > 0.5, are shown in **bold**.

	pH	TDS	TH	Ca^{2+}	Mg^{2+}	Na^{+}	K^{+}	Cl^{-}	HCO_3^{-}	SO_4^{2-}	NO_3^{-}
pH	1										
TDS	−0.098	1									
TH	−0.203	**0.873**	1								
Ca^{2+}	−0.198	**0.837**	**0.961**	1							
Mg^{2+}	−0.182	**0.803**	**0.918**	**0.771**	1						
Na^{+}	0.014	**0.906**	**0.586**	**0.567**	**0.533**	1					
K^{+}	−0.004	−0.045	0.003	−0.004	0.012	−0.108	1				
Cl^{-}	−0.053	**0.977**	**0.847**	**0.818**	**0.770**	**0.891**	−0.056	1			
HCO_3^{-}	−0.220	0.007	0.021	0.011	0.033	−0.004	0.045	−0.102	1		
SO_4^{2-}	−0.140	**0.753**	**0.675**	**0.629**	**0.648**	**0.668**	0.008	**0.610**	0.091	1	
NO_3^{-}	−0.273	0.152	0.261	0.289	0.185	0.030	0.009	0.091	0.235	0.253	1

3.3. Chemical Analysis, Spatial Distribution, and Correlation Coefficients

The results of the chemical analysis indicated that the dominant cations are Na^{+}, followed by Ca^{2+} and Mg^{2+}, while the dominant anions are Cl^{-} followed by SO_4^{2-} and HCO_3^{-}, with a minor contribution of NO_3^{-}. Cations and anions reflect sodium chloride water type (Table 1). The pH of the groundwater ranges between 6.33 and 8.68, with an average of 7.66 indicating more or less neutral groundwater that is generally suitable for drinking. TDS is an important parameter for assessing salinity hazards and suitability for drinking and irrigation. The TDS ranges from 128 mg/L in the boreholes located further inland to 8320 mg/L close to the coastline, with an average of 1709 mg/L; thus, a wide range of variation was detected (Figure 6). The higher TDS values are recorded in groundwater wells near the Red Sea coast, indicating significant groundwater salinisation due to seawater intrusion. This seawater intrusion in the coastal aquifer of Jazan was confirmed by Abdalla [42], Abdalla et al. [2], Al-Bassam and Hussein [57].

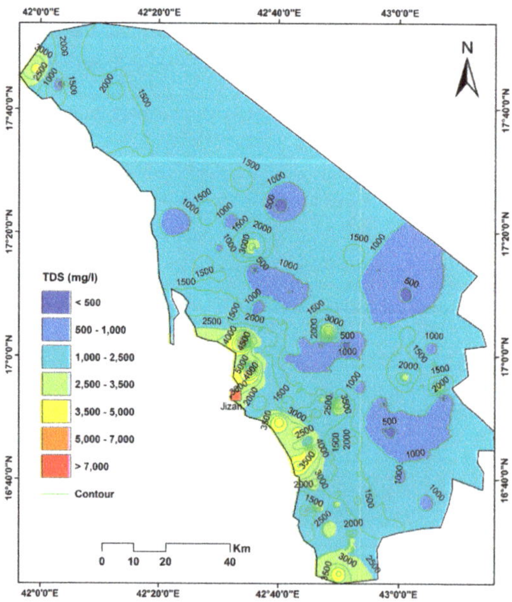

Figure 6. Total dissolved solids (TDS) zonation map for the study area.

The increase of Ca^{2+} and Mg^{2+} concentrations with the increasing salinity could indicate reverse ion exchange in the aquifer. High Cl^- and $SO_4{}^{2-}$ concentrations were recorded in places close to the coastline and indicate seawater intrusion.

The spatial distribution of the major cations (K^+, Na^+, Mg^{2+}, Ca^{2+}) and major anions ($SO_4{}^{2-}$, Cl^-, $HCO_3{}^-$, $NO_3{}^-$) is shown in Figures 7 and 8, respectively.

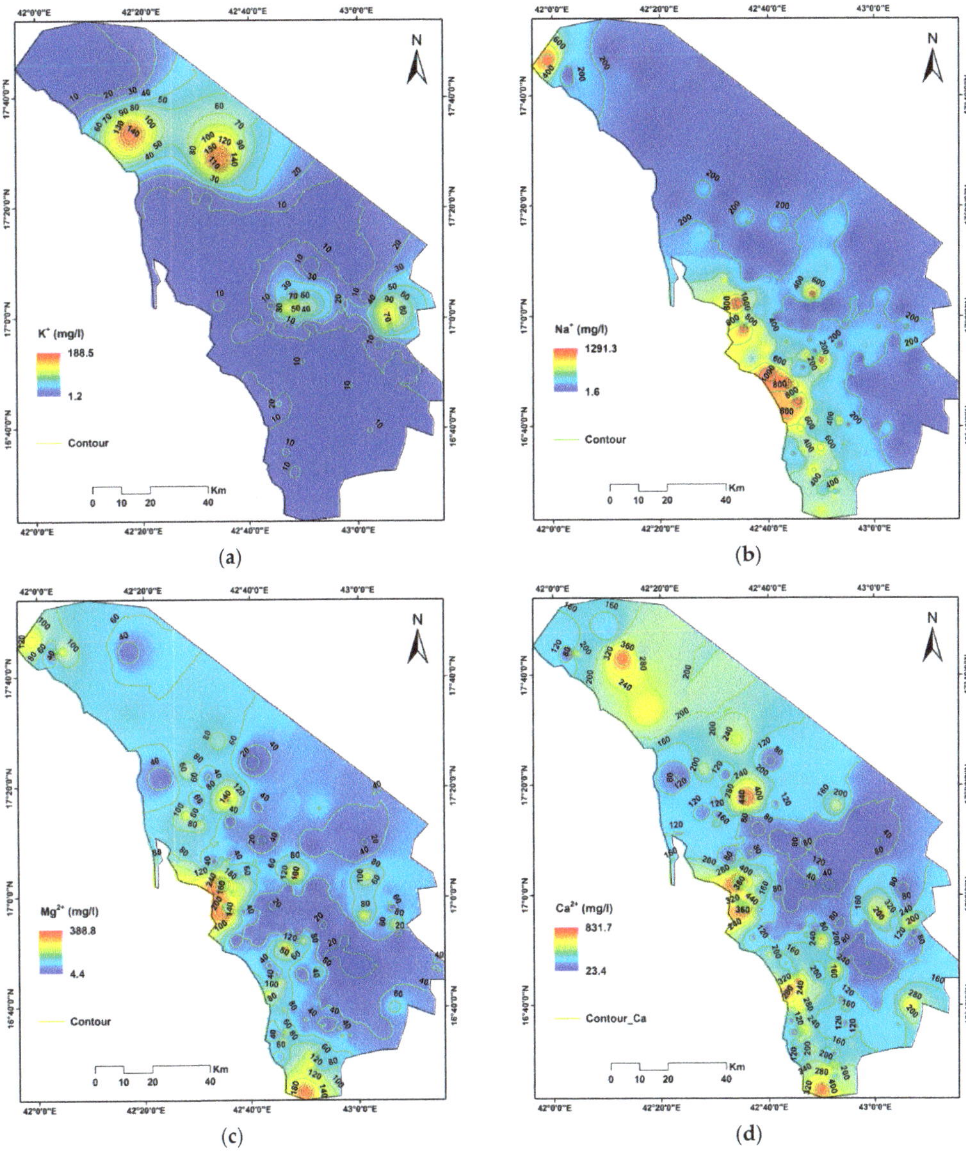

Figure 7. K^+ (**a**) Na^+ (**b**) Ca^{2+} (**c**) Mg^{2+} (**d**) distribution (zonation map) for the study area.

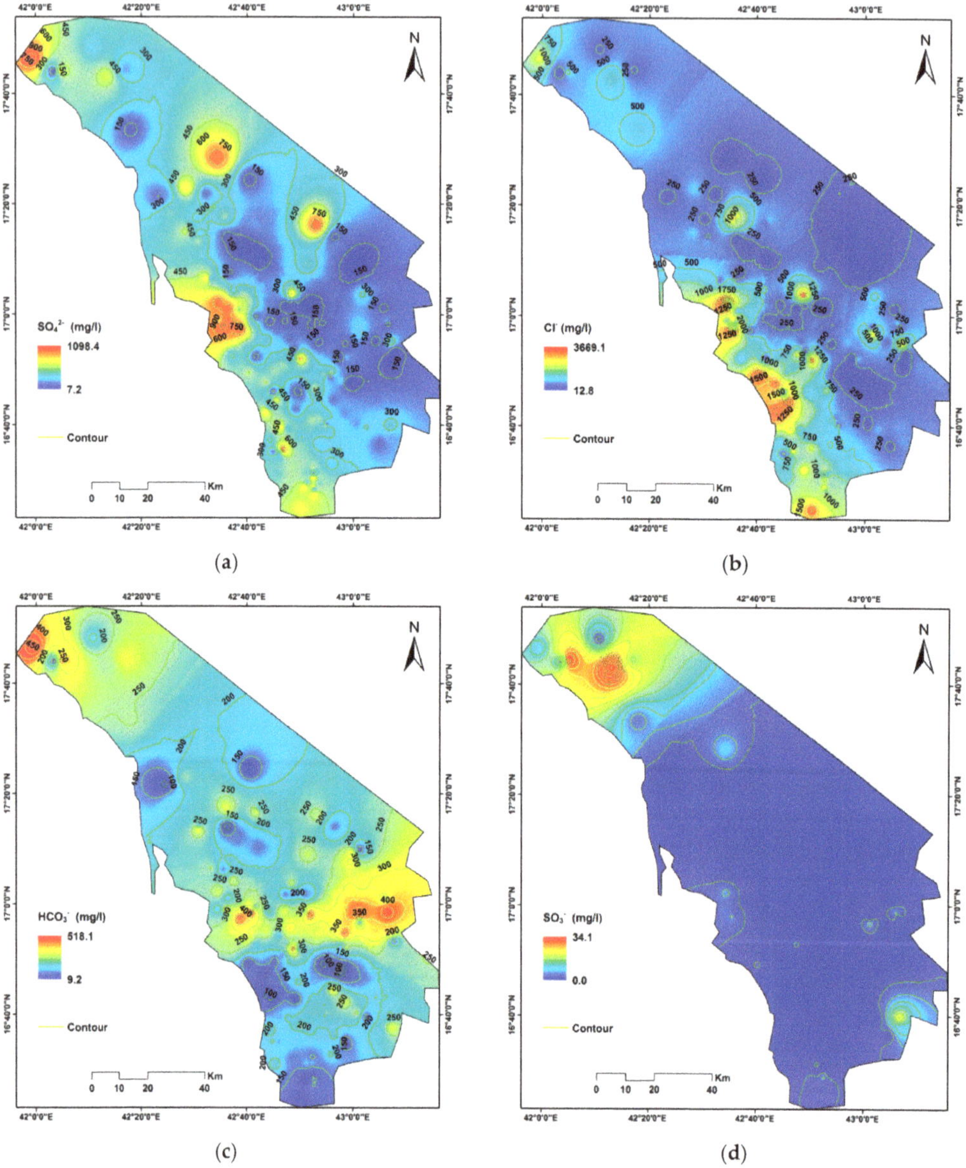

Figure 8. SO_4^{2-} (**a**) Cl^- (**b**) HCO_3^- (**c**) NO_3^- (**d**) distribution (zonation map) for the study area.

3.4. Water Quality Index Distribution and Classification

The calculated WAWQI shows that 15 wells have a score of less than 50, indicating excellent water quality, according to Kumar et al. [45] (Table 3). The WAWQI of 58 wells ranges from 50.1 to 100, indicating good quality water. Figure 9 depicts the WQI classification: 52 wells have a WQI of 100.1–200, indicating water of poor quality, and 13 wells have a WQI of 200.1–300, indicating water of extremely poor quality. Finally, seven wells have a WQI of greater than 300, indicating that the water is unfit for consumption. The spatial distribution of the WQI over the region is depicted using inverse distance weighted (IDW) interpolation in Figure 10.

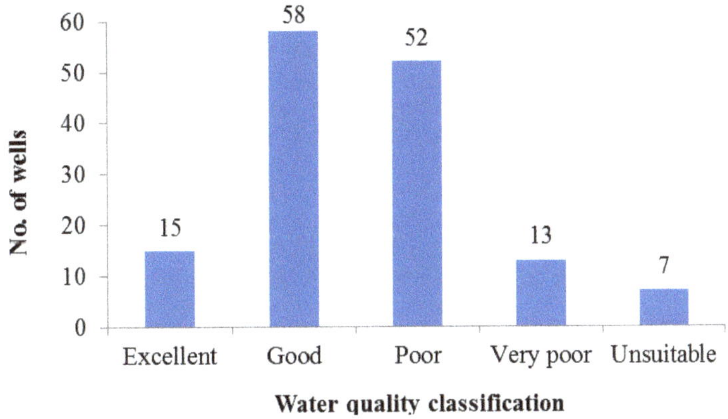

Figure 9. Classification of water quality according to the WAWQI.

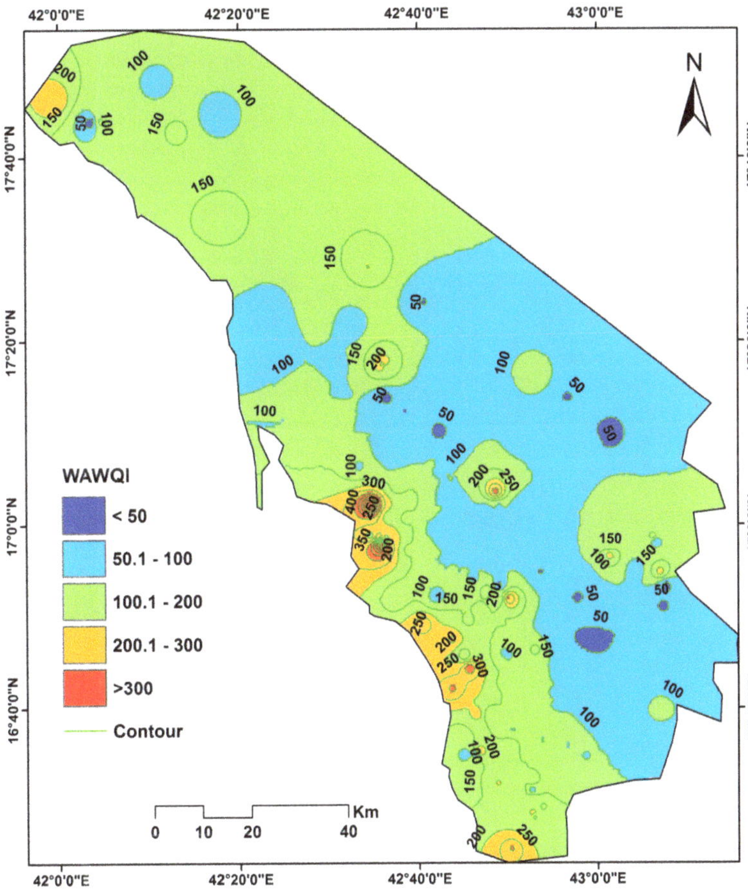

Figure 10. WAWQIs spatial distribution in the study area.

3.5. Evaluation of Data-Driven Models in WQ Prediction

The data were fitted using stepwise regression for the linear regression model, which produced a model that closely matched the observed and predicted WAWQI values. The input data were first structured as a dataset array, and the response data (WAWQI) were then arranged as a column vector. Each row of input data represents one observation. The regression model is then developed. The stepwise fit model begins with a single model, such as a constant, and then adds or subtracts terms one by one. Then, in a greedy manner, it selects an ideal parameter each time until it can no longer improve. The data should then be checked for outliers. The model coefficients obtained from the linear regression model are presented in Table 6. The model findings for this case study show that pH, K^+, and NO_3^- are the most significant variables and HCO_3^- is considered an insignificant variable. The model appears as in Equation (14). $R^2 = 1$ indicates that the data fit the model well. The MSE is 0.003, and the RMSE is 0.0023. Figure 11 shows the comparison of predicted and measured WAWQI.

$$\mathbf{WAWQI} = -0.0057308 + 1.4292\,\mathrm{pH} + 0.069523\,\mathrm{TH} - 0.098639\,Ca^{2+} + 0.025649\,Mg^{2+} + 0.050002\,\mathrm{Na}^+ + 0.41667\,\mathrm{K}^+ + 0.050032\,\mathrm{Cl}^- + 0.0071646\,HCO_3^- + 0.03574\,SO_4^{2-} + 0.24998\,NO_3^- \quad (14)$$

Table 6. Model coefficients from multiple regressions.

	Estimate	SE	*p* Value
Intercept	−0.005731	0.0070781	0.41958
pH	1.4292	0.0008888	<0.00000
TDS	0.012481	1.18×10^{-5}	<0.00000
TH	0.069523	0.018668	0.0002884
Ca^{2+}	−0.098639	0.046613	0.036195
Mg^{2+}	0.025649	0.076817	0.73898
Na^+	0.050002	1.91×10^{-5}	<0.00000
K^+	0.41667	1.09×10^{-5}	0
Cl^-	0.050032	1.73×10^{-5}	<0.00000
HCO_3^-	0.007165	1.02×10^{-5}	<0.00000
SO_4^{2-}	0.03574	1.30×10^{-5}	<0.00000
NO_3^-	0.24998	6.28×10^{-5}	0

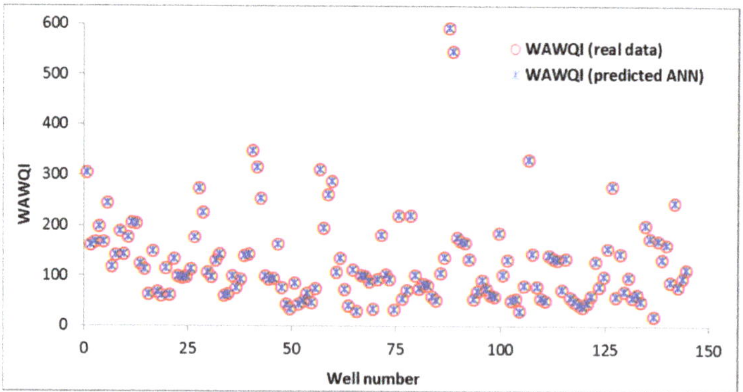

Figure 11. Observed and predicted WAWQI values.

The best result for the ANN modelling was obtained for the network with nine neurons after 19 iterations, while the best validation result was displayed for the network after 13 iterations (Figure 12).

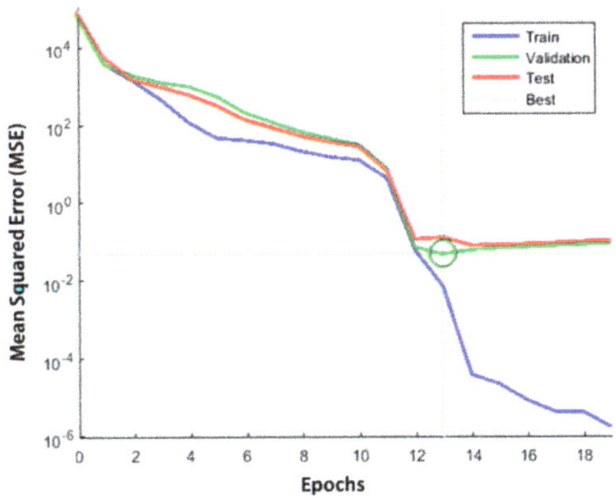

Figure 12. ANN modelling validation with a best performance MSE of 0.048016 at iteration 13.

The ME, MAE, RMSE, MAPE, regression factor R, and R^2 were used to validate the applied models' performance (Table 7). The scatterplots of the predicted and calculated WAWQI for the applied models in the training stage (right) and testing stage (left) are shown in Figure 13.

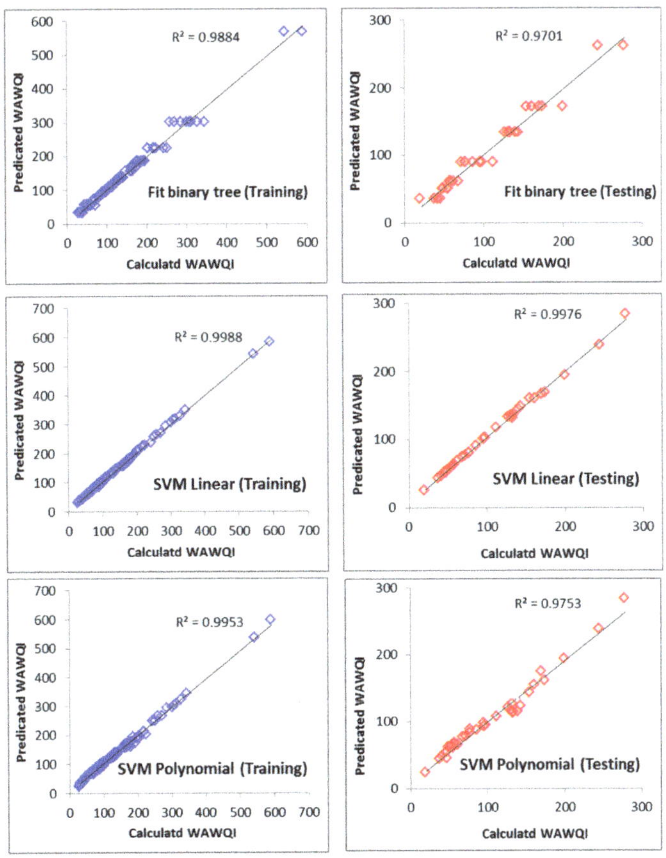

Figure 13. *Cont.*

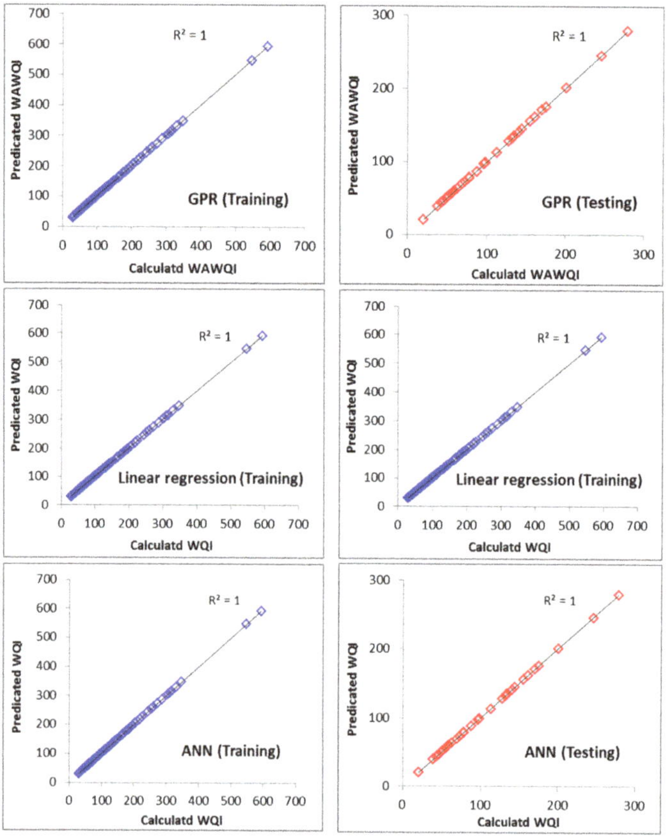

Figure 13. Predicted vs. calculated WAWQI for the applied models in the training stage (**right**) and testing stage (**left**).

Table 7. The performance indices of the developed models for the WAWQI during the training and testing stages.

Model/Indices		ME	MAE	RMSE	MAPE %	R	R^2
Fit binary tree	Training	0.00	6.066	10.085	4.920	0.9884	0.9942
	Testing	0.00	7.723	10.222	8.772	0.9701	0.9849
SVM (linear)	Training	1.920	3.388	3.825	3.380	0.9988	0.9994
	Testing	3.185	4.469	4.688	5.642	0.9976	0.9988
SVM (polynomial kernel)	Training	−0.234	6.424	8.144	5.453	0.9953	0.9976
	Testing	−1.078	8.302	10.040	9.343	0.9753	0.9876
Gaussian process regression (GPR)	Training	0.00	0.1935	0.2552	0.234	1.00	1.00
	Testing	0.00	0.2529	0.326	0.391	1.00	1.00
Linear regression (stepwise)	Training	0.00	0.0023	0.0028	0.0025	1.00	1.00
	Testing	0.00	0.0024	0.0029	0.0014	1.00	1.00
ANN	Training	0.088	0.088	0.0884	0.0969	1.00	1.00
	Testing	0.074	0.074	0.075	0.0982	1.00	1.00

The GPR, linear regression (stepwise), and ANN models all worked perfectly in the training and testing phases, as shown in Table 7. In both phases, these models had a high correlation, nearly to one, and small statistical errors. The linear regression (stepwise) model produced the best results, with MAE = 0.0023, RMSE = 0.0028, MAPE = 0.0025%,

R = 1.0, and R^2 = 1.0. The ANN model came in second with MAE = 0.088, RMSE = 0.0884, MAPE = 0.0969%, R = 1.0, and R^2 = 1.0. At the same time, the GPR model came in third with MAE = 0.194, RMSE = 0.255, MAPE = 0.234%, R = 1.0, and R^2 = 1.0. This was followed by the SVM (Linear) model with MAE = 3.39, RMSE = 3.83, MAPE = 3.38%, R = 0.9988, and R^2 = 0.9994. The SVM (Gaussian kernel) and Fit binary tree models performed the worst in the training and testing phases of the prediction procedure. For the SVM (polynomial kernel) model, the MAE, RMSE, MAPE, R, and R^2 were 6.42, 8.14, 5.45%, 0.9953, and 0.9976, respectively. For the Fit binary tree, the MAE, RMSE, MAPE, R, and R^2 were 6.07, 10.09, 4.92%, 0.9884, and 0.9942, respectively (Table 7).

Overall results showed that the proposed methods generated satisfactory outputs for estimating WAWQI close to observed data. The results obtained are highly satisfactory compared with the findings from Sakaa et al. [59], who showed that a combination of all input parameters attained a best predictive performance of R^2 testing = 0.82, RMSE testing = 5.17, while a couple of five input parameters, such as pH, EC, TDS, T, and saturation, achieved the second-best predictive precision (R^2 testing = 0.81, RMSE testing = 5.55). In addition, the current findings are in agreement with the results of Mokhtar et al. [27], who used SVM, extreme gradient boosting, Random Forest and stepwise regression, principal components regression, partial least squares regression, and ordinary least squares regression for WQI modelling and stated that all models used with values less than 0.1 show good prediction ability for all indices. These findings were extremely acceptable and agreed with those suggested by Elbeltagi et al. [60], who applied additive regression (AR), M5P tree model (M5P), random subspace (RSS), and SVM in WQI modelling and found that AR outperformed the other data-driven models (R^2 = 0.9993, MAE = 0.5243, RMSE = 0.06356, RAE% = 3.8449, and RRSE% = 3.9925). The AR was offered as an optimal model with good outcomes due to improved prediction precision with the fewest input parameters. Moreover, eight artificial intelligence algorithms, e.g., multi-linear regression (MLR), random forest (RF), M5P tree (M5P), random subspace (RSS), additive regression (AR), artificial neural network (ANN), support vector regression (SVR), and locally weighted linear regression (LWLR) have been applied by Kouadri et al. [35]. Their results stated that the MLR model performed better than the other models, whereas the RF model performed better. Also, the model results coincide with Kouadri et al. [36], who implemented long short-term memory (LSTM), multi-linear regression (MLR), and ANN and stated that the results are highly accurately predicted using ANN and MLR models compared to LSTM model. These models also generated more favourable outcomes than those achieved by Iqbal et al. [61], who used the WASP8 for water quality simulations. Their results clarified that Pearson correlation coefficient values are around 0.66, 0.68, and 0.58, respectively.

3.6. Best Subset Regression for Selecting the Most Important Parameters

Feature selection is one of the most important stages in a soft computing model to forecast and predict phenomena with many input variables. There are several approaches to specify the best combinations among all possible, including best subset regression, mutual information, and forward stepwise selection. The best subset regression analysis was performed in the current study to determine the best input combinations for the WAWQI model. For this purpose, six statistical criteria, including MSE, determination coefficients (R^2), adjusted R^2, Mallows' Cp, Akaike's AIC, and Amemiya's prediction criterion (PC), were computed, and the results are shown in Table 8. As can be seen, the eight bold input combinations have the lowest values of Amemiya's PC (from 0.00 to 0.027) among all input combinations. These eight input combinations have a high R^2 (from 0.975 to 1.00) and high Adj-R^2 (from 0.974 to 1.00) and were identified as the best input combination for the prediction of the WAWQI model. It is noteworthy that a total of 145 datasets were randomly split into two training and testing subsets. Moreover, 75% of the datasets were allocated for training the models, and the remaining were considered for testing and validating the models.

Table 8. The best subset regression analysis for determining the best input combinations to model WQI.

No. of Variables	Variables	MSE	R^2	Adjusted R^2	Akaike's AIC	Schwarz's SBC	Amemiya's PC
1	NO_3	7467.34	0.02	0.023	1295.139	1301.092	0.984
2	SO_4/NO_3	3356.68	0.56	0.561	1180.181	1189.112	0.445
3	$HCO_3/SO_4/NO_3$	3342.15	0.57	0.563	1180.527	1192.434	0.446
4	**$Cl/HCO_3/SO_4/NO_3$**	**197.359**	**0.975**	**0.974**	**771.240**	**786.124**	**0.027**
5	**$K/Cl/HCO_3/SO_4/NO_3$**	**60.957**	**0.992**	**0.992**	**601.848**	**619.708**	**0.008**
6	**$Na/K/Cl/HCO_3/SO_4/NO_3$**	**17.220**	**0.998**	**0.998**	**419.503**	**440.340**	**0.002**
7	**$Mg/Na/K/Cl/HCO_3/SO_4/NO_3$**	**1.773**	**1.000**	**1.000**	**90.825**	**114.639**	**0.000**
8	**$Ca/Mg/Na/K/Cl/HCO_3/SO_4/NO_3$**	**0.184**	**1.000**	**1.000**	**−236.474**	**−209.684**	**0.000**
9	**$TH/Ca/Na/K/Cl/HCO_3/SO_4/NO_3$**	**0.184**	**1.000**	**1.000**	**−236.474**	**−209.684**	**0.000**
10	**$TDS/TH/Ca/Na/K/Cl/HCO_3/SO_4/NO_3$**	**0.161**	**1.000**	**1.000**	**−255.097**	**−225.330**	**0.000**
11	**$pH/TDS/TH/Ca/Na/K/Cl/HCO_3/SO_4/NO_3$**	**0.000**	**1.000**	**1.000**	**0.000**	**0.000**	**0.000**

The best models for the selection criteria are displayed in **bold**.

4. Conclusions and Outlook

This study analysed the ability of six different AI techniques and regressions, such as linear regression (stepwise), support vector regression SVM (linear and polynomial kernels), Gaussian process regression (GPR), Fit binary tree, and artificial neural network ANN (Bayesian) for forecasting a WQI based on 11 physicochemical parameters (pH, TDS, Ca^{2+}, Mg^{2+}, Na^+, K^+, Cl^-, SO_4^{2-}, HCO_3^-, NO_3^-, and TH) collected from 145 groundwater wells in Jizan, Saudi Arabia.

The outcome of the resultant WQI model clearly identified (forecasted and predicted) the best input combination for the prediction of the WAWQI model. This might contribute significantly to the knowledge and understanding of the groundwater quality within the study area and its impact on any agricultural investments and sustainable development, as the study area has high importance to national and regional economic development especially agricultural and industrial activities.

In addition, ArcGIS was used to create maps of the spatial distribution of groundwater quality parameters. The best subset regression analysis was used to find the optimum input combinations for the WQI model.

The following findings have been obtained:

- Higher levels of Cl^- and SO_4^{2-} were found near the coast, which is indicative of seawater intrusion and serves as a proxy for salinisation. Furthermore, seven wells had a WAWQI of more than 300, suggesting that the water is unsafe for human consumption.
- The results of the stepwise fit model revealed that pH, K^+, and NO_3^- are the most important variables, while HCO_3^- is a non-significant variable. The best results were obtained from the simulated ANN modeling for the nine-neuron network after 19 iterations, whereas the best validation performance was 0.048016 at iteration 13.
- The GPR, linear regression (Stepwise), and ANN models worked flawlessly during the training and testing stages, with a high correlation of 1 and low statistical errors.
- The linear regression (stepwise) model generated the best results, with MAE = 0.0023, RMSE = 0.0028, and R = 1.0. This good performance is due to its special mechanism with repeated regressions, each time deleting the weakest associated variable until the observed and measured values fully match. The ANN model came in second with MAE = 0.088, RMSE = 0.0884, and R = 1.0. The GPR model finished in third with MAE = 0.194, RMSE = 0.255, and R = 1.0. The SVM (Linear) model was next, with MAE = 3.39, RMSE = 3.83, R = 0.9988.
- The SVM (polynomial kernel) and Fit binary tree models performed the worst during the training and testing phases of the prediction procedure.
- The optimum input combination for WAWQI model prediction was the eight input combinations with high R^2 (from 0.975 to 1.00) and high Adj-R^2 (from 0.974 to 1.00).

These findings are of importance to water planners in terms of WQI for enhancing sustainable groundwater resource management policies.

In conclusion, the best subset regression analysis is useful, and when only a portion of the relevant data are available, we can use the best subset regression model to determine which input parameters will best match the ML model for WQ prediction.

This study recommends not using SVM (polynomial kernel) and Fit binary tree models because of performing the worst during the training and testing phases of the prediction procedure. It can be recommended, in future works, standalone and hybrid artificial intelligence models for predicting WQIs in several regions under different conditions should be developed to recommend which model is most suitable for all these regions based on limited input variables. Future research can also incorporate depth to groundwater variation data into AI/ML methods to investigate its effects on groundwater quality. It is also recommended that seawater intrusion be controlled in the study area by implementing one of the following techniques: decreasing pumping rates, hydraulic barriers, artificial recharge using treated wastewater [62–64], using a freshwater surface recharge canal [65], cutoff walls [66,67], and brackish water pumping [68].

Author Contributions: Conceptualization, M.E.-R.; methodology, M.E.-R., O.B. and A.E.; software, M.E.-R.; validation, M.E.-R. and A.E.; formal analysis, M.E.-R. and A.E.; investigation, M.E.-R.; resources, M.E.-R.; data curation, M.E.-R.; writing—original draft preparation, M.E.-R. writing—review and editing, M.E.-R., O.B., F.A., S.A., M.S.A. and A.E.; visualization, M.E.-R. and A.E. All authors have read and agreed to the published version of the manuscript.

Funding: This research was funded by the Deputyship for Research and Innovation, Ministry of Education in Saudi Arabia, project no. (IFKSURC-1-7314).

Data Availability Statement: Data available based on request from first author.

Acknowledgments: The authors extend their appreciation to the Deputyship for Research & Innovation, Ministry of Education in Saudi Arabia, for funding this research. (IFKSURC-1-7314).

Conflicts of Interest: The authors declare no conflict of interest.

Abbreviations

AdaBoost	Adaptive Boosting
AI	Artificial Intelligence
ANN	Artificial Neural Network
APHA	American Public Health Association
AR	Additive Regression
BRBP	Bayesian Regularisation Back Propagation
BST	A Binary Search Tree
Ca^{2+}	Calcium Ion
CCMEWQI	Canadian Council of Ministers of the Environment Water Quality Index
Cl^{-}	Chlorine Ion
GA	Genetic Algorithms
GIS	Geographic Information System
GPR	Gaussian Process Regression
$HCO_3{}^{-}$	Bicarbonate Ion
IDW	Inverse Distance Weighted
IWQ	Irrigation Water Quality
K^{+}	Potassium Ion
kNN	K-Nearest Neighbour
LSTM	Long Short-Term Memory
LWLR	Locally Weighted Linear Regression
MAE	Mean Absolute Error
MAPE	Mean Absolute Percentage Error
ME	Mean Error

Mg^{2+}	Magnesium Ion
ML	Machine Learning
MLR	Multiple Linear Regression
MLR	Multi-Linear Regression
Na^{+}	Sodium Ion
NO_3^{-}	Nitrate Ion
NSFWQI	National Sanitation Foundation Water Quality Index
OWQI	Oregon Water Quality Index
pH	Potential Of Hydrogen
Poly kernel	Polynomial Kernel
R	Coefficient of Correlation
RF	Random Forest
RMSE	Root Mean Square Error
R-squared	Coefficient of Determination
RSS	Random Subspace
RSS	Random Subspace
SGD	Stochastic Gradient Descent
SO_4^{2-}	Sulfate Ion
St. Dev.	Standard Deviation
SVM	Support Vector Machine
TDS	Total Dissolved Solids
TH	Total Hardness
WAWQI	Weighted Arithmetic Water Quality Index
WHO	World Health Organization
WQI	Water Quality Index

References

1. Al-Turki, A.; Abdel-Nasser, G.; Al-Faraj, A.; Shahwan, A.; Al-Malik, A. Evaluation of well water quality in Southern Tihama plain, Saudi Arabia. *Resour. Bull.* **2011**, *172*, 5–47.
2. Abdalla, F.; Al-Turki, A.; Al Amri, A. Evaluation of groundwater resources in the Southern Tihama plain, Saudi Arabia. *Arab. J. Geosci.* **2015**, *8*, 3299–3310. [CrossRef]
3. Warner, K.L.; Barataud, F.; Hunt, R.J.; Benoit, M.; Anglade, J.; Borchardt, M.A. Interactions of water quality and integrated groundwater management: Examples from the United States and Europe. In *Integrated Groundwater Management*; Jakeman, A.J., Barreteau, O., Hunt, R.J., Rinaudo, J.-D., Ross, A., Eds.; Springer: Cham, Switzerland, 2016; pp. 347–376.
4. Akinbile, C.O.; Yusoff, M.S. Environmental Impact of Leachate Pollution on Groundwater Supplies in Akure, Nigeria. *Int. J. Environ. Sci. Dev.* **2011**, *2*, 81. [CrossRef]
5. Fernández, N.; Ramírez, A.; Solano, F. Physico-chemical Water Quality indices a comparative review. *Bistua Rev. De La Fac. De Cienc. Básicas* **2004**, *2*, 19–30.
6. Amer, R.; Ripperdan, R.; Wang, T.; Encarnación, J. Groundwater quality and management in arid and semi-arid regions: Case study, Central Eastern Desert of Egypt. *J. Afr. Earth Sci.* **2012**, *69*, 13–25. [CrossRef]
7. Ismail, E.; El-Rawy, M. Assessment of groundwater quality in West Sohag, Egypt. *Desalination Water Treat.* **2018**, *123*, 101–108. [CrossRef]
8. El-Rawy, M.; Ismail, E.; Abdalla, O. Assessment of groundwater quality using GIS, hydrogeochemistry, and factor statistical analysis in Qena Governorate, Egypt. *Desalin. Water Treat.* **2019**, *162*, 14–29. [CrossRef]
9. Sadat-Noori, S.M.; Ebrahimi, K.; Liaghat, A.M. Groundwater quality assessment using the Water Quality Index and GIS in Saveh-Nobaran aquifer, Iran. *Environ. Earth Sci.* **2014**, *71*, 3827–3843. [CrossRef]
10. Masoud, A.A. Groundwater quality assessment of the shallow aquifers west of the Nile Delta (Egypt) using multivariate statistical and geostatistical techniques. *J. Afr. Earth Sci.* **2014**, *95*, 123–137. [CrossRef]
11. El-Rawy, M.; Abdalla, F.; Negm, A.M. Groundwater Characterisation and Quality Assessment in Nubian Sandstone Aquifer, Kharga Oasis, Egypt. In *Groundwater in Egypt's Deserts*; Springer: Cham, Switzerland, 2021; pp. 177–199.
12. Gundaz, O.; Simsek, C. Assessment of three wastewater treatment plants in Turkey. In *Wastewater Reuse–Risk Assessment, Decision-Making and Environmental Security*; Springer: Dordrecht, The Netherlands, 2007; pp. 159–167.
13. Usali, N.; Ismail, M.H. Use of Remote Sensing and GIS in Monitoring Water Quality. *J. Sustain. Dev.* **2010**, *3*, 228. [CrossRef]
14. Akter, T.; Jhohura, F.T.; Akter, F.; Chowdhury, T.R.; Mistry, S.K.; Dey, D.; Barua, M.K.; Islam, A.; Rahman, M. Water Quality Index for measuring drinking water quality in rural Bangladesh: A cross-sectional study. *J. Health Popul. Nutr.* **2016**, *35*, 4. [CrossRef]
15. Sharma, D.; Kansal, A. Water quality analysis of River Yamuna using water quality index in the national capital territory, India (2000–2009). *Appl. Water Sci.* **2011**, *1*, 147–157. [CrossRef]
16. Fathi, H.; El-Rawy, M. GIS-based evaluation of water quality index for groundwater resources nearby wastewater treatment plants, Egypt. *Poll. Res.* **2018**, *37*, 105–116.

17. Lumb, A.; Halliwell, D.; Sharma, T. Application of CCME Water Quality Index to Monitor Water Quality: A Case Study of the Mackenzie River Basin, Canada. *Environ. Monit. Assess.* **2006**, *113*, 411–429. [CrossRef]
18. Chaturvedi, M.K.; Bassin, J.K. Assessing the water quality index of water treatment plant and bore wells, in Delhi, India. *Environ. Monit. Assess.* **2010**, *163*, 449–453. [CrossRef]
19. Sharma, P.; Meher, P.K.; Kumar, A.; Gautam, Y.P.; Mishra, K.P. Changes in water quality index of Ganges river at different locations in Allahabad. *Sustain. Water Qual. Ecol.* **2014**, *3*, 67–76. [CrossRef]
20. Hameed, M.; Sharqi, S.S.; Yaseen, Z.M.; Afan, H.A.; Hussain, A.; Elshafie, A. Application of artificial intelligence (AI) techniques in water quality index prediction: A case study in tropical region, Malaysia. *Neural Comput. Appl.* **2017**, *28*, 893–905. [CrossRef]
21. Aldhyani, T.H.H.; Al-Yaari, M.; Alkahtani, H.; Maashi, M. Water Quality Prediction Using Artificial Intelligence Algorithms. *Appl. Bionics Biomech.* **2020**, *2020*, 6659314. [CrossRef]
22. Agrawal, P.; Sinha, A.; Kumar, S.; Agarwal, A.; Banerjee, A.; Villuri, V.G.K.; Annavarapu, C.S.R.; Dwivedi, R.; Dera, V.V.R.; Sinha, J.; et al. Exploring Artificial Intelligence Techniques for Groundwater Quality Assessment. *Water* **2021**, *13*, 1172. [CrossRef]
23. Prasad, D.V.V.; Kumar, P.S.; Venkataramana, L.Y.; Prasannamedha, G.; Harshana, S.; Srividya, S.J.; Indraganti, S. Automating water quality analysis using ML and auto ML techniques. *Environ. Res.* **2021**, *202*, 111720. [CrossRef]
24. Ubah, J.I.; Orakwe, L.C.; Ogbu, K.N.; Awu, J.I.; Ahaneku, I.E.; Chukwuma, E.C. Forecasting water quality parameters using artificial neural network for irrigation purposes. *Sci. Rep.* **2021**, *11*, 24438. [CrossRef]
25. Abduljaleel, H.Y.; Schüttrumpf, H.; Azzam, R. *A GIS-Based Water Quality Management for Shatt Al-Arab River System, South of Iraq*; No. RWTH-2020-09237; Lehrstuhl für Ingenieurgeologie und Hydrogeologie: Aachen, Germany, 2020.
26. Setshedi, K.J.; Mutingwende, N.; Ngqwala, N.P. The Use of Artificial Neural Networks to Predict the Physicochemical Characteristics of Water Quality in Three District Municipalities, Eastern Cape Province, South Africa. *Int. J. Environ. Res. Public Health* **2021**, *18*, 5248. [CrossRef]
27. Mokhtar, A.; Elbeltagi, A.; Gyasi-Agyei, Y.; Al-Ansari, N.; Abdel-Fattah, M.K. Prediction of irrigation water quality indices based on machine learning and regression models. *Appl. Water Sci.* **2022**, *12*, 76. [CrossRef]
28. Wang, W.; Men, C.; Lu, W. Online prediction model based on support vector machine. *Neurocomputing* **2008**, *71*, 550–558. [CrossRef]
29. Noori, R.; Safavi, S.; Shahrokni, S.A.N. A reduced-order adaptive neuro-fuzzy inference system model as a software sensor for rapid estimation of five-day biochemical oxygen demand. *J. Hydrol.* **2013**, *495*, 175–185. [CrossRef]
30. Gazzaz, N.M.; Yusoff, M.K.; Aris, A.Z.; Juahir, H.; Ramli, M.F. Artificial neural network modeling of the water quality index for Kinta River (Malaysia) using water quality variables as predictors. *Mar. Pollut. Bull.* **2012**, *64*, 2409–2420. [CrossRef]
31. El Bilali, A.; Taleb, A. Prediction of irrigation water quality parameters using machine learning models in a semi-arid environment. *J. Saudi Soc. Agric. Sci.* **2020**, *19*, 439–451. [CrossRef]
32. Gupta, A.N.; Kumar, D.; Singh, A. Evaluation of Water Quality Based on a Machine Learning Algorithm and Water Quality Index for Mid Gangetic Region (South Bihar plain), India. *J. Geol. Soc. India* **2021**, *97*, 1063–1072. [CrossRef]
33. Kulisz, M.; Kujawska, J.; Przysucha, B.; Cel, W. Forecasting Water Quality Index in Groundwater Using Artificial Neural Network. *Energies* **2021**, *14*, 5875. [CrossRef]
34. El-Rawy, M.; Abd-Ellah, M.K.; Fathi, H.; Ahmed, A.K.A. Forecasting effluent and performance of wastewater treatment plant using different machine learning techniques. *J. Water Process. Eng.* **2021**, *44*, 102380. [CrossRef]
35. Kouadri, S.; Elbeltagi, A.; Islam, A.R.M.T.; Kateb, S. Performance of machine learning methods in predicting water quality index based on irregular data set: Application on Illizi region (Algerian southeast). *Appl. Water Sci.* **2021**, *11*, 190. [CrossRef]
36. Kouadri, S.; Pande, C.B.; Panneerselvam, B.; Moharir, K.N.; Elbeltagi, A. Prediction of irrigation groundwater quality parameters using ANN, LSTM, and MLR models. *Environ. Sci. Pollut. Res.* **2022**, *29*, 21067–21091. [CrossRef] [PubMed]
37. Horton, R.K. An Index Number System for Rating Water Quality. *J Water Pollut. Control Fed.* **1965**, *37*, 300–306.
38. The Climate Change Knowledge Portal (CCKP). Current Climate: Climatology. Saudi Arabia. Available online: https://climateknowledgeportal.worldbank.org/country/saudi-arabia/climate-data-historical (accessed on 13 January 2023).
39. Sulaiman, A.; Elawadi, E.; Mogren, S. Gravity interpretation to image the geologic structures of the coastal zone in al Qunfudhah area, southwest Saudi Arabia. *Geophys. J. Int.* **2018**, *214*, 1623–1632. [CrossRef]
40. Alshehri, F.; Sultan, M.; Karki, S.; Alwagdani, E.; Alsefry, S.; Alharbi, H.; Sahour, H.; Sturchio, N. Mapping the Distribution of Shallow Groundwater Occurrences Using Remote Sensing-Based Statistical Modeling over Southwest Saudi Arabia. *Remote Sens.* **2020**, *12*, 1361. [CrossRef]
41. Alarifi, S.S.; Abdelkareem, M.; Abdalla, F.; Alotaibi, M. Flash Flood Hazard Mapping Using Remote Sensing and GIS Techniques in Southwestern Saudi Arabia. *Sustainability* **2022**, *14*, 14145. [CrossRef]
42. Abdalla, F. Ionic ratios as tracers to assess seawater intrusion and to identify salinity sources in Jazan coastal aquifer, Saudi Arabia. *Arab. J. Geosci.* **2018**, *9*, 40. [CrossRef]
43. APHA. *Standard Methods for the Examination of Water and Wastewater*, 21st ed.; American Public Health Association: Washington, DC, USA, 2005.
44. WHO. *Guidelines or Drinking-Water Quality*, 4th ed.; World Health Organization: Geneva, Switzerland, 2011.
45. Kumar, S.K.; Bharani, R.; Magesh, N.S.; Godson, P.S.; Chandrasekar, N. Hydrogeochemistry and groundwater quality appraisal of part of south Chennai coastal aquifers, Tamil Nadu, India using WQI and fuzzy logic method. *Appl. Water Sci.* **2014**, *4*, 341–350. [CrossRef]

46. Chatterjee, S.; Hadi, A.S. *Regression Analysis by Example*; John Wiley & Sons: Hoboken, NJ, USA, 2013.
47. Sarkar, A.; Pandey, P. River Water Quality Modelling Using Artificial Neural Network Technique. *Aquat. Procedia* **2015**, *4*, 1070–1077. [CrossRef]
48. MacKay, D.J. Bayesian interpolation. *Neural Comput.* **1992**, *4*, 415–447. [CrossRef]
49. Foresee, F.D.; Hagan, M.T. Gauss-Newton approximation to Bayesian learning. In Proceedings of the International Conference on Neural Networks (ICNN'97), Houston, TX, USA, 9–12 June 1997; pp. 1930–1935.
50. Vapnik, V. *The Nature of Statistical Learning Theory*; Springer: New York, NY, USA, 1995.
51. Cantillo-Luna, S.; Moreno-Chuquen, R.; Chamorro, H.R.; Riquelme-Dominguez, J.M.; Gonzalez-Longatt, F. Locational Marginal Price Forecasting Using SVR-Based Multi-Output Regression in Electricity Markets. *Energies* **2022**, *15*, 293. [CrossRef]
52. Awad, M.; Khanna, R. *Efficient Learning Machines: Theories, Concepts, and Applications for Engineers and System Designers*; Springer nature: Berlin/Heidelberg, Germany, 2015; p. 268.
53. Rasmussen, C.E.; Williams, C.K.I. *Gaussian Processes for Machine Learning*; The MIT Press: Boston, MA, USA, 2006; pp. 69–106.
54. Markonis, Y.; Koutsoyiannis, D. Scale-dependence of persistence in precipitation records. *Nat. Clim. Chang.* **2015**, *6*, 399–401. [CrossRef]
55. Goldberg, Y.; Elhadad, M. SplitSVM: Fast, Space-Efficient, non-Heuristic, Polynomial Kernel Computation for NLP Applications. In Proceedings of the 46th Annual Meeting of the Association for Computational Linguistics on Human Language Technologies: Short Papers, Columbus, OH, USA, 16–17 June 2008.
56. Hussein, M.; Bazuhair, A. Groundwater in Haddat Al Sham-Al Bayada area, western Saudi Arabia. *Arab. Gulf J. Sci. Res.* **1992**, *1*, 23–43.
57. Al-Bassam, A.; Hussein, M. Combined geo-electrical and hydrochemical methods to detect salt-water intrusions: A case study from southwestern Saudi Arabia. *Manag. Environ. Qual.* **2008**, *19*, 179–193. [CrossRef]
58. Al Trbag, A.; Al-Amri, A.; El Derby, A. *Assessment of Groundwater at the Sites of the Jazan for Agricultural Development—Report Prepared for the Benefit of Jazan Development Co. Agricultural, Jazan—Saudi Arabia*, College of Science, King Saud University: Riyadh, Saudi Arabia, 1997; *unpublished*. (In Arabic)
59. Sakaa, B.; Elbeltagi, A.; Boudibi, S.; Chaffaï, H.; Islam, A.R.M.T.; Kulimushi, L.C.; Choudhari, P.; Hani, A.; Brouziyne, Y.; Wong, Y.J. Water quality index modeling using random forest and improved SMO algorithm for support vector machine in Saf-Saf river basin. *Environ. Sci. Pollut. Res.* **2022**, *29*, 48491–48508. [CrossRef]
60. Elbeltagi, A.; Pande, C.B.; Kouadri, S.; Islam, A.R.M.T. Applications of various data-driven models for the prediction of groundwater quality index in the Akot basin, Maharashtra, India. *Environ. Sci. Pollut. Res.* **2021**, *29*, 17591–17605. [CrossRef]
61. Iqbal, M.M.; Li, L.; Hussain, S.; Lee, J.L.; Mumtaz, F.; Elbeltagi, A.; Waqas, M.S.; Dilawar, A. Analysis of Seasonal Variations in Surface Water Quality over Wet and Dry Regions. *Water* **2022**, *14*, 1058. [CrossRef]
62. Al-Maktoumi, A.; El-Rawy, M.; Zekri, S. Management options for a multipurpose coastal aquifer in Oman. *Arab. J. Geosci.* **2016**, *9*, 636. [CrossRef]
63. Al-Maktoumi, A.; Zekri, S.; El-Rawy, M.; Abdalla, O.; Al-Wardy, M.; Al-Rawas, G.; Charabi, Y. Assessment of the impact of climate change on coastal aquifers in Oman. *Arab. J. Geosci.* **2018**, *11*, 501. [CrossRef]
64. El-Rawy, M.; Al-Maktoumi, A.; Zekri, S.; Abdalla, O.; Al-Abri, R. Hydrological and economic feasibility of mitigating a stressed coastal aquifer using managed aquifer recharge: A case study of Jamma aquifer, Oman. *J. Arid. Land* **2019**, *11*, 148–159. [CrossRef]
65. Motallebian, M.; Ahmadi, H.; Raoof, A.; Cartwright, N. An alternative approach to control saltwater intrusion in coastal aquifers using a freshwater surface recharge canal. *J. Contam. Hydrol.* **2019**, *222*, 56–64. [CrossRef] [PubMed]
66. Abdoulhalik, A.; Ahmed, A.A. The effectiveness of cutoff walls to control saltwater intrusion in multi-layered coastal aquifers: Experimental and numerical study. *J. Environ. Manag.* **2017**, *199*, 62–73. [CrossRef] [PubMed]
67. Laabidi, E.; Bouhlila, R. A new technique of seawater intrusion control: Development of geochemical cutoff wall. *Environ. Sci. Pollut. Res.* **2021**, *28*, 41794–41806. [CrossRef] [PubMed]
68. Sherif, M.M.; Hamza, K. Mitigation of Seawater Intrusion by Pumping Brackish Water. *Transp. Porous Media* **2001**, *43*, 29–44. [CrossRef]

water

MDPI

Article

Mapping Specific Constituents of an Ochre-Coloured Watercourse Based on In Situ and Airborne Hyperspectral Remote Sensing Data

Christoph Ulrich [1], Michael Hupfer [2,3], Robert Schwefel [2], Lutz Bannehr [1] and Angela Lausch [1,4,5,*]

1 Department of Architecture, Facility Management and Geoinformation, Institute for Geoinformation and Surveying, Bauhausstraße 8, D-06846 Dessau, Germany
2 Department of Ecohydrology and Biogeochemistry, Leibniz Institute of Freshwater Ecology and Inland Fisheries, Müggelseedamm 301, D-12587 Berlin, Germany,
3 Department of Aquatic Ecology, Brandenburg Technical University Cottbus-Senftenberg, Seestr. 45, D-15526 Bad Saarow, Germany
4 Department of Computational Landscape Ecology, Helmholtz Centre for Environmental Research–UFZ, Permoserstr. 15, D-04318 Leipzig, Germany
5 Department of Geography and Geoecology, Martin Luther University Halle-Wittenberg, Von-Seckendorff-Platz 4, D-06120 Halle, Germany
* Correspondence: angela.lausch@ufz.de; Tel.: +49-341-235-1961; Fax: +49-341-235-1939

Abstract: It is a well-known fact that water bodies are crucial for human life, ecosystems and biodiversity. Therefore, they are subject to regulatory monitoring in terms of water quality. However, land-use intensification, such as open-cast mining activities, can have a direct impact on water quality. Unfortunately, in situ measurements of water quality parameters are spatially limited, costly and time-consuming, which is why we proposed a combination of hyperspectral data, in situ data and simple regression models in this study to estimate and thus monitor various water quality parameters. We focused on the variables of total iron, ferrous iron, ferric iron, sulphate and chlorophyll-a. Unlike other studies, we used a combination of airborne hyperspectral and RGB data to ensure a very high spatial resolution of the data. To investigate the potential of our approach, we conducted simultaneous in situ measurements and airborne hyperspectral/RGB aircraft campaigns at different sites of the Spree River in Germany to monitor the impact of pyrite weathering on water bodies after open-cast mining activities. Appropriate regression models were developed to estimate the five variables mentioned above. The model with the best performance for each variable gave a coefficient of determination R^2 of 64% to 79%. This clearly shows the potential of airborne hyperspectral/RGB data for water quality monitoring. In further investigations, we focused on the use of machine learning techniques, as well as transferability to other water bodies. The approach presented here has great potential for the development of a monitoring method for the continuous monitoring of still waters and large watercourses, especially given the freely available space-based hyperspectral missions via EnMAP.

Keywords: remote sensing; hyperspectral data; RGB; in situ; ochre-coloured rivers; pyrite weathering; water constituents; water quality

Citation: Ulrich, C.; Hupfer, M.; Schwefel, R.; Bannehr, L.; Lausch, A. Mapping Specific Constituents of an Ochre-Coloured Watercourse Based on In Situ and Airborne Hyperspectral Remote Sensing Data. *Water* **2023**, *15*, 1532. https://doi.org/10.3390/w15081532

Academic Editor: Imokhai Theophilus Tenebe

Received: 3 March 2023
Revised: 7 April 2023
Accepted: 11 April 2023
Published: 13 April 2023

1. Introduction

The abundance, distribution and use of water are becoming ever more critical on a global scale [1]. Nevertheless, water is a decisive factor in the evolution, distribution and maintenance of ecosystems and biodiversity. Severe anthropogenic influences (e.g., eutrophication, extraction, contamination, sewage, toxins, coal mines and climate change) have already led to a partially irreversible deterioration and disturbance of freshwater ecosystems, their associated biodiversity and their loss of ecosystem services.

Lignite and coal are important energy sources in many countries. However, mining directly and indirectly affects surrounding ecosystems and nearby water bodies [2]. Before contaminated water is discharged into local surface waters, it is often collected in sedimentation ponds and treated with lime [3]. The liming process generates copious amounts of chemically stable sludge, which contains heavy metals. Furthermore, the groundwater level is also lowered during open-cast mining activities. In areas with soils containing iron disulphide minerals, e.g., marcasite or pyrite, low groundwater levels enable the reaction between these minerals, atmospheric oxygen and precipitation water, which leads to the decomposition of the minerals into sulphate, ferrous iron (Fe(II)) ions and hydrogen [4,5]. The low groundwater level also has a negative impact on the surrounding flora and fauna, as well as stream biodiversity [6–9]. When the open-cast mining activities are over, renaturation occurs and the groundwater rises naturally. Through this process, the products of pyrite weathering reach the local surface waters through aquifers [6]. In the oxygen-rich surface waters, the Fe(II) ions oxidise to ferrous iron (Fe(III)) ions and flocculate, leading to an ochre-like colouring of the water, which is consequently called ochre-coloured water. The ochre causes high turbidity in the water, and the deposition of iron precipitates, thus attenuating the living conditions for sediment dwellers. As a result, biodiversity in the affected areas decreases [10].

Monitoring of the effects of open-cast mining and subsequent renaturation using in situ measurement methods has been carried out for some time now. However, extensive temporal and spatial studies are time- and cost-intensive [11]. Airborne and spaceborne multispectral [12] and hyperspectral remote sensing (RS) [13] provide an alternative approach for monitoring various water quality indicators [14,15]. RS approaches provide continuous spatial and temporal maps of specific water quality parameters. Furthermore, they enable efficient, repeatable and standardised monitoring of a specific area and the identification and quantification of water constituents [12,16].

Owing to technical and methodological developments, hyperspectral RS is being increasingly used to derive water quality parameters, such as chlorophyll a (Chl-a) content [17], turbidity and visibility [18], depth [19] or coloured dissolved organic matter (CDOM) [17]. The quality of RS approaches and algorithms was greatly improved by the use of in situ monitoring data for calibration and validation. The use of hyperspectral RS approaches to detect water characteristics will play a crucial role in the regional and global monitoring of water quality indicators in the context of existing and future hyperspectral satellite missions, such as the DLR Earth Sensing Imaging Spectrometer (DESIS) [20], the Environmental Mapping and Analysis Program (EnMAP) [21], the PRecursore IperSpettrale della Missione Applicativa (PRISMA) [22] or the Hyperspectral Imager Suite (HISUI [23]).

Knowledge of the traits of water and its dissolved and particulate components forms the basis for numerous RS algorithms [24,25]. Consequently, all optically active water constituents determine the reflectance (RRS) of the water through the specific absorption and backscattering behaviour per wavelength [26]. On the other hand, the RS of water constituents often relies on empirical models [16,25,27]. These models try to establish statistical relationships between remote sensing signals and in situ data from locally collected water samples. In empirical models, mechanistic explanations based on the radiative transfer equation are considered from statistical correlations. For waterbodies with similar optical properties, empirical models based on a limited amount of local field data may provide fast and accurate results with a high spatial resolution for a specific aquatic environment [28].

The relevant spectral range to derive optically active water constituents, i.e., CDOM, phytoplankton and total suspended matter (TSM) [25,29], ranges between 400 and 1000 nm. Water absorbs a large amount of sunlight and exhibits low backscattering behaviour, which only increases within the wavelength range of 400 to 500 nm [30]. CDOM mainly absorbs light within the wavelength range of 400 and 500 nm [31,32]. Therefore, most CDOM-derivation models are based on this wavelength range [33]. Humic substances and other dissolved organic carbon (DOC) are often assigned to this group. Phytoplankton is one of the living and particulate components of waters measured using RS and Chl-a, which is a key molecule of photosynthetically active organisms that is often used as an indicator of biomass production. The spectral characteristics of Chl-a include the low reflectance between 400 and 500 nm [32] and maximum reflectance at 580 nm [27,30]. The non-living particulate matter can be assigned to TSM. This group includes particulate organic deceased matter or small sand particles [30]. Water bodies with a high TSM concentration exhibit increased backscattering behaviour. In addition to the abovementioned active water constituents, TFe, Fe(II), Fe(III) and sulphate also affect the water colouring.

Repic et al. [34] already used multispectral video data with three spectral channels in the wavelength range from 400 to 1500 nm to detect the iron ions of two abandoned open pit lakes. Furthermore, Anderson and Robbins [35] developed a method to spectrally discriminate acid mine drainage (AMD) and natural streams using water samples, in situ spectral data and multispectral RS data. They found that a $pH < 4$ may have prevented the oxidation of Fe(II) to Fe(III). As a result, iron could not precipitate in the form of Fe(III). On this basis, Anderson and Robbins [35] found wavelengths at 650 nm and 750 nm to be very sensitive to iron oxides. Williams et al. [36] used the same airborne multispectral video system to identify and map AMDs. At a pH of 3.2 and 6.9, they were able to show that the 650 and 750 nm wavelengths were sensitive enough to visualize the ochre precipitates of the mine water.

Following the initial studies on AMDs, in subsequent years, research focused on open-cast mining residual lakes [30,37,38].

In recent decades, significant progress was achieved in monitoring and assessing the quality of inland waters using RS [12,15,25]. Through improved computer-based interpretation techniques and spectrally (up to 3.25 nm per channel) plus spatially (up to 50 cm ground sampling distance) high-resolution satellites, coastal zones and lakes were increasingly monitored [39–41]. Owing to their small spatial extent and complex chemical composition, small watercourses have received relatively little attention [42] and the focus of previous analyses was on very wide rivers, such as the Mississippi, the Gironde or the Rhein, and river estuaries [13,43–46].

By using the institute's own gyrocopter with extensive sensor technology (hyperspectral/RGB/TIR), Fe(II) and Fe(III) were thus recorded, modelled and derived separately for the first time with low spatial RS data resolution (50 cm), even with lower concentrations of iron for smaller flowing watercourses with a neutral pH.

The objectives of this study were as follows:

- To show that airborne hyperspectral/RGB RS technologies are suitable for monitoring water quality parameters for small streams.
- To propose simple linear modelling approaches for modelling and predicting TFe, Fe (II), Fe(III), sulphates and Chl-a in ochre streams.
- To develop and test a robust procedure to derive TFe, Fe(II), Fe(III), sulphates and Chl-a based on airborne hyperspectral RS data and simultaneous field sampling in a river section influenced by mining activities.
- To transfer the point results from the in situ field sampling to the area.
- To discuss the framework conditions as well as the limitations of the presented approach.

2. Materials and Methods

2.1. Study Area

The study area was located southeast of Berlin in Lusatia along the River Spree and extended from the village Ruhlmühle to the end of the Spremberg reservoir (Figure 1). The flow direction of the Spree was from south to north.

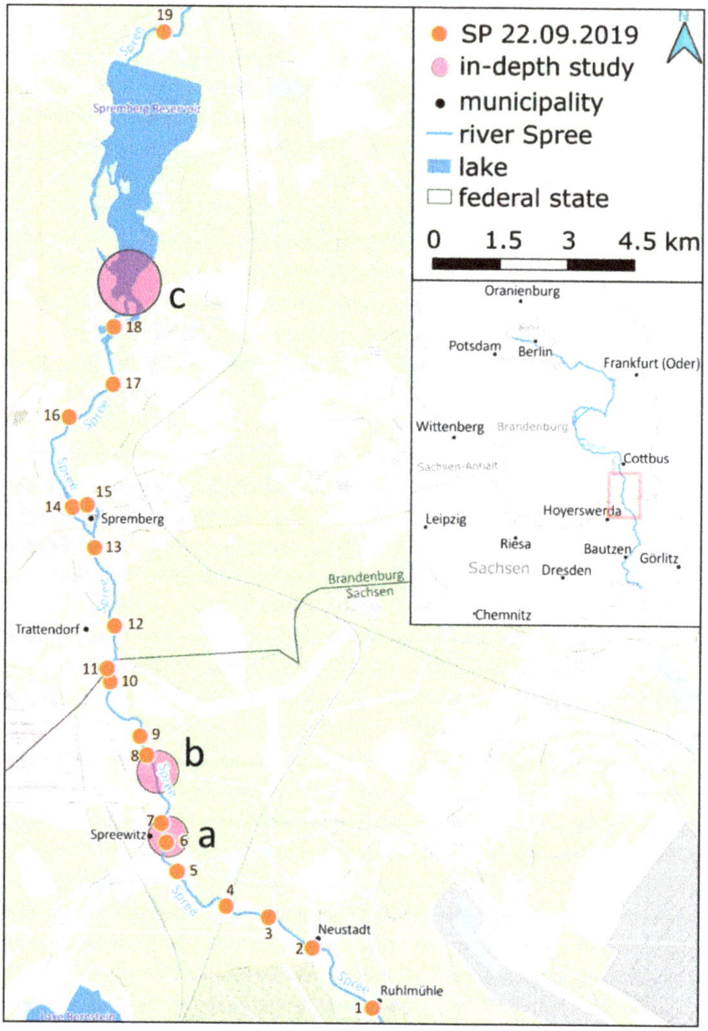

Figure 1. Study site in the south-east of Berlin in Lusatia along the Spree River from the Ruhlmühle village to the Spremberg reservoir. The flow direction of the Spree from south to north. (a–c) shows the sites for more in-depth studies.

Open-cast lignite mining in Lusatia has the greatest impact on this section of the river [47]. Increased loads from the products of pyrite weathering into the Spree River caused (1) an increased concentration of sulphate, (2) an increased concentration of Fe(II) and Fe(III), and (3) increased turbidity of the water body owing to sludge formation.

In the mid-20th century, the groundwater level in the study area was lowered significantly due to open-cast mining. Consequently, the pyrite in the soil reacted with oxygen in the air and precipitation water (Figure 1) and subsequently disintegrated into Fe(II), sulphate and hydrogen ions. Following the implementation of recovery measures in 1990, many open-cast mines were closed and flooded via the natural process. Thus, the groundwater level rose in many parts of the mining area [48]. The products of pyrite weathering in the hydrosphere reached surrounding surface waters via aquifers. The Fe(II) ions were oxidized in the oxygen-rich water to Fe(III) ions [5].

In pH-neutral water bodies, Fe(III) ions were hydrolyzed to ferrous-oxyhydroxide (Fe(III)-oxyhydroxide) and flocculated [5,49]. A Fe(III)-hydroxide concentration of 2 mg/L causes the water to appear yellow-to-reddish-brown in colour [50,51]. The process of iron clogging is schematically summarized in Figure 2.

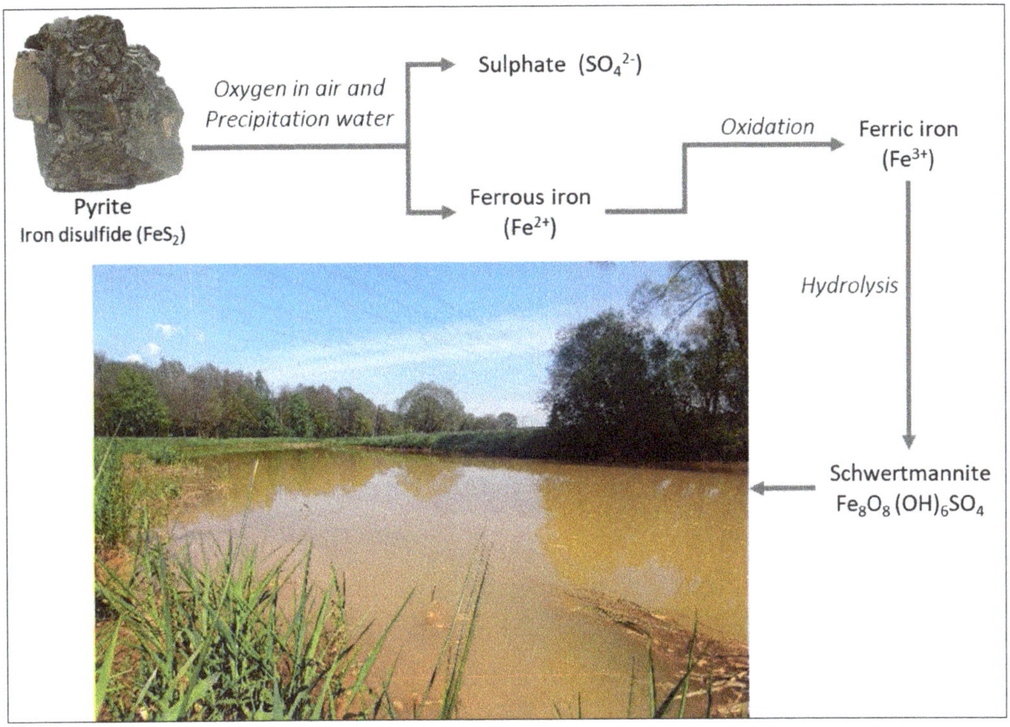

Figure 2. Schematic diagram of the weathering of pyrite from the source material (pyrite) to the clogging in the flowing water (Spree), Fe(III)-oxyhydroxide is represented by Schwertmannite.

The municipalities of Ruhlmühle and Spreewitz, the mouth of the industrial canal of Schwarze Pumpe industrial park, the pre-dam Bühlow and the Spremberg reservoir were significant spots in the study area (Figure 1). Analyses conducted by Bilek and Koch [6] showed the discharge of Fe(II) loads from anoxic groundwater into the oxygen-rich river water. Uhlmann et al. [52] observed the input of large amounts of Fe(II) into the river at the village Ruhlmühle (SP 1 in (Figure 1). In the oxygen-rich water and under natural pH conditions, the Fe(II)-oxide flocculated out as Fe(III)-oxyhydroxide, causing an ochre colour in the river water.

At Spreewitz (SP 7 in Figure 1), the Kleine Spree River flows into the Spree River and is enriched with iron. In the south of Trattendorf town (SP 8 in Figure 1), the canal from the Schwarze Pumpe industrial park increases the sulphate concentration in the Spree River. The treatment of the water in a pre-dam of the Spremberg reservoir (Bühlow) is intended to increase iron retention [53] (SP 18 in Figure 1). In the Spremberg reservoir, the reduced flow velocity leads to the deposition of Fe(III)-oxyhydroxide, as well as reduced turbidity (between SP 18 and 19 in Figure 1). This process significantly reduces the brown colouring of the Spree below the Spremberg reservoir.

2.2. In Situ Data

On 22 September 2019, a field campaign was conducted to simultaneously collect RS data and field samples to develop empirical derivation models. Water samples were taken between 10 a.m. and 2 p.m. at the same time that a gyrocopter flew over (see Section 2.3) using a scoop down to a depth of 30 cm. Thus, every sample represented a composite sample of the top water column. The in situ data served as reference data for the development of additional empirical models in order to derive the distribution of water constituents over an extensive area from airborne hyperspectral data.

The total iron (TFe) was measured following digestion with 2 mL of H_2O_2 (5%) and 2 mL of H_2SO_4 (5 M) according to DIN 38406 using flame-atomic absorbance spectroscopy (AAS PinAAcle 900T, PerkinElmer, Waltham, MA, USA). The dissolved Fe(II) analysis was prepared on site according to DIN 38406 by stabilizing 15 mL of the filtrated water sample with 0.15 mL of 5 M sulphuric acid and storing it free of air bubbles. On the same day, the concentration of Fe(II) was determined photometrically after complexation with 1.10 phenanthroline (Spekol 1500, Analytik Jena, Jena, Germany). The total Fe(III) concentration was calculated using the difference between the TFe concentration and the Fe(II) concentration. Sulphate was determined according to DIN EN ISO 10304-1 by filtering the water sample on site. The sulphate concentration was quantified using ion chromatography (Compact IC Flex 930, Metrohm, Switzerland). To determine the dissolved substances, the sample was filtered using a syringe with a 0.45 μm filter (cellulose acetate; Whatman GmbH, Dassel, Germany). The unfiltered water samples for determining the Chl-a concentration were stored on site in 1 L vessels in a dark and cool place and processed further within 24 h. In the laboratory, the analysis was performed via HPLC (Waters Alliance, Milford, MA, USA) according to the procedure reported by Mehnert et al. [54]. The uncertainty of the analytical replicates in these analyses was typically <3%.

2.3. Remote Sensing Data

The gyrocopter at the Institute for Geoinformation and Surveying of the Anhalt University (see Figure 3) was used to record the hyperspectral RS data (HySpex VNIR, VSWIR) in the research area within the wavelength range of 400–2500 nm. The flight altitude was 600 m. Usually, hyperspectral data within the wavelength range of 400–1000 nm are used to derive water constituents. Because of the high turbidity and high TFe concentration throughout the study area, the wavelength range of 1000–2500 nm will be investigated further.

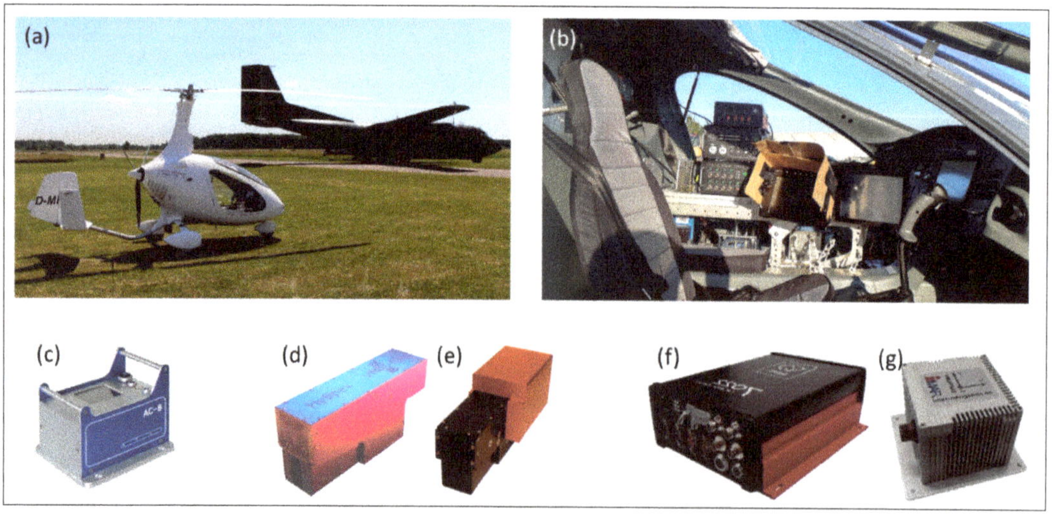

Figure 3. (**a**) Image of the gyrocopter from the Institute for Geoinformation and Surveying of HSA and (**b**) the measurement technology on the co-pilot's seat. Integrated sensor technology: (**c**) Nikon D800e modified by ATS, (**d**) HySpex VNIR, (**e**) HySpex SWIR from Neo, (**f**) GNSS receiver from PPM and (**g**) IMU from iMAR.

For the monitoring, the digital camera NIKON D800e (see Figure 3c) and the hyperspectral sensors HySpex VNIR 1600 (400–1000 nm) and HySpex SWIR 384 (1000–2500 nm) from Norsk Elektro Optikk AS (Neo) were used as optical sensors (see Figure 3d,e). An inertial navigation system (INS) was used for the high-precision acquisition of position and orientation data. The INS consisted of a GNSS receiver from PPM and an inertial measurement unit (IMU) from iMAR (see Figure 3f,g). The sensors are triggered automatically by the flight management software Aviatrix.

To detect small but significant spectral differences, the spectral sampling rate of the HySpex VNIR was set to a bandwidth of 7.4 nm. This resulted in 80 channels within the spectral wavelength range of 400–1000 nm. At an altitude of 600 m, the ground-sampling distance (GSD) was 20 cm. The data from the HySpex SWIR had a spectral sampling rate of 5.45 nm with 288 channels within the wavelength range of 1000–2500 nm. The GSD of the HySpex SWIR was 60 cm at a flight altitude of 600 m.

As a basis for processing the hyperspectral data, a near-true orthophoto with a GSD of 6 cm was calculated from the RGB images using the structure from motion (SfM) algorithm [55]. This served as the basis for determining the boresight angle for processing the hyperspectral data. These data were rectified, georeferenced and atmospherically corrected using PARGE Version 3.3 [56] and ATCOR-4 Version 7.2 [57] software. Owing to the flight altitude of 600 m and a field of view of the hyperspectral sensors of 34° and 32°, respectively, adjacent and atmospheric effects significantly influenced the hyperspectral images.

The atmospheric correction was carried out with the ATCOR software using an urban model based on the presence of open-cast mines, industry and urban areas. Specific absorptions and reflectances of the atmosphere influence a large part of the measurable radiance, which requires correction for reliable derivation [58]. For this reason, according to Ray [59], the determined pixel values represent an apparent reflectance (aRRS), which is comparable with the RRS in most cases.

Although aRRS is comparable to RRS, adjacent effects cannot be compensated for using an atmospheric correction because the information on the simultaneous illumination conditions is not available. To exclude these influences for subsequent calculations, the hyperspectral image data of the surrounding river environment were masked out, meaning that only the relevant spectral information of the river forms the basis of all subsequent model calculations. Adjacent effects can impair the measurable radiance through reflections and lead to saturation of the sensor [60]. The adjacent effects can be assigned, for example, to reflections of solar panels or waves from the water body or even shadows from vegetation on the water's surface. If these influences could also be seen on the water's surface, they were also masked.

2.4. Model Development for the Area-Wide Derivation

The empirical approach links the in situ sampling points (SPs) and airborne hyperspectral RS data via regression to determine the coefficient of determination (R^2) between them [26,61]. For the area coverage derivation of the water constituents, the workflow can be divided into six tasks: (1) image pre-processing, (2) data linking in two steps, (3) development of spectral indices, (4) data filtering using the standard deviation (SD), (5) regression determination and (6) derivation. Due to the two separate steps for linking the SP data with the RS data, an identification number (ID) was assigned to each SP (SP-ID) (see Figure 4).

Initially (task 1), the hyperspectral data was processed as described in Section 2.3. This was followed (task 2) by linking the in situ data with the hyperspectral data in two steps. First, the link was made through the GNSS position. This allowed for the extraction of specific spectral information for each sampling point and the assignment of the particular information to the sampling point identification number (SP-ID). Second, we linked the spectral indices based on specific wavelength bands with the in situ data. As described by Ulrich et al. [62] the spectral indices are based on previous studies on the determination of water quality [27,30,32,45,63–66], spectral simulations and analyses. Based on this, the following spectral indices were developed as a template (Equations (1)–(5); task 3):

$$\text{Type 1 single Wavelength: } SI\,1 = R_{RS}(\text{b1}) \tag{1}$$

$$\text{Type 2 Ratio}: \; SI\,2 = \frac{R_{RS}(\text{b1}) - R_{RS}(\text{b2})}{R_{RS}(\text{b1}) + R_{RS}(\text{b2})} \tag{2}$$

$$\text{Type 3 Ratio}: \; SI\,3 = \frac{R_{RS}(\text{b1}) - R_{RS}(\text{b3})}{R_{RS}(\text{b2}) - R_{RS}(\text{b3})}\text{t} \tag{3}$$

$$\text{Type 4 Ratio}: \; SI\,4 = \frac{R_{RS}(\text{b1}) - R_{RS}(\text{b3})}{R_{RS}(\text{b2}) + R_{RS}(\text{b3})} \tag{4}$$

$$\text{Type 5 Slope}: \; SI\,5 = \frac{R_{RS}(b2) - R_{RS}(b1)}{b2 - b1} \tag{5}$$

where R_{RS}: reflectance and b: band of the hyperspectral image.

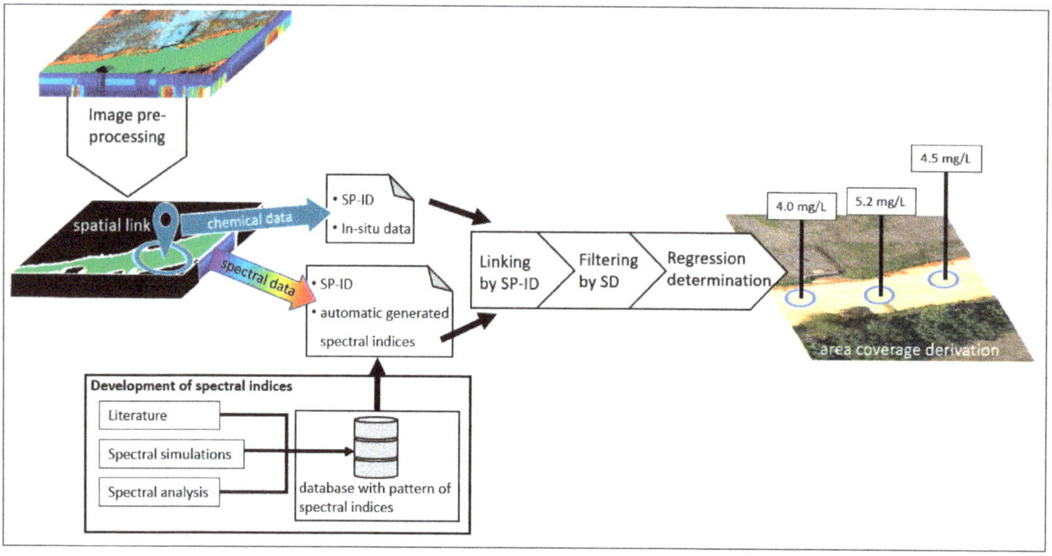

Figure 4. Flowchart of the methodical approach used to derive specific constituents of an ochre-coloured watercourse based on sampling points (SP) and airborne hyperspectral remote sensing data. Image pre-processing, development of spectral indices, data linking in two steps, data filtering using the standard deviation (SD), regression determination and derivation. Due to the two separate steps for linking the SP data with the RS data, an identification number (ID) was assigned to each SP (SP-ID).

To improve the derivation accuracy, the water samples were filtered using 2.5 times the SD to eliminate incorrect samples or analyses (task 4).

Next (task 5), suitable regression models were determined for each water constituent: TFe, Fe(II), Fe(III) and sulphate. For this purpose, each chemical parameter was related to 100 spectral indices and tested for linear, logarithmic, exponential and potential regressions (Equations (6)–(9)) following the general regression model presented by the International Ocean-Colour Coordinating Group [67].

$$\text{Linear regression}: \ \hat{p} = \alpha \times (SI) + \beta \tag{6}$$

$$\text{Logarithmic regression}: \ \hat{p} = \alpha \times log(SI) + \beta \tag{7}$$

$$\text{Exponential regression}: \ \hat{p} = \beta \times e^{\alpha*(SI)} \tag{8}$$

$$\text{Potential regression}: \ \hat{p} = \beta \times (SI)^{\alpha} \tag{9}$$

where $\hat{p}$: parameter to be estimated in the quantitative unit; SI: spectral index; R_i: spectral band; and α and β: regression parameters between the spectral and in situ data.

The most suitable type was determined using the coefficient of determination (R^2) value and stored with the corresponding regression parameters. Consequently, this procedure resulted in a specific regression model for each water constituent per spectral index for the area coverage derivation.

Finally, for the accuracy analysis, the RMSE and relative RMSE (rRMSE) values were calculated for 10 derivation models with the highest R^2 (task 6). For this purpose, the derived concentrations at the sampling site were linked to the filtered limnic data. The rRMSE was calculated according to the procedure reported by Gao et al. [68] and Wang and Lu [69] (Equation (10)). This allowed for the most appropriate spectral index to be selected for each water body constituent.

$$rRMSE\ [\%] = \frac{RMSE}{\overline{Obs}} \times 100\% \tag{10}$$

where $\overline{Obs}$: mean value of the observations.

3. Results and Discussion

The limnological characteristics of watercourses influenced by mining differ from those of natural watercourses [37]. This is because of the geological conditions, the geohydrological processes of the mining areas and a rise in the groundwater level [38]. In this study, a framework consisting of simultaneous monitoring of in situ data and airborne hyperspectral/RGB data was presented, enabling the modelling of water quality indicators (TFe, Fe(II), Fe(III), sulphate and Chl-A).

3.1. In Situ Measurements of Water Quality

In situ data from 19 sampling locations were analysed for TFe, Fe(II), Fe(III), sulphate and Chl-A (see Table 1). Only 15 of the sampling locations were covered by the gyrocopter survey and used to calibrate the remote sensing data. Table 1 gives an overview of the monitored limnic parameters and their mean values, standard deviations, minima and maxima.

Table 1. List of the in situ monitored chemical parameters. The SP-IDs 3-14 and 16-18 were used for the derivation models. Thus, the total quantity (N) ranged between 13 and 15 values. This was due to suitable spatial overlaying with the RS data.

SP ID	TFe [mg/L] (1)	Fe(II) [mg/L]	Fe(III) [mg/L]	Sulphate [mg/L] (2)	Chl-a [μg/L]
1	0.7	0.04	0.66	356	11.9
2	2.8	2.19	0.61	362	12.31
3	3.8	2.45	1.35	364	9.76
4	5.7	4.53	1.17	369	10.64
5	6.3	4.47	1.83	367	9.96
6	6.3	5.17	1.13	366	11.04
7	7.0	5.28	1.72	336	10.20
8	4.9	3.1	1.8	436	7.41
9	4.7	2.74	1.96	434	7.06

Table 1. *Cont.*

SP ID	TFe [mg/L] [(1)]	Fe(II) [mg/L]	Fe(III) [mg/L]	Sulphate [mg/L] [(2)]	Chl-a [µg/L]
10	3.9	2.00	1.9	422	6.92
11	3.9	1.89	2.01	420	7.18
12	4.00	1.52	2.48	421	7.03
13	3.2	0.94	2.26	421	6.63
14	2.3	0.58	1.72	426	6.15
15	3.1	0.64	2.46	419	6.32
16	3.2	0.31	2.89	398	8.66
17	1.8	<0.05	-	399	6.43
18	2.9	<0.05	-	399	6.35
19	0.6	<0.05	-	430	10.39
Mean	**3.74**	**2.37**	**1.75**	**3.74**	**2.37**
Standard deviation	**1.74**	**1.68**	**0.62**	**31**	**2.05**
Maximum	**7.00**	**5.28**	**2.89**	**436**	**12.32**
Minimum	**0.60**	**0.04**	**0.61**	**336**	**6.15**

Note: The regulatory limits according to German law are (https://www.gesetze-im-internet.de/trinkwv_2001/BJNR095910001.html, accessed on 4 December 2023): [(1)] 0.2 mg/L for (total) iron and [(2)] 250 mg/L for sulphate.

Figure 5 shows the plot of the distribution of the recorded in situ water parameter measurements for Chl-a, TFe, Fe(II), Fe(III) (Figure 5a) and sulphate (Figure 5b) based on a probability function. The value range of the sulphate concentration deviated strongly from the concentrations of the other water constituents.

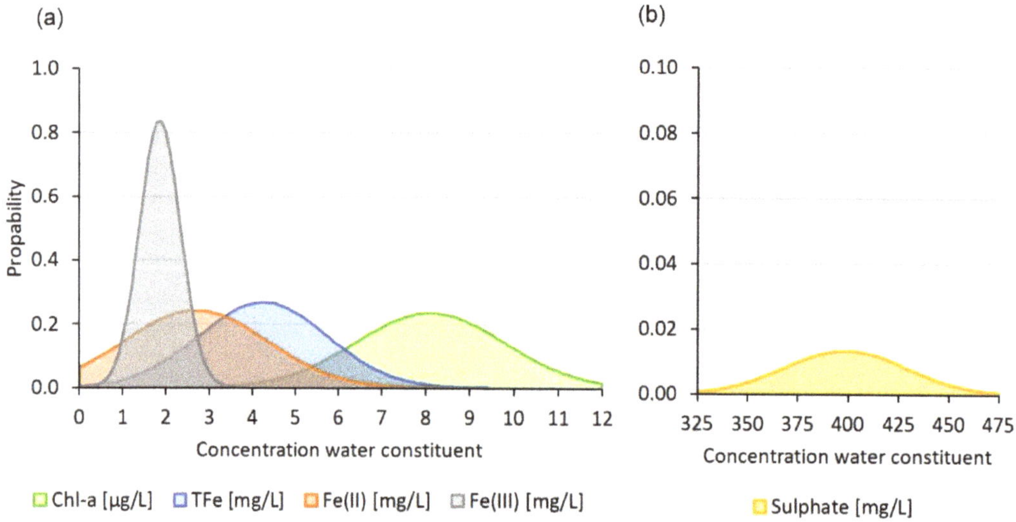

Figure 5. Plot of the distribution of the recorded in situ water parameter measurements: (**a**) Chl-a, TFe, Fe(II), Fe(III) and (**b**) sulphate based on a probability function in a concentration range from 0 to 475 mg/L. The value range of the sulphate concentration deviated strongly from the concentrations of the other water constituents.

Likewise, the correlation matrix (Table 2) shows that the in situ water parameters Chl-a, TFe, Fe(II), Fe(III) and sulphate were correlated with each other, making modelling and a clear delineation of the individual water parameters using hyperspectral data difficult. TFe and Fe(II) showed the highest correlations with r = 0.97. This indicates a high proportion of dissolved Fe(II) in the measured water depth of 30 cm. Furthermore, the water body might not have been completely saturated with oxygen on the day of the measurement. The correlation between Fe(III) and TFe was r = −0.50 and between Fe(III) and Fe(II) was r = −0.70. Both values indicated that the Fe(III) concentration increased or decreased inversely to the concentrations of TFe and Fe(II) (see Table 2).

Table 2. Correlations matrix with the recorded in situ water parameters: TFe, Fe(II), Fe(III), sulphate and Chl-a.

	TFe [mg/L]	Fe(II) [mg/L]	Fe(III) [mg/L]	Sulphate [mg/L]	Chl-a [µg/L]
TFe [mg/L]	**1.00**	**0.97**	−0.50	−0.58	**0.79**
Fe(II) [mg/L]	**0.97**	**1.00**	−0.70	−0.70	**0.79**
Fe(III) [mg/L]	−0.50	−0.70	**1.00**	0.48	−0.56
Sulphate [mg/L]	−0.58	−0.70	0.48	**1.00**	**−0.84**
Chl-a [µg/L]	**0.79**	**0.79**	−0.56	**−0.84**	**1.00**

Similarly, sulphate also showed a correlation of r = −0.58 with TFe and r =−0.70 with Fe(II). This was due to additional inputs of sulphate from the Schwarze Pumpe industrial park at sample point 8 (see Figure 1). Chl-a indicated a correlation of r = 0.79 with TFe and Fe(II). This indicated a uniform decrease of the Chl-a concentration in relation to the TFe and Fe(II) concentrations, resulting from the increasing ocherous colour due to the flocculation of the Fe(III) ions. Accordingly, the correlation between Chl-a and Fe(III) was r = −0.56. Moreover, Chl-a also showed a correlation of r = −0.84 with sulphate.

Figure 6 shows five spectra within the wavelength range of 415 to 1615 nm. The corresponding chemical parameters are listed in Table 1. These spectra were extracted from the hyperspectral images at sampling points 4, 7, 11, 16 and 19.

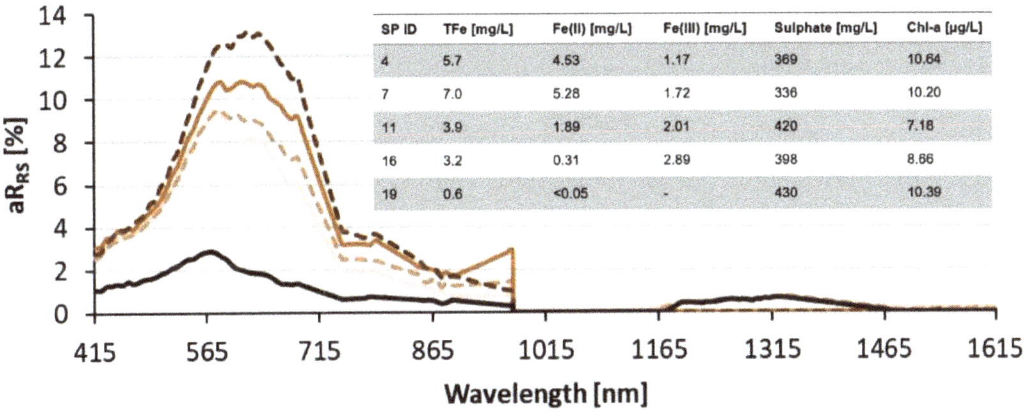

Figure 6. Representation of five airborne hyperspectral spectra in the wavelength range of 420 to 1500 nm with different TFe concentrations at the sampling points. The jump at 1000 nm resulted from the inaccuracies of the internal calibration of the two hyperspectral data sets of HySpex VNIR and HySpex SWIR to the one final hyperspectral data set (400–2500 nm). These overlapped in the wavelength range of approximately 1000 nm and could occur with differences of up to 4%. The table contains the in situ measurements of the five measuring points (see also Table 1).

3.2. *Airborne Hyperspectral RS and Modelling*

Thanks to the institute's own gyrocopter with the hyperspectral sensor technology HySpex, it was possible to record airborne hyperspectral images in the wavelength range from 420 to 1500 nm at the same time as the in situ measurements. Figure 6 shows the results of the refluxes in the wavelength range from 440 to 1400 nm at five in situ measuring points for which airborne hyperspectral RS was conducted.

Figure 6 clearly shows that the reflectance in the wavelength range from 560 to 700 nm increased with increasing iron(III) concentration. This can also be seen from the correlation of $R^2 = 0.78$ with the wavelength at 675 nm (see Table 3). Within the wavelength range of 400–560 nm, the aRRS was low owing to absorption by CDOM and Chl-a [32]. In addition, Weyhenmeyer et al. [70] observed that Fe(II), like DOC, correlates with absorption at 420 nm. Moreover, according to Asmala et al. [31], the influence of Fe(II) on the absorption behaviour of the water body is visible beyond 520 nm. However, based on the acquired data, the spectral influence of Fe(II) and sulphate was not visible within this wavelength range. This was particularly due to the rapid oxidation of Fe(II) to Fe(III). In addition, in the spectral range of 560–700 nm, most maxima exceeding 6% could be determined at the measuring points with a TFe concentration exceeding 3 mg/L. The highest maximum at 580 nm shows the spectrum of SP 16 with the highest Fe(III) concentration of 2.89 mg/L (see Figure 6a). Owing to the high concentration of Fe(III) and the resulting flocculation, the water became very turbid, which, according to Frauendorf [30], makes Fe(III) classifiable as TSM.

Table 3. List of the area coverage derivation results showing the number of samples used (N), the ID to link to the regression models in Figure 1, the spectral index, the regression equation, the coefficient of determination (R^2), the RMSE and the rRMSE. The models displayed in **bold font** were used in Figure 6 for the specific concentration derivation.

Parameter	N	ID	Spectral Index	Equation	R^2	RMSE	rRMSE
TFe [mg/L]	15	(a)	$(R_{RS}(580)-R_{RS}(455))/(R_{RS}(580)+R_{RS}(455))$	$\hat{p} = 51.04 \times \left(\frac{R_1}{R_2}\right) + 27.69$	0.70	0.95	22.19%
		(b)	$\mathbf{(R_{RS}(455)-R_{RS}(580))/(R_{RS}(480)+R_{RS}(580))}$	$\hat{p} = \mathbf{55.86} \times \left(\frac{R_1}{R_2}\right) + \mathbf{30.14}$	**0.64**	**0.93**	**21.91%**
Fe(II) [mg/L]	13	**(c)**	$\mathbf{(R_{RS}(580)-R_{RS}(455))/(R_{RS}(580)+R_{RS}(455))}$	$\hat{p} = \mathbf{59.43} \times \left(\frac{R_1}{R_2}\right) + \mathbf{29.78}$	**0.78**	**0.95**	**35.33%**
Fe(III) [mg/L]	13	(d)	$(R_{RS}(580)-R_{RS}(1250))/(R_{RS}(650)-R_{RS}(1250))$	$\hat{p} = 2.17 \times (SI)^{-2.8}$	0.84	0.27	14.53%
		(e)	$\mathbf{R_{RS}(701)/R_{RS}(563)}$	$\hat{p} = \mathbf{0.24}\, e^{\mathbf{2.81}(SI)}$	**0.79**	**0.22**	**11.86%**
Sulphate [mg/L]	15	**(f)**	$\mathbf{(R_{RS}(580)-R_{RS}(480))/(R_{RS}(580)+R_{RS}(480))}$	$\hat{p} = \mathbf{1.03} \times (SI)^{\mathbf{1.07}}$	**0.53**	**21**	**5.31%**
Chl-a [µg/L]	15	(g)	$(R_{RS}(580)-R_{RS}(455))/(R_{RS}(580)+R_{RS}(455))$	$\hat{p} = 201.69\, e^{-7.05(SI)}$	0.72	1.09	13.48%
		(h)	$\mathbf{(R_{RS}(580)-R_{RS}(480))/(R_{RS}(580)+R_{RS}(480))}$	$\hat{p} = \mathbf{26.3} \times ln(SI) - \mathbf{15.48}$	**0.67**	**0.98**	**12.09%**

By means of the multi-spectrometer Ocean and Land Colour Instrument (OLCI) on board the Sentinel 3, Knaeps et al. [71] established that with increasing TSM concentration up to 402 mg/L, additional information on the short-wave infrared (SWIR) range contributes to more accurate modelling. Furthermore, iron compounds in the soil show specific spectral characteristics, especially in the SWIR range [72]. The spectra with a TFe concentration under 3 mg/L are similar to the aRRS of natural rivers (SP 19 in Figure 6). Absorption by water is clearly visible between 700–1000 nm. The aRRS between 1000–1615 nm do not show additional information.

The most suitable regression models to link particular spectral indices to water constituents are shown in Figure 7. They are also displayed in Table 3, which additionally shows the number of samples used, the spectral index, the ID to link to the regression models, the regression equation, R^2, the RMSE and the rRMSE. The number of samples for modelling was lower than indicated because not every sampling location was within the coverage area of the RS data. The empirical models achieved a derivation accuracy of 0.93 mg/L for the total iron concentration, 0.95 mg/L for ferrous iron, 0.22 mg/L for ferric iron and 21 mg/L for sulphate.

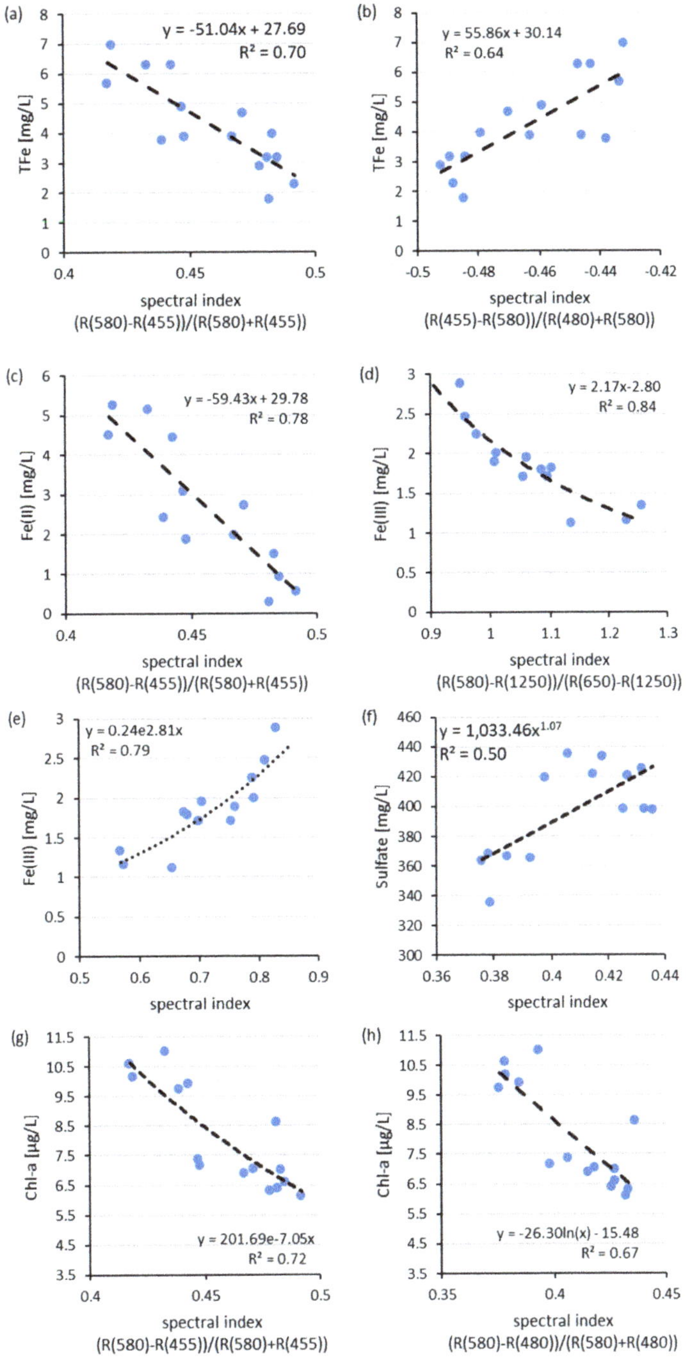

Figure 7. Representation of the regression models (in Table 3) showing the regression equivalence and the coefficient of determination. (**a–h**) The most suitable regression models to link particular spectral indices to water constituents.

Both indices for the derivation of TFe were based on the wavelengths of 455 and 580 nm and can be explained based on spectral analysis. The wavelength range between 450–500 nm was characterized by a broad absorption band of iron oxides. This was also confirmed by Rowan and Goetz [72]. As shown in Figure 7, there was a maximum in the reflectance at 580 nm, which was caused by an increased scattering by TSM and a low absorption by Chl-a [32,73]. Since TFe and Chl-a are correlated, it is difficult to distinguish between these constituents based on regression models. A linear regression of the single wavelength at 580 nm and TFe and Chl-a (Equations (1) and (6)) shows that the reflection decreased with increasing TFe (r^2 = 0.5) and Chl-a (r^2 = 0.3) concentrations, indicating that iron was mainly responsible for the optical properties within this wavelength range.

As a result, there was a steep slope from 450–580 nm in the spectral curve. The most suitable model for deriving the TFe content obtained an rRMSE of 21.91% and an R^2 of 0.64. The regression model for determining the Fe(II) concentration with the highest correlation of 0.78 was based on the same spectral analysis as TFe but the validation showed a higher rRMSE of 35.33%. The use of the same spectral index as the basis for the derivation model can be explained by the high correlation r = 0.97 between TFe and Fe(II). The first model for the derivation of Fe(III) was based on the study conducted by Knaeps et al. [46] and is suitable for inland waters with a TSM concentration of 30 to 1400 mg/L. The index showed an R^2 of 0.80 and an rRMSE of 14.53%. The second model used to derive Fe(III) was based on the spectral index RRS(701)/RRS(563) and yielded an accuracy of 11.86% and an R^2 of 0.76. According to Gitelson [32], this index is suitable for deriving Chl-a. However, spectral analyses with the software package WASI [74] showed that the reflectance increases around 580 nm with increasing TSM concentration. Furthermore, the absorption by water increased steadily from 700 nm, resulting in a strong reduction in reflectance. As a result, there was a strong gradient between 580 and 700 nm.

The model used to derive sulphate yielded an R^2 = 0.53 and an rRMSE of 5.31%. The derivation algorithm used a potential regression model based on the spectral index (RRS(580)-RRS(480))/(RRS(580)+RRS(480)). This spectral index was based on the analyses carried out. As described, there was an increased scattering at 580 nm owing to the TSM concentration [32]. Furthermore, the aforementioned absorption of iron oxides at the 480 nm range contributed to a strong gradient in the 480–580 nm range.

The spectral indices obtained for the derivation of Chl-a were similar to those of the derivation of TFe. This was probably related to the high correlation of r = 0.79 between Chl-a and TFe, as well as Fe(II). Nevertheless, the best fitting regression model was (RRS(580) − RRS(480))/(RRS(580) + RRS(480)) based on a logarithmic approach and a derivative accuracy of RMSE = 0.98 μg/L.

To summarize, it was concluded that the derivative models, with the exception of Fe(III), mainly referred to the spectral range from 450 to 580 nm. Based on the day of the investigation, this spectral range had the highest derivation accuracy. Wavelengths beyond 560 nm were only considered for the derivation of Fe(III).

Figure 8 shows the derivation results of TFe, Fe(II), Fe(III), sulphate and Chl-a using three different locations as examples (see Figure 1). The background of the derivation results consisted of channels 580, 680 and 780 nm of the hyperspectral image. Therefore, the River Spree and the Spremberg Dam appeared blue. Masked areas are shown in black. Areas outside of the water body are displayed in grey. Data-free areas are presented in white. For the derivation of the water constituents TFe, Fe(II), Fe(III), sulphate and Chl-a, the models displayed in bold in Table 3 were used.

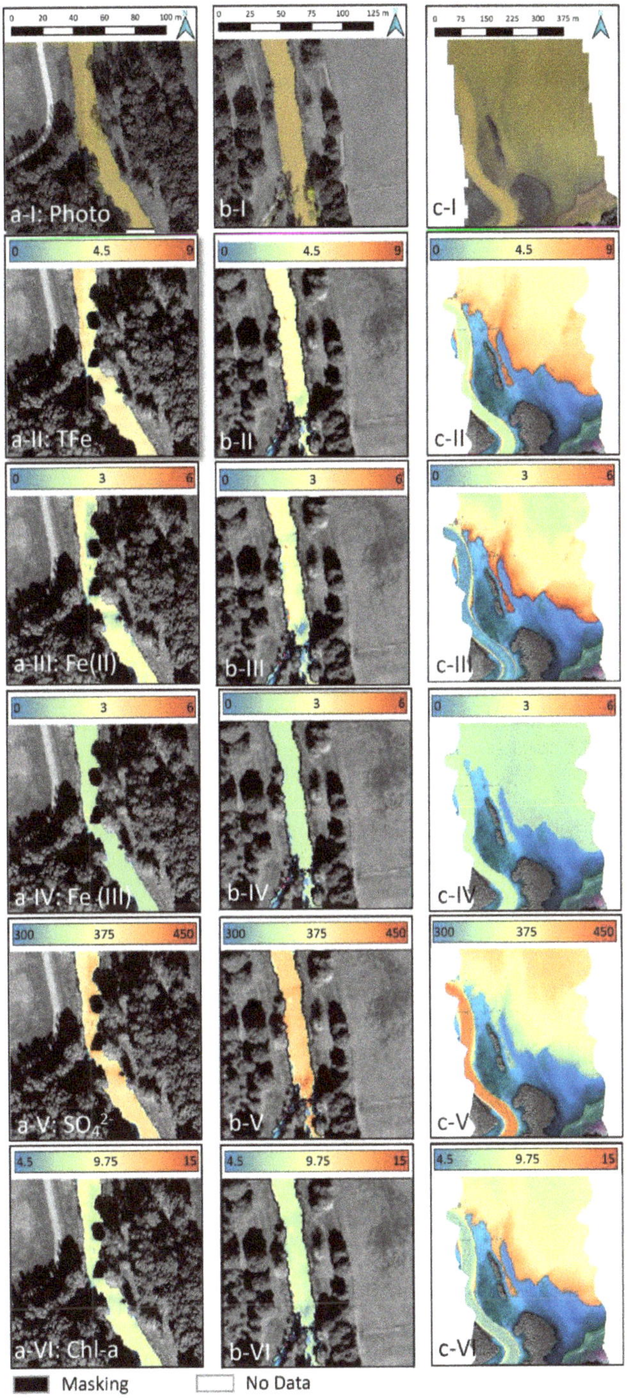

Figure 8. Three different regions with the corresponding derivation results of (**II**) TFe [mg/L], (**III**) Fe(II) [mg/L], (**IV**) Fe(III) [mg/L], (**V**) sulphate [mg/L] and (**VI**) Chl-a [µg/L] are shown in columns: (**a**) the estuary of the Kleine Spree River, (**b**) an industrial canal, and (**c**) the estuary of the Spree River

into the Spremberg Dam. (**I**) represents the study area based on the high-resolution true orthophoto. The models shown in bold in Table 3 were used for the area coverage derivation. The background of the derivation results consisted of channels 580, 680 and 780 nm of the hyperspectral image, whereby the water body appeared blue. Masked areas, e.g., due to shadows or adjacent effects, are shown in black, and areas outside the water body are shown in grey. Data-free areas are shown in white. The locations of the in-depth studies are shown in Figure 1a–c.

Figure 8 shows the mouth of the Kleine Spree into the River Spree. Based on the derivation results of TFe (see Figure 8(a-II)), a difference was observed between the concentration in the Kleine Spree and that in the River Spree. In the case of Fe(II) derivation (see Figure 8(a-III)), the influence of the Kleine Spree was comparatively low. The influence of the Kleine Spree River can only be observed directly at the river mouth based on Chl-a derivatives (see Figure 8(a-VI)). The derived Fe(III) (see Figure 8(a-IV)) and sulphate (see Figure 8(a-V)) concentrations showed no visible change. In the case of Fe(III), this could be attributed to the low oxygen saturation in the Kleine Spree River, as observed by Uhlmann et al. [52], whereby only a small amount of Fe(II) was oxidised.

Figure 8 column b shows the mouth area of the industrial canal from the Schwarze Pumpe industrial park. A clear increase in the sulphate concentration can be seen there (see Figure 8(b-V)). Based on the derived TFe (see Figure 8(b-II)), Fe(II) (see Figure 8(b-III)) and Chl-a (see Figure 8(b-VI)) concentrations, a decrease in the mouth area of the industrial canal was first identified. After that, the concentration rose. From the Fe(III) concentration shown (see Figure 8(b-IV)), no change was detected.

Figure 8 column c displays the mouth of the Spree into the Spremberg reservoir. Owing to chemical and spectral differences between the River Spree and the southern part of the Spremberg reservoir, no suitable concentrations could be derived. Nevertheless, clear differences between the River Spree and the Spremberg reservoir were observed, except for the derivation result of Fe(III) (see Figure 8(c-IV)). Thus, the derived TFe concentration clearly showed that the River Spree only had a low TFe concentration after the upstream pre-dam Bühlow (see Figure 8(c-II)). This concentration increased in the Spremberg reservoir. Similar observations were made with respect to the Fe(II) (see Figure 8(c-III)) and Chl-a (see Figure 8(c-IV)) concentrations. However, the derived sulphate concentration was different (see Figure 8(c-V)). In the resulting image, the River Spree shows a very high concentration and the reservoir initially exhibited a low concentration. This raises the possibility that another source of iron input was located south of the Spremberg reservoir, such as swamps that were drained during open-cast mining activities.

4. Conclusions and Outlook

In this study, a combination of airborne imaging hyperspectral/RGB-RS was used to monitor water quality parameters of smaller rivers that were altered by open-cast mining processes with subsequent renaturation and groundwater recharge. The focus of monitoring, modelling and prediction was on the water quality variables TFe, Fe(II), Fe(III), sulphate and Chl-a. Due to the availability of a university-owned gyrocopter with corresponding RS sensors (HySpex, RGB, TIR), simultaneous imaging airborne RS data and in situ measurement data could be collected on site in the Spree.

The acquisition of simultaneous RGB data guaranteed the creation of a highly accurate terrain model, which was urgently needed for processing (geometric correction of) airborne hyperspectral data. With this approach, airborne hyperspectral RS and in situ data could be used for the first time to separately derive the water quality indicators Fe(II) and Fe(III) for smaller streams with neutral pH by means of regression. They were then modelled and the point measurements were transferred into space. The determination of the total iron concentration in open-cast mining lakes was already carried out by numerous studies, but the differentiation between the types Fe(II) and Fe(III) had not yet been made in any study available so far. This was partly due to the very low concentration of Fe(III) in the open-cast mining lakes. In addition, the examples on AMDs showed that iron could be identified, but no concentrations have been derived directly; therefore, the results should only be understood as an index. Furthermore, the advantage of our study was that in situ measurements could be conducted at the same time as hyperspectral measurements (400–2500 nm). Especially in rapidly changing aquatic systems, this method was found to be useful. The empirical models achieved a derivation accuracy of 0.93 mg/L for the total iron concentration, 0.95 mg/L (R^2 = 64%) for iron, 0.22 mg/L (R^2 = 79%) for ferric iron and 21 mg/L (R^2 = 64%) for sulphate (R^2 = 53%). An empirical approach was chosen for the study because it was robust; could be implemented quickly; and, with the focus on sulphate, was particularly target-oriented.

The obvious disadvantages, such as its low transferability and high statistical dependence, were taken into account by investigating a large number of spectral indices and using a valid statistical evaluation from the coefficient of determination and RMSE value. This also avoided an overspecialisation of the algorithm. The derivation results shown in Figure 8 demonstrate how inflowing river water could be clearly distinguished from reservoir water and how local phenomena, such as inflowing channels or rivers, could be reliably detected. This underlines the applicability of airborne RS for smaller flowing waters for continuous water quality monitoring.

The general conditions and requirements were as follows:

- The present results were only achieved by combining airborne hyperspectral RS data with simultaneous in situ measurements.
- Airborne hyperspectral sensors acquire very high-resolution and continuous spectra that allow detailed analyses to be carried out. The high spatial resolution offers significant advantages over satellite data (multispectral and hyperspectral) with a low spatial resolution for the derivation of water constituents from inland waters.
- Machine learning methods must be applied rather than simple regression models for modelling and prediction to achieve a better generalisation and transferability of the results.
- Spectral databases need to be in place for the quantification of water quality indicators.
- Scale dependencies have to be undertaken to transfer from high-resolution airborne hyperspectral RS data to the now freely available spaceborne hyperspectral data (EnMAP, DESIS, Prisma).

The method has tremendous potential to be practically established and transferred into official monitoring procedures in order to combine official in situ measurements with airborne and spaceborne RS information and to achieve added value for the monitoring of water quality indicators.

Author Contributions: Conceptualisation, C.U., L.B., M.H. and A.L.; methodology, C.U., L.B. and A.L.; statistical analysis and validation, C.U., L.B., M.H. and R.S. All authors have read and agreed to the published version of the manuscript.

Funding: Financial support was provided by the Institute of Geoinformation and Surveying of the Anhalt University of Applied Sciences. Further, this research was part of the Graduate Research School (GRS) 'Signatures of severely disturbed landscapes- case study mining landscapes' of the BTU Cottbus-Senftenberg. The study was partly funded by the Leibniz Institute of Freshwater Ecology and Inland Fisheries.

Data Availability Statement: Not applicable.

Acknowledgments: The authors would like to thank the many supporters of the measurement campaign, especially in the chemical laboratories of the IGB and HSA. We would especially like to thank Sylvia Jordan (IGB) for organising and assisting with sampling. We are grateful to Barbara Stein (IGB) for the carefully conducted Chl-a analyses. We acknowledge the German Research Foundation (DFG) for their contributions to the hyperspectral sensors HySpex VNIR and HySpex SWIR. Further, we would like to thank Milan Geoservice GmbH for their support and expertise.

Conflicts of Interest: The authors declare no conflict of interest.

References

1. Van Dijk, A.I.J.M.; Beck, H.E.; de Jeu, R.A.M.; Dorigo, W.A.; Hou, J.; Preimesberger, W.; Rahman, J.; Rozas Larraondo, P.R.; Van Der, R.S. Global Water Monitor 2022, Summary Report. Global Water Monitor. Available online: www.globalwater.online (accessed on 3 March 2023).
2. Tiwary, R.K. Environmental Impact of Coal Mining Onwater Regime and Its Management. *Water. Air. Soil Pollut.* **2001**, *132*, 185–199. [CrossRef]
3. Margareta, M.; Kaartinen, T.; Mäkinen, J.; Punkkinen, H.; Häkkinen, A.; Mamelkina, M.; Tuunila, R.; Lamberg, P.; Gonzales, M.S.; Sandru, M.; et al. *Water Conscious Mining (WASCIOUS)*; TemaNord; Nordic Council of Ministers: Copenhagen, Denmark, 2017; ISBN 9789289349628.
4. España, J.S. Acid Mine Drainage in the Iberian Pyrite Belt: An Overview with Special Emphasis on Generation Mechanisms, Aqueous Composition and Associated Mineral Phases. In Proceedings of the Conferencia Invitada: Sánchez España, Macla, Spain, Madrid; 2008; Volume 10, pp. 34–43. Available online: https://www.researchgate.net/publication/235355094_Acid_Mine_Drainage_in_the_Iberian_Pyrite_Belt_an_Overviewwith_Special_Emphasis_on_Generation_MechanismsAqueous_Composition_and_Associated_Mineral_Phases (accessed on 1 February 2023).
5. Schultze, M.; Pokrandt, K.-H.; Hille, W. Pit lakes of the Central German lignite mining district: Creation, morphometry and water quality aspects. *Limnologica* **2010**, *40*, 148–155. [CrossRef]
6. Bilek, F.; Koch, C. Eisenretention in der Talsperre Spremberg, 2012. Available online: https://lfu.brandenburg.de/lfu/de/ueber-uns/veroeffentlichungen/detail/~31-10-2012-eisenretention-in-der-talsperre-spremberg (accessed on 1 February 2023).
7. Giam, X.; Olden, J.D.; Simberloff, D. Impact of coal mining on stream biodiversity in the US and its regulatory implications. *Nat. Sustain.* **2018**, *1*, 176–183. [CrossRef]
8. Hüttl, R.F. Ecology of post strip-mining landscapes in Lusatia, Germany. *Environ. Sci. Policy* **1998**, *1*, 129–135. [CrossRef]
9. Zerbe, S.; Wiegleb, G. *Renaturierung von Ökosystemen in Mitteleuropa*; Zerbe, S., Wiegleb, G., Eds.; Spektrum Akademischer Verlag: Heidelberg, Germany, 2009; ISBN 978-3-8274-1901-9.
10. Lunt, J.; Smee, D.L. Turbidity alters estuarine biodiversity and species composition. *ICES J. Mar. Sci.* **2020**, *77*, 379–387. [CrossRef]
11. Ramadas, M.; Samantaray, A.K. Applications of Remote Sensing and GIS in Water Quality Monitoring and Remediation: A State-of-the-Art Review. In *Water Remediation*; Springer Nature: Singapore, 2018; pp. 225–246.
12. Sagan, V.; Peterson, K.T.; Maimaitijiang, M.; Sidike, P.; Sloan, J.; Greeling, B.A.; Maalouf, S.; Adams, C. Monitoring inland water quality using remote sensing: Potential and limitations of spectral indices, bio-optical simulations, machine learning, and cloud computing. *Earth-Sci. Rev.* **2020**, *205*, 103187. [CrossRef]
13. Olmanson, L.G.; Brezonik, P.L.; Bauer, M.E. Airborne hyperspectral remote sensing to assess spatial distribution of water quality characteristics in large rivers: The Mississippi River and its tributaries in Minnesota. *Remote Sens. Environ.* **2013**, *130*, 254–265. [CrossRef]
14. Chawla, I.; Karthikeyan, L.; Mishra, A.K. A review of remote sensing applications for water security: Quantity, quality, and extremes. *J. Hydrol.* **2020**, *585*, 124826. [CrossRef]
15. Gholizadeh, M.; Melesse, A.; Reddi, L. A Comprehensive Review on Water Quality Parameters Estimation Using Remote Sensing Techniques. *Sensors* **2016**, *16*, 1298. [CrossRef]
16. Ritchie, J.C.; Zimba, P.V.; Everitt, J.H. Remote Sensing Techniques to Assess Water Quality. *Photogramm. Eng. Remote Sens.* **2003**, *69*, 695–704. [CrossRef]
17. Göritz, A.; Berger, S.; Gege, P.; Grossart, H.-P.; Nejstgaard, J.; Riedel, S.; Röttgers, R.; Utschig, C. Retrieval of Water Constituents from Hyperspectral In-Situ Measurements under Variable Cloud Cover—A Case Study at Lake Stechlin (Germany). *Remote Sens.* **2018**, *10*, 181. [CrossRef]
18. Pan, X.; Wang, Z.; Ullah, H.; Chen, C.; Wang, X.; Li, X.; Li, H.; Zhuang, Q.; Xue, B.; Yu, Y. Evaluation of Eutrophication in Jiaozhou Bay via Water Color Parameters Determination with UAV-Borne Hyperspectral Imagery. *Atmosphere* **2023**, *14*, 387. [CrossRef]
19. Kakuta, S.; Ariyasu, E.; Takeda, T. Shallow Water Bathymetry Mapping Using Hyperspectral Data. In Proceedings of the IGARSS 2018—2018 IEEE International Geoscience and Remote Sensing Symposium, Valencia, Spain, 22–27 July 2018; pp. 1539–1542.
20. Alonso, K.; Bachmann, M.; Burch, K.; Carmona, E.; Cerra, D.; de los Reyes, R.; Dietrich, D.; Heiden, U.; Hölderlin, A.; Ickes, J.; et al. Data Products, Quality and Validation of the DLR Earth Sensing Imaging Spectrometer (DESIS). *Sensors* **2019**, *19*, 4471. [CrossRef] [PubMed]
21. Guanter, L.; Kaufmann, H.; Segl, K.; Foerster, S.; Rogass, C.; Chabrillat, S.; Kuester, T.; Hollstein, A.; Rossner, G.; Chlebek, C.; et al. The EnMAP Spaceborne Imaging Spectroscopy Mission for Earth Observation. *Remote Sens.* **2015**, *7*, 8830–8857. [CrossRef]
22. Lopinto, E.; Fasano, L.; Longo, F.; Varacalli, G.; Sacco, P.; Chiarantini, L.; Sarti, F.; Agrimano, L.; Santoro, F.; Cogliati, S.; et al. Current Status and Future Perspectives of the PRISMA Mission at the Turn of One Year in Operational Usage. In Proceedings of the 2021 IEEE International Geoscience and Remote Sensing Symposium IGARSS, Brussels, Belgium, 11–16 July 2021; pp. 1380–1383.
23. Imatani, R.; Ito, Y.; Ikehara, K.; Iwasaki, A.; Inada, H.; Tanii, J.; Kashimura, O. The flight model performances of Hyperspectral Imager Suite (HISUI). In Proceedings of the Sensors, Systems, and Next-Generation Satellites XXV, Online Conference, Spain; 2021; p. 3. Available online: https://ui.adsabs.harvard.edu/abs/2021SPIE11858E..08I/abstract (accessed on 1 February 2023).
24. Gege, P. Radiative Transfer Theory for Inland Waters. In *Bio-Optical Modeling and Remote Sensing of Inland Waters*; Elsevier: Amsterdam, The Netherlands, 2017; pp. 25–67, ISBN 978-0-12-804644-9.

25. Ogashawara, I.; Mishra, D.R.; Gitelson, A.A. Remote Sensing of Inland Waters. In *Bio-Optical Modeling and Remote Sensing of Inland Waters*; Elsevier: Amsterdam, The Netherlands, 2017; pp. 1–24, ISBN 978-0-12-804644-9.
26. Dörnhöfer, K.; Oppelt, N. Remote sensing for lake research and monitoring—Recent advances. *Ecol. Indic.* **2016**, *64*, 105–122. [CrossRef]
27. Dekker, A.G. Detection of Optical Water Quality Parameters for Eutrophic Waters by High Resolution Remote Sensing. Ph.D. Thesis, Free University, Berlin, Germany, 1993.
28. Matthews, M.W. A current review of empirical procedures of remote sensing in inland and near-coastal transitional waters. *Int. J. Remote Sens.* **2011**, *32*, 6855–6899. [CrossRef]
29. Kirk, J.T.O. *Light and Photosynthesis in Aquatic Ecosystems*; 3. Auflage.; Cambridge University Press: Canberra, Australia, 2010; ISBN 978-0-521-15175-7.
30. Frauendorf, J. *Entwicklung und Anwendung von Fernerkundungsmethoden zur Ableitung von Wasserqualitätsparametern verschiedener Restseen des Braunkohlentagebaus in Mitteldeutschland*; Martin-Luther-Universität Halle-Wittenberg: Halle, Germany, 2002. [CrossRef]
31. Asmala, E.; Stedmon, C.A.; Thomas, D.N. Linking CDOM spectral absorption to dissolved organic carbon concentrations and loadings in boreal estuaries. *Estuar. Coast. Shelf Sci.* **2012**, *111*, 107–117. [CrossRef]
32. Gitelson, A.A. The peak near 700 nm on radiance spectra of algae and water: Relationships of its magnitude and position with chlorophyll concentration. *Int. J. Remote Sens.* **1992**, *13*, 3367–3373. [CrossRef]
33. Kopačková, V.; Hladíková, L. Applying Spectral Unmixing to Determine Surface Water Parameters in a Mining Environment. *Remote Sens.* **2014**, *6*, 11204–11224. [CrossRef]
34. Repic, R.L.; Lee, J.K.; Mausel, P.W. An Analysis of Selected Water Parameters in Surface Coal Mines Using Multispectral Videography. *Photogramm. Eng.* **1991**, *4*, 1589–1596.
35. Anderson, J.E.; Robbins, E. Spectral Reflectance and Detection of Iron-Oxide Precipitates Associated with Acidic Mine Drainage. *Photogramm. Eng. Remote Sens.* **1998**, *64*, 1201–1208.
36. Williams, D.J.; Bigham, J.M.; Cravotta III, C.A.; Trainaa, S.J.; Anderson, J.E.; Lyon, J.G. Assessing mine drainage pH from the color and spectral reflectance of chemical precipitates. *Appl. Geochem.* **2002**, *17*, 1273–1286. [CrossRef]
37. Gläßer, C.; Groth, D.; Frauendorf, J. Monitoring of hydrochemical parameters of lignite mining lakes in Central Germany using airborne hyperspectral casi-scanner data. *Int. J. Coal Geol.* **2011**, *86*, 40–53. [CrossRef]
38. Schroeter, L.; Gläβer, C. Analyses and monitoring of lignite mining lakes in Eastern Germany with spectral signatures of Landsat TM satellite data. *Int. J. Coal Geol.* **2011**, *86*, 27–39. [CrossRef]
39. Brando, V.E.; Dekker, A.G. Satellite hyperspectral remote sensing for estimating estuarine and coastal water quality. *IEEE Trans. Geosci. Remote Sens.* **2003**, *41*, 1378–1387. [CrossRef]
40. Bukata, R.P.; Jerome, J.H.; Kondratyev, K.Y.; Pozdnyakov, D.V. *Optical Properties and Remote Sensing of Inland and Coastal Waters*; CRC Press: Boca Raton, FL, USA, 2018; ISBN 9780203744956.
41. Kutser, T.; Hedley, J.; Giardino, C.; Roelfsema, C.; Brando, V.E. Remote sensing of shallow waters—A 50 year retrospective and future directions. *Remote Sens. Environ.* **2020**, *240*, 111619. [CrossRef]
42. Kuhn, C.; de Matos Valerio, A.; Ward, N.; Loken, L.; Sawakuchi, H.O.; Kampel, M.; Richey, J.; Stadler, P.; Crawford, J.; Striegl, R.; et al. Performance of Landsat-8 and Sentinel-2 surface reflectance products for river remote sensing retrievals of chlorophyll-a and turbidity. *Remote Sens. Environ.* **2019**, *224*, 104–118. [CrossRef]
43. Balasubramanian, S.V.; Pahlevan, N.; Smith, B.; Binding, C.; Schalles, J.; Loisel, H.; Gurlin, D.; Greb, S.; Alikas, K.; Randla, M.; et al. Robust algorithm for estimating total suspended solids (TSS) in inland and nearshore coastal waters. *Remote Sens. Environ.* **2020**, *246*, 111768. [CrossRef]
44. Baschek, B.; Fricke, K.; Dörnhöfer, K.; Oppelt, N. Grundlagen und Möglichkeiten der passiven Fernerkundung von Binnengewässern. In *Handbuch Angewandte Limnologie: Grundlagen—Gewässerbelastung—Restaurierung—Aquatische Ökotoxikologie—Bewertung—Gewässerschutz*; Wiley: Hoboken, NJ, USA, 2018; pp. 1–28. ISBN 978-3-527-67848-8.
45. Doxaran, D.; Froidefond, J.-M.; Castaing, P. Remote-sensing reflectance of turbid sediment-dominated waters Reduction of sediment type variations and changing illumination conditions effects by use of reflectance ratios. *Appl. Opt.* **2003**, *42*, 2623. [CrossRef]
46. Knaeps, E.; Ruddick, K.G.; Doxaran, D.; Dogliotti, A.I.; Nechad, B.; Raymaekers, D.; Sterckx, S. A SWIR based algorithm to retrieve total suspended matter in extremely turbid waters. *Remote Sens. Environ.* **2015**, *168*, 66–79. [CrossRef]
47. Friedland, G.; Grüneberg, B.; Hupfer, M. Geochemical signatures of lignite mining products in sediments downstream a fluvial-lacustrine system. *Sci. Total Environ.* **2021**, *760*, 143942. [CrossRef]
48. Bilek, F.; Moritz, F.; Albinus, S. Iron-Hydroxide-Removal from Mining Affected Rivers. In *Proceedings of the Mining Meets Water—Conflicts and Solutions*; International Mine Water Association (IMWA): Freiberg, Germany, 2016; pp. 151–158.
49. Gleisner, M.; Herbert, R.B. Sulfide mineral oxidation in freshly processed tailings: Batch experiments. *J. Geochem. Explor.* **2002**, *76*, 139–153. [CrossRef]
50. Uhlig, U.; Radigk, S.; Uhlmann, W.; Preuß, V.; Koch, T. Iron removal from the Spree River in the Bühlow pre-impoundment basin of the Spremberg reservoir IMWA 2016. Available online: https://www.imwa.info/docs/imwa_2016/IMWA2016_Uhlig_75.pdf (accessed on 1 February 2023).

51. Uhlmann, W.; Theiss, S.; Nestler, W.; Claus, T. Fortführung der Studie zur Talsperre Spremberg—Abschlussbericht (Dezember 2013). Available online: https://docplayer.org/83470767-Fortfuehrung-der-studie-zur-talsperre-spremberg-abschlussbericht-dezember-2013.html (accessed on 1 February 2023).
52. Uhlmann, W.; Theiss, S.; Nestler, W.; Zimmermann, K.; Claus, T. *Weiterführende Untersuchungen zu den Hydrochemischen und Ökologischen Auswirkungen der Exfiltration von Eisenhaltigem, Saurem Grundwasser in die Kleine Spree und in Die Spree Projektphase 2: Präzisierung der Ursachen und Quellstärken für die Hohe Eisenbel*; IWB: Dresden, Germany, 2012.
53. LMBV LMBV: Spree bei Wilhelmsthal bekommt wieder Kalk—Übergang in Spätsommer-Fahrweise. Available online: https://www.lmbv.de/index.php/pressemitteilung/lmbv-spree-bei-wilhelmsthal-bekommt-wieder-kalk-uebergang-in-spaetsommer-fahrweise-4576.html (accessed on 1 February 2023).
54. Mehnert, G.; Rücker, J.; Nicklisch, A.; Leunert, F.; Wiedner, C. Effects of thermal acclimation and photoacclimation on lipophilic pigments in an invasive and a native cyanobacterium of temperate regions. *Eur. J. Phycol.* **2012**, *47*, 182–192. [CrossRef]
55. Schonberger, J.L.; Frahm, J.-M. Structure-from-Motion Revisited. In Proceedings of the 2016 IEEE Conference on Computer Vision and Pattern Recognition (CVPR), Las Vegas, NV, USA, 27–30 June 2016; pp. 4104–4113.
56. Schläpfer, D. PARGE Airborne Image Rectification PARGE®Image Rectification. 2022. Available online: https://www.rese-apps.com/software/parge/index.html (accessed on 1 February 2023).
57. Schläpfer, D. ATCOR for Airborne Remote Sensing. ATCOR 4—For Airborne Remote Sensing Systems. 2022. Available online: https://www.rese-apps.com/software/atcor-4-airborne/index.html (accessed on 1 February 2023).
58. Hovis, W.A.; Leung, K.C. Remote Sensing of Ocean Color. *Opt. Eng.* **1977**, *16*, 439–472. [CrossRef]
59. Ray, T.W. A FAQ on Vegetation in Remote Sensing, 1994, IEEE/ACM Third International Conference on Cyber-Physical Systems. Pasadena. 1994. Available online: http://www.remote-sensing.info/wp-content/uploads/2012/07/A_FAQ_on_Vegetation_in_Remote_Sensing.pdf (accessed on 1 February 2023).
60. Moses, W.J.; Sterckx, S.; Montes, M.J.; De Keukelaere, L.; Knaeps, E. Atmospheric Correction for Inland Waters. In *Bio-Optical Modeling and Remote Sensing of Inland Waters*; Elsevier: Amsterdam, The Netherlands, 2017; pp. 69–100, ISBN 978-0-12-804644-9.
61. Håkanson, L.; Bryhn, A.C.; Blenckner, T. Operational Effect Variables and Functional Ecosystem Classifications—A Review on Empirical Models for Aquatic Systems along a Salinity Gradient. *Int. Rev. Hydrobiol.* **2007**, *92*, 326–357. [CrossRef]
62. Ulrich, C.; Bannehr, L.; Lausch, A. Ableitung von Eisen(II, III)oxid in Fließgewässern mittels Multispektraldaten. In Proceedings of the Gesellschaft für Photogrammetrie, Fernerkundung und Geoinformation, Bern, Switzerland; 2016; Volume 25, pp. 34–43. Available online: https://www.google.com.hk/search?q=Ableitung+von+Eisen%28II%2C+III%29oxid+in+Flie%C3%9Fgew%C3%A4ssern+mittels+Multispektraldaten&ei=JL83ZOaWDb-m2roP7qah8Ao&ved=0ahUKEwjmuNmCuqb-AhU_k1YBHW5TCK4Q4dUDCA4&uact=5&oq=Ableitung+von+Eisen%28II%2C+III%29oxid+in+Flie%C3%9Fgew%C3%A4ssern+mittels+Multispektraldaten&gs_lcp=Cgxnd3Mtd2l6LXNlcnAQA0oECEEYAFAAWABgiQNoAHABeACAAXmIAXmSAQMwLjGYAQCgAQKgAQHAAQE&sclient=gws-wiz-serp (accessed on 1 February 2023).
63. Durning, W.P.; Polis, S.R.; Frost, E.G.; Kaiser, J.V. Integrated Use of Remote Sensing and GIS for Mineral Exploration-Final Report. Affiliated Research Center, San Diego State University: San Diego, CA, USA, 1998.
64. Eloheimo, K.; Hannonen, T.; Härmä, P.; Pyhälahti, T.; Koponen, S.; Pulliainen, J.; Servomaa, H.; Kutser, T. Coastal monitoring using satellite, airborne and in situ data in the archipelago of Baltic Sea. In Proceedings of the 5th International Conference on Remote Sensing for Marine and Coastal Environments, San Diego, CA, USA, 5–7 October 1998; pp. 306–331.
65. Robinson, M.-C.; Morris, K.P.; Dyer, K.R. Deriving Fluxes of Suspended Particulate Matter in the Humber Estuary, UK, Using Airborne Remote Sensing. *Mar. Pollut. Bull.* **1999**, *37*, 155–163. [CrossRef]
66. van der Meer, F.D.; van der Werff, H.M.A.; van Ruitenbeek, F.J.A. Potential of ESA's Sentinel-2 for geological applications. *Remote Sens. Environ.* **2014**, *148*, 124–133. [CrossRef]
67. IOCCG. Remote Sensing of Ocean Colour in Coastal, and Other Optically-Complex, Waters, Reports of the International Ocean-Colour Coordinating Group. Reports of the International Ocean-Colour Coordinating Group, Dartmouth, Kanada. 2020. Available online: http://ioccg.org/wp-content/uploads/2015/10/ioccg-report-03.pdf (accessed on 1 February 2023).
68. Gao, H.; Birkett, C.; Lettenmaier, D.P. Global monitoring of large reservoir storage from satellite remote sensing. *Water Resour. Res.* **2012**, *48*, 2012WR012063. [CrossRef]
69. Wang, J.-J.; Lu, X.X. Estimation of suspended sediment concentrations using Terra MODIS: An example from the Lower Yangtze River, China. *Sci. Total Environ.* **2010**, *408*, 1131–1138. [CrossRef]
70. Weyhenmeyer, G.A.; Prairie, Y.T.; Tranvik, L.J. Browning of Boreal Freshwaters Coupled to Carbon-Iron Interactions along the Aquatic Continuum. *PLoS ONE* **2014**, *9*, e88104. [CrossRef]
71. Knaeps, E.; Dogliotti, A.I.; Raymaekers, D.; Ruddick, K.; Sterckx, S. In situ evidence of non-zero reflectance in the OLCI 1020 nm band for a turbid estuary. *Remote Sens. Environ.* **2012**, *120*, 133–144. [CrossRef]
72. Rowan, L.C.; Goetz, A.F.H.; Ashley, R.P. Discrimination of hydrothermallyy altered and unaltered rocks in visible and near infrared multispectral images. *Geophysics* **1977**, *42*, 522–535. [CrossRef]

73. Haardt, H.; Maske, H. Specific in vivo absorption coefficient of chlorophyll a at 675 nm1. *Limnol. Oceanogr.* **1987**, *32*, 608–619. [CrossRef]
74. Gege, P. The water colour simulator WASI: A software tool for forward and inverse modeling of optical in-situ spectra. In Proceedings of the IGARSS 2001. Scanning the Present and Resolving the Future. Proceedings. IEEE 2001 International Geoscience and Remote Sensing Symposium (Cat. No.01CH37217), Sydney, Australia, 9–13 July 2001; Volume 6, pp. 2743–2745.

water

Article

Assessing the Hydrochemistry, Groundwater Drinking Quality, and Possible Hazard to Human Health in Shizuishan Area, Northwest China

Shan Xiao [1,2], Yuan Fang [1,3], Jie Chen [1,2,*], Zonghua Zou [1,2], Yanyan Gao [1,2], Panpan Xu [1,2], Xueke Jiao [1,2] and Miaoyue Ren [1,2]

1 School of Water and Environment, Chang'an University, Xi'an 710064, China
2 Key Laboratory of Subsurface Hydrology and Ecological Effects in Arid Region of the Ministry of Education, Chang'an University, Xi'an 710064, China
3 Ningxia Survey and Monitor Institute of Land and Resources, Yinchuan 750002, China
* Correspondence: jiechen17@chd.edu.cn

Abstract: Groundwater is an important source of drinking water, particularly in arid regions. In this study, a total of 66 groundwater samples were collected from the phreatic aquifer in the Shizuishan area, a traditional irrigation region of Ningxia. The results showed that the TDS values were above the drinking water standards for nearly 50% of the groundwater samples. The ions followed the order of $Na^+ > Ca^{2+} > Mg^{2+} > K^+$ and $SO_4^{2-} > Cl^- > HCO_3^-$ in the groundwater. There were four dominant factors in controlling groundwater chemistry based on principal component analysis: the salinity factor, alkalinity factor, carbonate factor, and pollution factor. The high concentration of NH_4-N in groundwater was attributed to agricultural activities, but the high NO_3-N levels were mainly due to sewage or wastewater. F and As were derived from geogenic sources. Based on the result of the WQI assessment, about 40% of the samples in the central part of the study region showed unacceptable water quality for drinking, which was mainly associated with high NH_4-N, TDS, and As concentrations. The total non-carcinogenic risks of drinking the groundwater were 0.05–10.62 for adults and 0.09–20.65 for children, respectively. The order of pollutants in the groundwater in terms of their hazard to residents was: As > F^- > NO_3-N > NH_4-N. The carcinogenic risk values of As through oral ingestion for children and adults were $0–7.37 \times 10^{-4}$ and $0–1.89 \times 10^{-4}$, respectively. Chronic exposure by oral ingestion presented as the main source of susceptibility to exposure to groundwater contaminants for children.

Keywords: groundwater; hydrochemistry; quality; health risk; China

Citation: Xiao, S.; Fang, Y.; Chen, J.; Zou, Z.; Gao, Y.; Xu, P.; Jiao, X.; Ren, M. Assessing the Hydrochemistry, Groundwater Drinking Quality, and Possible Hazard to Human Health in Shizuishan Area, Northwest China. *Water* **2023**, *15*, 1082. https://doi.org/10.3390/w15061082

Academic Editor: Imokhai Theophilus Tenebe

Received: 22 January 2023
Revised: 2 March 2023
Accepted: 8 March 2023
Published: 11 March 2023

1. Introduction

Groundwater plays an important role in supporting agricultural, domestic, and industrial water use because of its widespread distribution and relatively stable quality [1].

In many semi-arid and arid regions, groundwater is a major source of drinking water [2,3]. The impacts of human and geogenic activities on groundwater quality are of increasing concern. Geogenic sources, rapid urbanization, excessive withdrawals, improper disposal of waste, the overuse of fertilizer, etc., have caused major changes in the physical properties and chemical composition of groundwater to a great extent [4–6]. However, prolonged exposure to contaminants can have adverse effects on human health. Groundwater remediation is also a long and slow process.

Understanding the hydrochemistry of groundwater is vital for maintaining groundwater quality [7–9]. The chemistry of groundwater has evolved through the effects of precipitation, evaporation, weathering, sorption, and exchange reactions [10–12]. Many scholars have found by various methods that the chemical composition of groundwater is closely related to the geological environment and hydrogeological conditions [13]. For instance, Schot et al. [14] found that the hydrochemical composition of groundwater in the Gooi and Vechtstreek areas in the Netherlands was influenced by human activities, with urbanization and agricultural activities causing increased concentrations of nitrate, sulfate, and K^+ in the groundwater. Adams et al. [15] indicated that the chemical composition of the groundwater in the study area is dominated by salinization, mineral dissolution and precipitation, cation exchange, and human activities. Zhang et al. [16] showed that the high concentrations of Fe and Mn in the Songnen Plain were associated with the unique geomorphology and reducing environment. Güler et al. [17] showed that the main factors causing water chemistry changes in the Tarsus coastal plain were water-rock interaction and nitrate pollution, seawater intrusion and salinization, geological factors, and anthropogenic zinc pollution. In this context, the exposure through drinking water to high levels of nitrogen, fluoride, and arsenic is of increasing concern to humans. Nitrogen is one of the essential elements for plant growth [18,19]. Although nitrogen fertilizer provides certain nutrients for plant growth and improves crop yields, the use of large amounts of nitrogen fertilizer accelerates the leaching of nitrogen into groundwater, which will pose a serious threat to the environment and human health [20,21].

Groundwater contaminated with high levels of arsenic and fluoride is a serious threat to the safety of drinking water. It has become a global problem [22–24]. The chronic ingestion of high-fluoride (>1.0–1.5 mg/L) groundwater has been shown to contribute to fluorosis and various diseases, such as cardiovascular disease, osteosclerosis, endocrine disorders, and multi-organ lesions [10]. Chronic exposure to arsenic through contaminated food and drinking water can lead to arsenic poisoning, with symptoms including peripheral neuropathy, skin lesions, diabetes, cancer, and cardiovascular disease [25–27]. The Water Quality Index (WQI) is a practical approach to assessing the quality of groundwater by distilling a large amount of water quality data to an index representative of regional groundwater quality [28–31]. The United States Environmental Protection Agency's (USEPA) Human Health Risk Assessment (HHRA) (USEPA, 2004) is widely used to assess the risks of groundwater quality to human health. Integrating the HHRA and WQI can facilitate improved monitoring and groundwater quality conservation for sustainable groundwater management [32–35].

In northwestern China, water obtained from alluvial aquifers is used for drinking purposes, especially in most remote areas [36,37]. The Shizuishan area, located in northern Ningxia, is a typical arid and semi-arid region. As a traditionally agricultural region, a large amount of irrigation water infiltrated into the aquifer becomes the dominant groundwater inflow. Private hand-pumped wells are widely used in the rural part to support drinking due to their low cost and high efficiency. However, the phreatic aquifer is sensitive to human activities because of the shallow water table coupled with the high population densities. It is still a potential risk to human health to use groundwater as a source of drinking water in the study region. Therefore, the present study aims to interpret the complex groundwater hydrochemistry of the shallow aquifer in the Shizuishan area. In the present study, Principal Component Analysis (PCA) was applied to identify natural and anthropogenic factors with significant influence on groundwater hydrochemistry. The quality of groundwater and associated health risks were evaluated to provide more information for groundwater management.

2. Study Area

The Shizuishan area covers about 2241 km^2 and is located in the northern Yinchuan Plain, Ningxia (38°39′17″–39°23′16″ N, 106°8′14″–106°52′11″ E. Figure 1). The Yinchuan Plain has a traditional irrigation history dating back to more than 2000 years. According to the Ningxia Statistical Yearbook (2022) [38], Shizuishan has a population of about 800,000, and the main crops are rice, corn, and wheat. In the arid environment, the mean annual precipitation and potential evaporation are 179 mm and 1800 mm, respectively. Precipitation is mainly concentrated in May and October. The area has sufficient sunshine, a large temperature difference between day and night, and the annual average climate is about 10.6 °C [38,39]. Irrigation water in the area is mainly supplied by water transfer from the Yellow River, and there are many water diversion channels in the plain to transfer water through to ensure the needs of crops are met. Groundwater depth is impacted by irrigation and ranges from 1 to 3 m. Soil salinization has many major adverse effects on crop yield due to intense evaporation.

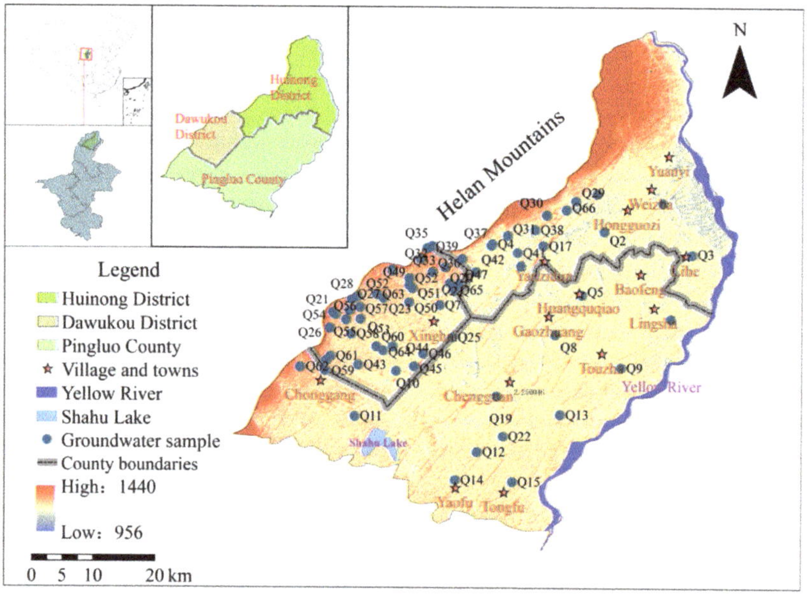

Figure 1. Groundwater sampling location of the Shizuishan area.

Groundwater serves as crucial support for drinking, irrigation, and industrial purposes. In general, the groundwater flow in the study area is from the southwest to the northeast. The aquifer in Shizuishan is mainly composed of Quaternary fine alluvial lacustrine deposits. From west to east, the landforms of the study area are leaning towards pluvial plain, pluvial alluvial plain, and alluvial lacustrine plain [40]. The lithology of aquifers in the pluvial plain becomes thinner from west to east. The lithology of the alluvial lacustrine plain is mainly medium fine sand [39]. From top to bottom, the aquifer system is composed of phreatic, upper confined, and lower confined aquifers with thicknesses of 20–30 m, 50–60 m, and 60–80 m, respectively. The aquitards consist mainly of clay between the adjacent aquifers. In this study, we focused primarily on the phreatic aquifer. The aquifer is recharged primarily by irrigation channels, irrigation infiltration, precipitation, and lateral inflow. Groundwater withdrawals include evaporation, discharge to drains, and artificial extraction. Many residents rely on groundwater for their drinking water.

3. Materials and Methods

3.1. Collection and Analyses of Samples

The present study collected 66 groundwater samples from private wells in April 2021 (Figure 1). The collection and processing of groundwater samples in the present study followed technical specifications for environmental monitoring. Sampling locations were recorded by using a handled GPS device. Before collecting groundwater samples, the well was flushed for 5 to 10 min to remove standing water. Water samples were stored in previously sterilized polyethylene bottles, and pH, total dissolved solids (TDS), and temperature were measured in the field using portable meters (DDBJ-350F, INESA Scientific

Instrument Co., Ltd., Shanghai, China). The samples were stored at 4 °C before analysis for physicochemical parameters.

The groundwater samples were analyzed according to the methods recommended by the Ministry of Health of the People's Republic of China (PRC) and the Standardization Administration of the PRC [41]. The laboratory of the Ningxia Geological Survey Institute analyzed all the samples. The samples were used for quality assurance/quality control (QA/QC) analysis. The sample analysis followed the Technical Guidelines for Groundwater Environmental Monitoring developed by the State Environmental Protection Administration (2004) [42].To ensure the accuracy and reliability of the results, the groundwater sample analysis was repeated three times. Flame atomic absorption spectrophotometry was used to measure the contents of K^+ and Na^+, while titration was used to measure those of Cl^-, Ca^{2+}, $SO_4{}^{2-}$, Mg^{2+}, and $HCO_3{}^-$. Ion chromatography was used to measure the content of F^-. Spectrophotometry was used to measure the contents of NH_4–N and NO_3–N. Arsenic content was determined through hydride generation atomic fluorescence spectrometry (HG-AFS).

In this study, SPSS 20.0 was used for the principal component analysis of the groundwater samples to analyze the main controlling factors in groundwater. Piper (1944) and Gibbs (1970) diagrams and ion ratios were drawn from to study the hydrochemical types and main control factors of the groundwater by using MATLAB (version 2016b) and the Origin (version 2020) software. An IDW-to-raster interpolation was applied to spatially map the groundwater quality and health risks from oral intake based on ArcGIS (version 10.7).

3.2. Groundwater Quality Assessment

The WQI provides a comprehensive assessment of drinking water quality. Weights (w_i; between 2 and 5) have been assigned to the different individual parameters making up the WQI according to their relative concentrations and importance for drinking water quality [43,44]. The WQI is calculated through:

$$W_i = \frac{w_i}{\sum\limits_{i}^{n} w_i} \tag{1}$$

$$Q_i = \frac{C_i - C_{ip}}{S_i - C_{ip}} \times 100 \tag{2}$$

$$SI_i = W_i \times Q_i \tag{3}$$

$$WQI = \sum_{i=1}^{n} SI_i \tag{4}$$

where W_i represents the weight of each parameter, w_i represents the weight assigned to each parameter, n represents the number of parameters, C_i denotes the concentration of a single parameter, C_{ip} represents the ideal distilled parameter value (a pH of 7 and zero concentration for the remaining parameters), Q_i represents the quality rating, S_i represents the WHO standard for a single parameter, and SI_i represents the sub-index of the "ith" parameter. The quality of groundwater in the study area could be classified into five classes according to the WQI: (1) excellent; (2) good; (3) medium; (4) poor; and (5) extremely poor.

3.3. Assessment of Hazards to Human Health

The present study applied the HHRA to estimate the adverse impacts of the ingestion and assimilation of toxicants on adults and children. The non-carcinogenic hazard posed by contaminants was determined by [45]:

$$D_i = \frac{C_i \times IR \times EF \times ED}{BW \times AT} \tag{5}$$

$$HQ_i = D_i / RfD_i \tag{6}$$

where D_i represents exposure dose by ingestion for the *i*th contaminant (mg/kg per day); C_i denotes contaminant concentration (mg/L); *IR* is the rate of ingestion (L/day); *EF* represents the frequency of exposure (days/year); *ED* represents the duration of exposure (years); *BW* means body mass (kg); *AT* is mean exposure duration (days); HQ_i represents the non-carcinogenic impact of an individual parameter; and *RfD* represents the non-carcinogenic reference dosage of a parameter (mg kg^{-1} day^{-1}). The *RfD* for NO_3-N, NH_4-N, F^-, and AS were 1.6, 0.9, 0.06, and 0.0003 mg/(kg·day), respectively [37,46,47].

The total hazard index (*HI*) represents the integrated hazard posed by contaminants in water used for drinking. A *HI* of > 1.0 and a *HI* of < 1.0 represent potential adverse human health impacts and acceptable levels of non-carcinogenic risk, respectively.

$$HI = \sum HQ \tag{7}$$

In this study, the potential health risks of ingesting NO_3^-, NH_4^+, F^-, and As in groundwater were considered. Because all residents in the study area rely on groundwater as a source of drinking water, *EF* was assigned a value of 365 days for both children and adults. *ED* was assigned to children and adults at ages 6 and 30. *AT* is equal to 365 days. Statistical data [48] indicated the *BW* of adults and children in Ningxia to be 62.5 kg and 15 kg, respectively. All parameters are listed in Table 1.

Table 1. Parameters of daily dose calculation models.

Parameters	Unit	Item	Children	Adults
IR	L/d	Oral	0.7	1.50
BW	kg	Oral	15	62.5
EF	d/a	Oral	365	365
ED	a	Oral	6	30
AT	d	Carcinogenic	74.68 × 365	74.68 × 365
		Non-carcinogenic	6 × 365	30 × 365

The carcinogenic risk of arsenic (R_{As}) can be estimated as:

$$R_{As} = q_{As} D_{As} \tag{8}$$

where q_{As} represents the carcinogenic coefficient of arsenic ingested through drinking water (1.5 kg day mg^{-1}). The USEPA usually applies a range of target reference risks of 10^{-4}–10^{-6} in the assimilation of carcinogenic toxins in drinking water, with 10^{-6} being generally recognized as an appropriate standard for drinking water [49]. R_{As} of > 10^{-4} represents the possibility of adverse impacts on human health posed by arsenic in groundwater.

4. Results and Discussion

4.1. Groundwater Chemistry

Table 2 provides a statistical summary of groundwater quality according to the analysis of groundwater samples (Table 2). The ranges of WHO and the Chinese standards for different chemicals [50,51] are considered in the assessment of the groundwater for use as drinking water.

The mean pH of the water samples was 7.59, indicating it was slightly alkaline. The TDS values ranged from 232 to 18,448 mg/L (mean of 1990 mg/L). The TDS value was above the drinking water standard (1000 mg/L) for nearly 50% of the groundwater samples [51]. The highest TDS concentrations in groundwater were found in Chengguan and Gaozhuang villages. The elevated TDS concentrations could be attributed to ion exchange, solubilization, and the extended groundwater residence time in the aquifer [52,53].

Table 2. Entropy weights for the parameters.

Parameters	Weight (Wi)	Relative Weight Wi
pH	3	0.063
TDS	5	0.104
K^+	2	0.042
Na^+	2	0.042
Ca^{2+}	3	0.063
Mg^{2+}	3	0.063
SO_4^{2-}	3	0.063
Cl^-	4	0.083
HCO_3^-	3	0.063
NO_3-N	5	0.104
NH_4-N	5	0.104
F^-	5	0.104
As	5	0.104

The cations in the groundwater could be ranked by mean content in the order $Na^+ > Ca^{2+} > Mg^{2+} > K^+$, with concentrations of 427, 135, 92, and 8.13 mg/L, respectively (Figure 2). The high concentration of sodium in groundwater may be associated with the dissolution of halite, Na-feldspar, and ion exchange [54]. Na^+ was the dominant groundwater cation with a concentration ranging from 30 to 5540 mg/L, and one-third of the samples had Na^+ concentrations higher than the WHO [51] standard of 200 mg/L. Groundwater Ca^{2+} and Mg^{2+} concentrations ranged from 24.02 to 1027 mg/L and 6.07 to 584 mg/L, respectively. The concentrations of Ca^{2+} and Mg^{2+} exceeded the WHO [51] limits (75 mg/L and 30 mg/L in three-quarters of the groundwater samples. Groundwater K^+ concentration was relatively low, with 16% of the samples exceeding the WHO [51] limit of 10 mg/L. The major ions in the groundwater followed the trend $SO_4^{2-} > Cl^- > HCO_3^-$ with mean concentrations of 663 mg/L, 487 mg/L, and 330 mg/L, respectively (Figure 2). The high concentrations may be due to the dissolution of sulfate minerals (such as mirabilite, gypsum, etc.) and human activities (such as domestic sewage discharge, industrial pollution, etc.) [55].

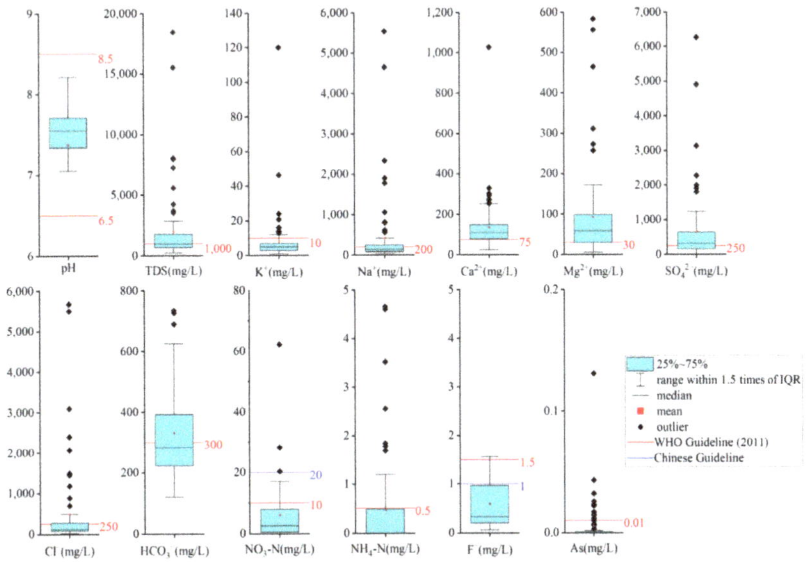

Figure 2. Box diagram of major chemical parameters in groundwater.

About 55% of the samples contained SO_4^{2-}, which exceeded the WHO [51] and Chinese permissible limit [50] (250 mg/L). The concentrations of Cl^- and $HCO3^-$ in 26% and 43% of the groundwater samples, respectively, exceeded the respective WHO [51] and Chinese permissible limits [50] (250 g/L and 300 mg/L). Most of the study region had high HCO_3^- concentrations in the groundwater, and only some villages (including Yuanyi, western parts of Chonggang, Xinghai, and Yanzidun; Figure 3i) along the western border had low HCO_3^- concentrations.

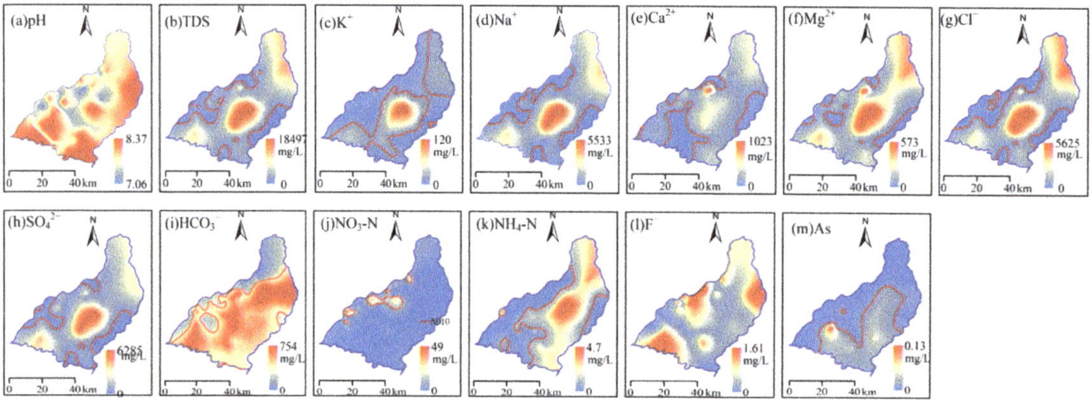

Figure 3. Geographic visualization of the hydrogeochemical variables: (**a**) pH; (**b**) TDS; (**c**) K^+; (**d**) Na^+;(**e**) Ca^{2+}; (**f**) Mg^{2+}; (**g**) Cl^-; (**h**) SO_4^{2-}; (**i**) HCO_3^-; (**j**) NO_3-N; (**k**) NH_4-N; (**l**) F^-; (**m**) As. The red line represents the WHO limit value.

In the traditional irrigated area, the nitrogen concentration in groundwater can better indicate the anthropogenic effects [56–58]. The concentrations of NO_3-N in the groundwater ranged from BDL (below the detection limit) to 62.15 mg/L (Figure 2). About 20% and 5% of the samples had NO_3-N concentrations higher than the WHO [51] and Chinese [50] drinking water standards of 10 mg/L and 20 mg/L, respectively. NH_4-N concentrations ranged between 0–4.7 mg/L (Figure 2). Nearly 23% of the samples had high NH_4-N levels, exceeding the threshold of the WHO [51] and Chinese standard [50] (0.5 mg/L). Nevertheless, NH_4-N was the dominant nitrogen pollutant in the study region due to its wide spatial distribution.

Fluoride is a dominant trace element in groundwater [59,60]. High concentrations of groundwater fluoride can result in fluorosis of the teeth and skeleton, including teeth discoloration and ligament deformation [61–63]. Groundwater fluoride concentrations ranged from 0.06 to 1.56 mg/L (mean of 0.58 mg/L). Among the 14 groundwater samples, 2 samples in the Lihe, Chonggang, and Xinghai villages (Figure 3l) contained higher fluoride levels than recommended by the WHO [51] and Chinese standards [50] of 1.5 and 1.0 mg/L, respectively. The high fluoride content may be due to the dissolution of fluorine-containing minerals, such as fluorite [48].

According to the WHO [51] and the Chinese standard [50], a concentration of As above 0.01 mg/L may hurt human health [62–64]. In the study region, the concentration of As was in the range of 0 to 0.131 mg/L. While the water quality parameters of most of the groundwater samples were within the drinking water standards, 10 samples from the central region contained levels of As that were higher than the permissible limit.

4.2. Hydrogeochemical Facies

The present study used a Piper [65] trilinear diagram (Figure 4) to illustrate the ion composition and chemical evolution of the groundwater. Groundwater cations were concentrated in the center and right corners of the cation triangle diagram. Groundwater cations indicated the groundwater types, namely a non-dominant type and a Na-dominant type. Groundwater anions showed a relatively dispersed distribution, with partial concentration in the SO_4^{2-} and HCO_3^- terminals. The facies of groundwater hydrochemistry in the study area were predominantly SO_4·Cl-Ca·Mg, and SO_4·Cl-Na. Some samples distributed along the Helan Mountains belonged to the HCO_3^- Ca·Mg type.

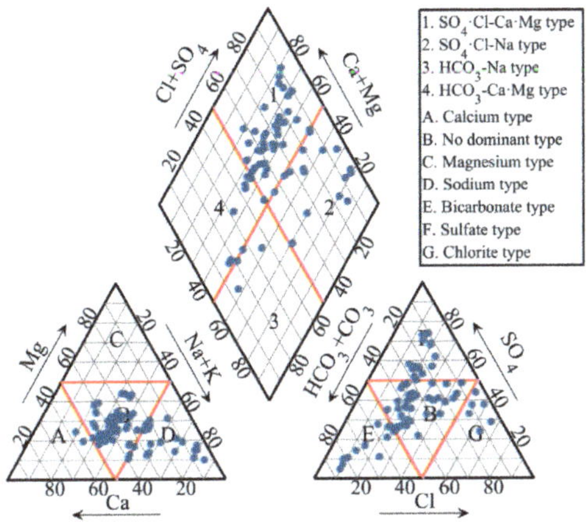

Figure 4. Piper diagram of groundwater chemical composition in the Shizuishan area.

4.3. Processes Regulating Groundwater Hydrochemistry

Principal Component Analysis (PCA) is a linear dimensionality reduction technique that reduces the number of variables and retains much more information [53]. The present study performed a PCA to identify the dominant factors regulating groundwater hydrochemistry in the study area. For the groundwater samples in this study, PC1 (43.40%), PC2 (13.63%), PC3 (13.18%), and PC4 (11.13%) could explain 81.34% of these 13 variables, which means that the principal component analysis was reliable (Figure 5).

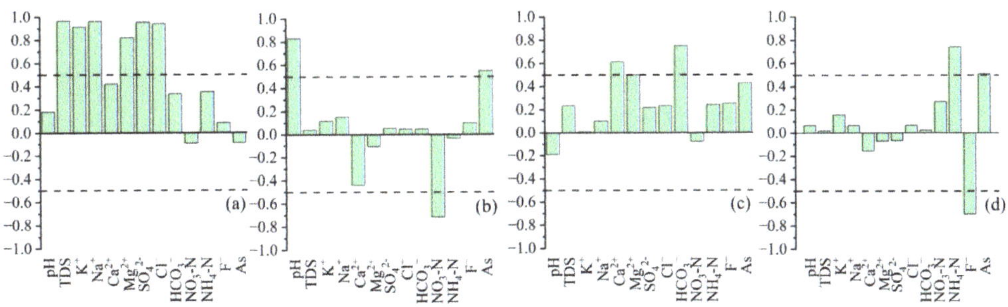

Figure 5. Load factor score of submersible principal component analysis. (**a**) salinity factor; (**b**) alkalinity factor; (**c**) carbonate factor; (**d**) pollution factor.

PC1 had a high contribution of TDS, K^+, Na^+, Mg^{2+}, Cl^-, and SO_4^{2-} (r = 0.97, 0.92, 0.96, 0.82, 0.94, and 0.96, respectively). These ions were the major constituents of the TDS. According to the correlation analysis, there were significant ($p < 0.05$) positive correlations between the TDS and K^+, Na^+, Mg^{2+}, Cl^-, and SO_4^{2-} (r^2 = 0.88, 0.97, 0.90, 0.98, and 0.98, respectively Figure 6). This indicates that the source of these ions may be derived from the dissolution of gypsum, halite, and dolomite [23,66]. The strong positive correlation demonstrated the dominant role of water-rock interaction in controlling groundwater hydrochemistry. Therefore, this component is referred to as the salinity factor.

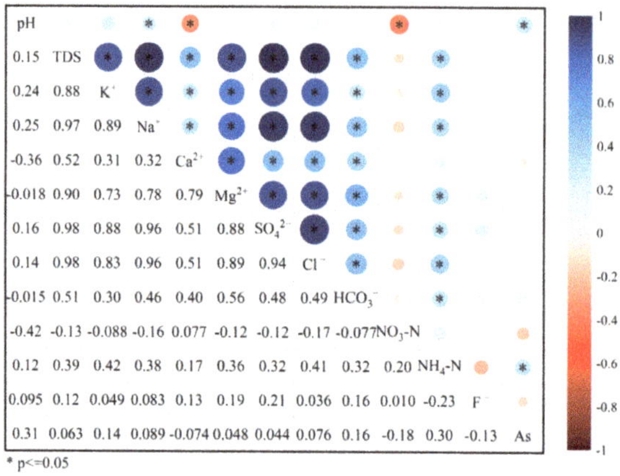

Figure 6. Correlation coefficient diagram of the groundwater hydrochemical indicators.

In PC2, there were strong positive loadings values of groundwater with pH and As (r = 0.83 and 0.55, respectively), and negative loadings of NO_3-N (r = −0.72). The level of the pH is an important factor for As accumulation in groundwater. Groundwater is weakly alkaline, which is helpful for As adsorption [67]. It has been shown that elevated groundwater pH affects the dissolution of As minerals due to chemical interactions between the underlying aquifer layer and the overlying water. This is due to the negative charge on the surface of sediment particles with increased pH, which forms electrostatic repulsion with arsenic-containing anions. As a result, arsenic in the adsorbed state can be desorbed and released into the groundwater [68]. Combined with the scattered distribution of high-NO_3-N groundwater and the negative relationship between pH and NO_3-N ($r^2 = -0.42$, $p < 0.05$, Figure 6), it is reasonable to assume that the possible source of NO_3-N may be derived from domestic and industrial wastewater infiltration. Therefore, this component is referred to as the alkalinity factor.

PC3 showed significant loadings for Ca^{2+}, Mg^{2+}, and HCO_3^- (r = 0.61, 0.50, and 0.75, respectively). These components were attributed to carbonate dissolution/precipitation. Positive significant ($p < 0.05$) correlations were found between Ca^{2+} and HCO_3^- and between Mg^{2+} and HCO_3^- ($r^2 = 0.40$ and 0.56, respectively, Figure 6), indicating that calcite and dolomite may be the major sources of those ions (Reaction 1 and Reaction 2). Therefore, this component is termed a carbonate factor.

$$CaCO_3 + CO_2 + H_2O \rightarrow 2HCO_3^- + Ca^{2+} \quad \text{(R1)}$$

$$CaMg(CO_3)_2 + 2CO_2 + 2H_2O \rightarrow 4HCO_3^- + Ca^{2+} + Mg^{2+} \quad \text{(R2)}$$

The PC4 had high positive loadings of NH_4-N (r = 0.71) and As (r = 0.51), and had negative loadings of F (r = −0.70). According to research [68–70], the occurrence of NH_4-N and As in groundwater is mostly detected in anaerobic environments in the northern part of the Yinchuan Plain. Positive relationships between NH_4-N and As ($r^2 = 0.3$, $p < 0.05$, Figure 6) in groundwater can demonstrate the controlling factor of redox conditions in hydrochemistry. Wu et al. [39] found that fluorite tends to dissolve in groundwater. Fluorite is the main source of fluoride in groundwater. PC4 contained the predominant groundwater contaminants in the study region. Therefore, this component is considered a pollution factor.

4.3.1. Water-Rock Interaction

Gibbs [71] proposed a model to better understand the mechanisms that regulate groundwater chemistry. This diagram reflects the influences of water-rock interactions, evaporation, and precipitation on groundwater hydrochemistry. The Gibbs curve shows that evaporation and rock weathering are the main driving forces, which are related to the complex geochemical mechanism of the study area. Other studies have made similar analyses [72]. The Gibbs diagram has been widely used in hydrogeochemical research [54,73]. Due to the hydrogeochemical facies shown in Figure 7, the majority of samples were positioned in the evaporation- and rock-dominant regions. The shallow depth of the groundwater facilitated the infiltration of irrigation water into the groundwater, resulting in evaporation, which is also the primary factor influencing groundwater chemistry. Since the study area is arid/semi-arid, groundwater sampling points in the present study were positioned along long distances from areas in which atmospheric precipitation regulates groundwater quality.

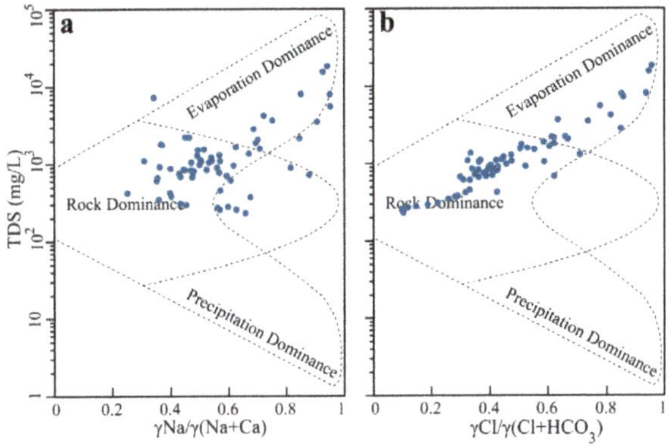

Figure 7. Gibbs diagram of groundwater chemical composition in the Shizuishan Area.

The Na^+: Cl^- (meq/L) ratio is widely used to identify the main sources of Na^+ [74]. Na^+: Cl^- = 1 when groundwater Na^+ is derived from halite dissolution. However, if the ratio is greater than 1.0, the Na^+ may have another source, such as the dissolution of silicate rocks or cation exchange [75]. As shown in Figure 8a, most of the sampling points were near or to the left of the 1:1 ratio line. Therefore, the groundwater Na^+ was mainly influenced by the dissolution of halite and also influenced by other sources (Reaction 3 and Reaction 4).

$$NaCl \rightarrow Na^{+} + Cl^{-} \quad \text{(R3)}$$

$$2NaAlSi_3O_8 + 2CO_2 + 11H_2O \rightarrow Al_2Si_2O_5(OH)_4 + 2Na + 4H_4SiO_4 + 2HCO_3^- \quad \text{(R4)}$$

Ca^{2+}/SO_4^{2-} is expected to be 1:1 when Ca^{2+} and SO_4^{2-} are derived from gypsum dissolution [76,77]. As shown in Figure 8b, most of the sampling points were positioned above the 1:1 line, indicating excess SO_4^{2-} and/or insufficient Ca^{2+} in groundwater. Cation exchange may reduce the Ca^{2+} concentration in the study region. Under those conditions, a moderate positive correlation was observed between concentrations of Ca^{2+} and SO_4^{2-} ($r^2 = 0.51$, $p < 0.05$).

In addition, the ratio ($Ca^{2+} + Mg^{2+}$): ($HCO_3^- + SO_4^{2-}$) can be helpful for determining groundwater sources of Ca^{2+} and Mg^{2+}. If the ratio is approximately 1.0, carbonate and sulfate rock dissolution is the main source of groundwater Ca^{2+} and Mg^{2+} [78]. In Figure 8c, most of the samples fell below the line. This demonstrates that the concentrations of HCO_3^- and SO_4^{2-} have obvious advantages for concentrations of $Ca^{2+} + Mg^{2+}$. The groundwater content of these ions may be regulated by the dissolution of both carbonate and gypsum. In addition, the cation exchange process involving the release of Na^+ and the uptake of Ca^{2+} or Mg^{2+} can cause a deficit in groundwater $Ca^{2+} + Mg^{2+}$.

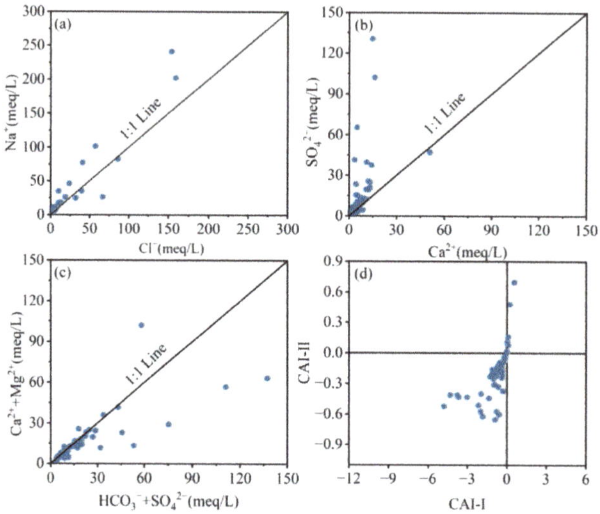

Figure 8. Ions correlation analysis picture of the Shizuishan Area.

Schoeller [79] proposed the chlor-alkali index (CAI-I and CAI-II) to measure the ion exchange reaction between groundwater and aquifers. If the chlor-alkali index is negative, it means that the amounts of Ca^{2+} and Mg^{2+} in the groundwater is decreasing, while those of Na+ and K^+ are increasing. Conversely, Ca^{2+} and Mg^{2+} increased, and Na^+ and K^+ decreased [53,72]. The above observations of the cation exchange process were further confirmed by the results of the chlor-alkali indices. As shown in Figure 8d, groundwater CAI-I and CAI-II were negative in most of the samples, ranging from −4.76 to 0.61 and −0.65 to 0.70, respectively. This result indicated the occurrence of positive cation exchange reactions in all the sampling areas. In the study region, groundwater Ca^{2+} displaces Na^+ from the host rock during positive cation exchange, increasing concentrations of groundwater Na^+ (Reaction 5).

$$CAI - \mathrm{I} = \left[Cl^- - (Na^+ + K^+)\right]/Cl^- \tag{9}$$

$$CAI - \mathrm{II} = \left[Cl^- - (Na^+ + K^+)\right]/(SO_4^{2-} + HCO_3^- + CO_3^{2-} + NO_3^-) \tag{10}$$

$$2\mathrm{NaX} + \mathrm{Ca}^{2+} \rightarrow \mathrm{CaX_2}{+}2\mathrm{Na}^+ \tag{R5}$$

4.3.2. Anthropogenic Activities

NO_3 and NH_4 are the main forms of nitrogen in groundwater. However, the spatial distributions of NO_3-N and NH_4-N (Figure 3j,k) indicate point and non-point contamination. The positive relationship between NH_4-N and K^+(r^2 = 0.42, $p < 0.05$) in Figure 6 can better explain the effect of fertilizer application. As a traditional irrigated region, nitrogen pollution in groundwater is caused by the overuse of fertilizer and flood irrigation. The average fertilizer application amount per hectare of arable land in Ningxia reached 821.81 kg, and the average fertilizer application rate was much higher than the national average (339 kg/ha) [38]. Rahman et al. [56] verified the positive correlation between NH4-N and K+ in groundwater due to intensive agricultural activities. However, when the groundwater is in a reducing environment, nitrate is reduced to nitrite nitrogen and then to ammonia nitrogen under the action of parthenogenic anaerobic bacteria [69]. The anaerobic conditions determined the predominance of NH_4-N in the groundwater. This observation is supported by the results of Chen et al. [40] in which high groundwater Fe and As concentrations were observed, which was further verified by the redox condition.

4.3.3. Geogenic Sources

Minerals containing arsenic and fluoride, such as fluorite, apatite, biotite, muscovite, hornblende, etc., are widely considered to be the sources of groundwater arsenic and fluoride [80]. The Shizuishan area contains a large amount of minerals including silica, coal, and pyrite. Scorodite ($FeAsO_4{\cdot}2H_2O$) is the main component of pyrite, and thus, it is considered a potential source of arsenic in groundwater [81]. Based on the chemical equilibrium in Reaction 6, scorodite may be a possible source of arsenic in groundwater [82]. The dissolution of CaF_2 (fluorite) is believed to be the main process leading to groundwater fluoride levels in Shizuishan (Reaction 7). This is consistent with previous research in Yinchuan Plain [36,39,76]. The high concentration of HCO_3^- in groundwater can control the solubility of fluoride, thereby releasing the concentration of fluoride into groundwater [83,84]. The arsenic and pH in the study region displayed an excellent positive correlation ($r^2 = 0.42$, $p < 0.05$). It made the alkaline environment the major controlling mechanism for the leaching of arsenic into groundwater. In the arid region, evaporation is beneficial to enhancing the concentration of arsenic in groundwater.

$$FeAsO_4 \bullet 2H_2O \rightarrow Fe^{3+} + AsO_4^{3-} + 2H_2O \tag{R6}$$

$$CaF_2 + 2HCO_3^- \rightarrow CaCO_3 + 2F^- + H_2O + CO_2 \tag{R7}$$

4.4. *Groundwater Quality Assessment*

Based on the weight given for each parameter, the relative weights were calculated (Table 2). The relative weights of TDS, Cl^-, NO_3-N, NH_4-N, F^-, and As were greater than other parameters, indicating that these were parameters with dominant effects on groundwater quality.

WQI is used to assess groundwater quality [43,44]. A WQI of < 50, 50 < WQI < 100, 100 < WQI < 200, 200 < WQI < 300, and > 300 indicate excellent, good, medium, poor, and extremely poor groundwater quality, respectively. Table 3 shows the distribution of groundwater sampling points in the Shizuishan area among these five categories. There are 15 groundwater samples of excellent quality, which are mainly distributed along the Helan Mountains. This may be related to the good quality of lateral inflow in the groundwater system.

Table 3. Quality classification of phreatic water and confined water.

	Categories	Ranges	NO. of Sample
WQI	Excellent	<50	15
	Good	50–100	24
	Medium	100–200	18
	Poor	200–300	3
	Extremely poor	>300	6

The aquifer along the Helan Mountains consists mainly of gravels, with much higher hydraulic gradients and faster groundwater flow rates. Considering the groundwater recharge sources, these samples may be associated with the good quality of lateral inflow in the groundwater system [39]. In total, 24 and 18 samples were classified as good and medium, respectively. This implies that among the sampling points, 60% of them contain groundwater that is suitable for drinking. In addition, there were three samples of poor quality. Six samples (Q1, Q5, Q8, Q11, Q19, and Q41) were of extremely poor quality. These samples were not suitable for direct drinking. For example, the maximum WQI value in Q8 was 965. The concentrations of TDS, NH4-N, and As in Q8 were 18448, 3.52, and 0.0253 mg/L, respectively, which far exceeded the WHO's [51] acceptable limits for drinking. The western and southern parts of the Shizuishan region had good groundwater quality, including Chengguan, Touzha, the northern part of Hongguozi, and the eastern part of Chonggang (Figure 9).

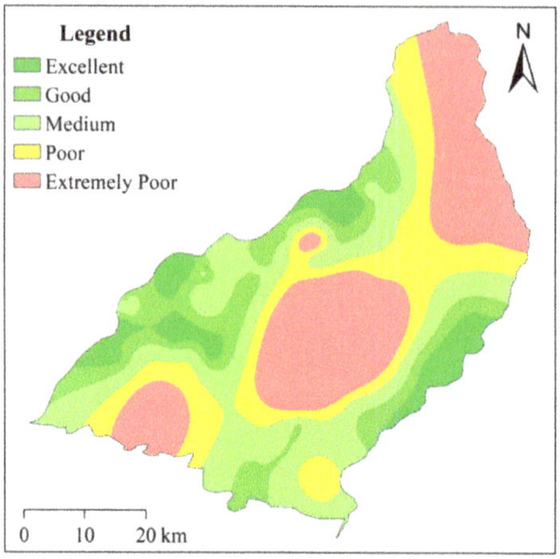

Figure 9. Groundwater EWQI zoning map of the Shizuishan area.

4.5. Assessment of the Hazard to Human Health

The non-carcinogenic health risk levels of groundwater oral intake were calculated. The health risks of NH_4-N, NO_3^-N, F^-, and As intake for children and adults are shown in Figure 10. The hazard quotients of NH_4-N were in the range of 0–0.22 for children and 0–0.11 for adults. The non-carcinogenic risks posed by NO_3-N and F^- were 0–1.81 and 0.05–1.81 for children, respectively. However, the corresponding risks in adults were acceptable in the range of 0–0.93 for NO_3-N and 0.02–0.62 for F^-. The non-carcinogenic risks associated with As in groundwater were predominant in the study region. The HQ values of As were between 0 and 20.38 for children, and between 0 and 10.48 for adults. Thus, for children, the cumulative hazard index (HI) through drinking water ranged from 0.09 to 20.65 (mean of 1.53), whereas the maximum was recorded at Q10 (20.65). The HI for adults ranged from 0.05 to 10.62, with a mean of 0.79, which is within the permissible risk level of 1.0. The level of non-carcinogenic hazards posed to children were 1.94 times higher than those posed to adults. In this regard, pollutants in groundwater posed higher non-carcinogenic risks to children. Similar results have been observed by other scholars. Adimalla et al. [43] identified the great potential risks in the agricultural area of Nanganur in southern India due to high concentrations of nitrate in groundwater. Qaiser et al. [47] also found that the non-carcinogenic index was higher for children than for adults in Pakistan.

As shown in Table 4, the contribution of HI was 57.07% for As, 29.91% for F^-, 11.42% for NO_3-N, and 1.61% for NH_4-N for both children and adults. The ranking of groundwater parameters according to their non-carcinogenic risks to human health was As > F^- > NO_3-N > NH_4-N. This also implies that concentrations of As and F have a great influence on the non-carcinogenic risks of groundwater. The spatial distribution of HI had similar characteristics to the spatial distribution of As and F^-. Groundwater is the only source of drinking water in the study region, and most of the residents, especially children, are exposed to adverse health risks due to the consumption of contaminated groundwater. In particular, the metabolic capacity of children is far lower than that of adults, and the harm from drinking contaminated water is much greater than that posed to adults. The long-term drinking of high-arsenic water can lead to chronic arsenic poisoning, including skin lesions, diabetes, cardiovascular diseases, and cancer [85,86]. Drinking high-fluoride groundwater will cause dental fluorosis and skeletal fluorosis [87]. Although the hydrochemical groundwater data showed that nitrogen contamination was more severe in the study region, the As concentration and the associated risks to residents should be given more attention by decision makers.

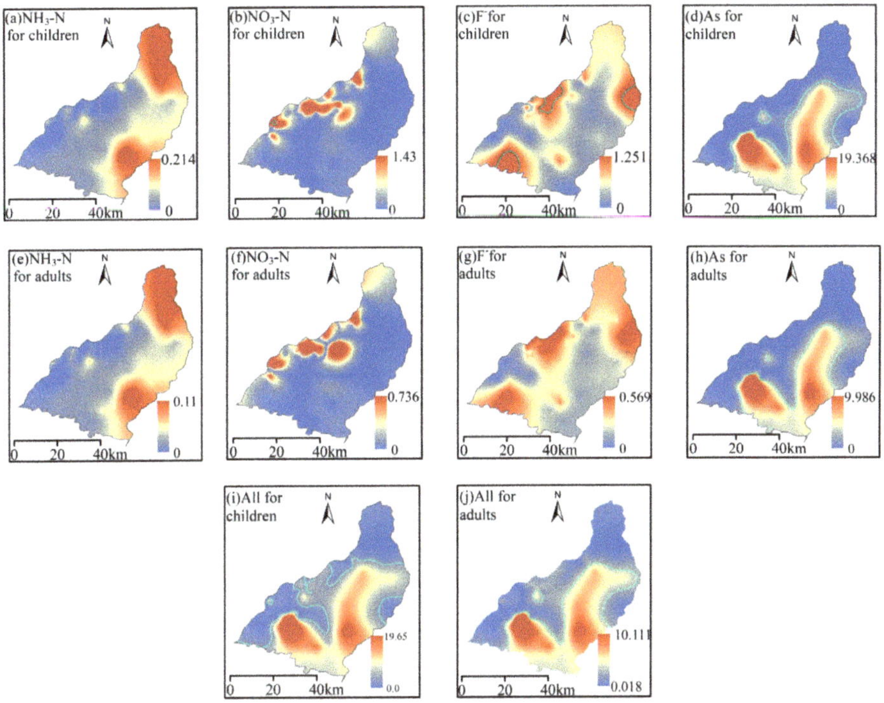

Figure 10. Non-carcinogenic risk zoning map in Shizuishan area. (**a**) NH_4-N for children; (**b**) NO_3-N for children; (**c**) F^- for children; (**d**) As for children; (**e**) NH_4-N for adults; (**f**) NO_3-N for adults; (**g**) F^- for adults; (**h**) As for adults; (**i**) all for children; (**j**) all for adults. The blue line represents the critical HI value.

Table 4. Non-carcinogenic and carcinogenic risk results of Shizuishan area through drinking water intake.

		HI			R_{As}	
		NO. of Samples	**Percentage of HQ > 1**	**Contribution rate of HI**	**NO. of Samples**	**Percentage of $R_{As} > 1 \times 10^{-4}$**
Children	NH_4-N	0	0%	1.61%	/	/
	NO_3-N	1	1.52%	11.42%	/	/
	F^-	13	19.70%	29.91%	/	/
	As	12	18.18%	57.07%	6	9.09%
	All	29	43.94%	/	/	/
Adults	NH_4-N	0	0%	1.61%	/	/
	NO_3-N	0	0%	11.42%	/	/
	F^-	0	0%	29.91%	/	/
	As	9	13.64%	57.07%	12	18.18%
	All	12	18.18%	/	/	/

For children (Figure 11a), the carcinogenic risk index of As by oral intake was 0–7.37×10^{-4}, with a mean value of 3.02×10^{-5}. For adults (Figure 11b), the carcinogenic risk index of As ranged between 0–1.89×10^{-4} with a mean level of 7.76×10^{-5}. There were 6 and 12 samples in the Middle East region that exceeded the acceptable level for children and adults, respectively. It may have adverse effects on human health.

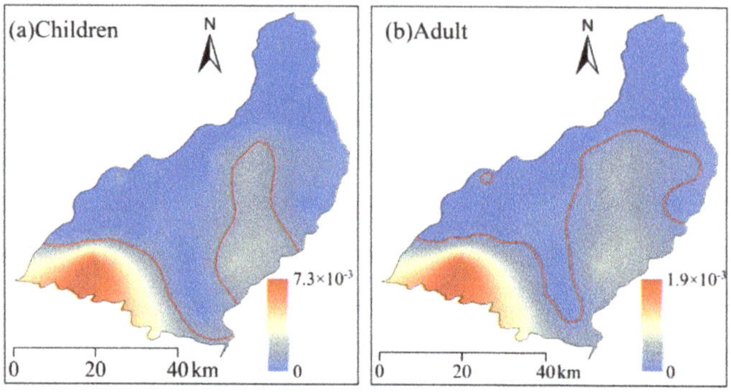

Figure 11. Carcinogenic risk zoning map of the Shizuishan area. The red line represents the risk threshold for R_{AS}.

4.6. Recommendation for Groundwater Management

Based on the above discussion, serious potential health risks exist in drinking groundwater from the study region. It is consistent with the previous findings [44,64] that the nitrogen pollution of groundwater contamination has become a serious challenge in the Yinchuan Plain [46,66]. The Jing Hui canal irrigation area is another old agricultural region with more than 2000 years of irrigation history. Similarly, nitrate pollution and the related health risks caused by intensive agricultural activities are very serious in the irrigation area [88]. In general, NO_3-N is a common component of nitrogen. When groundwater is in a reducing environment, nitrate will be reduced to nitrite nitrogen and then to ammonia nitrogen under the action of anaerobic bacteria [89]. In other words, NH_4-N in groundwater should also be paid more attention. However, there is an inconsistency between the dominant pollutants and the health-risk-induced pollutants in groundwater. In the study area, the exceedance rate of arsenic in groundwater was 15%, but it accounted for 57.7% of the total non-carcinogenic risk and was the main carcinogenic contaminant causing health risks. Therefore, arsenic contamination in groundwater should be of concern. These high-arsenic groundwaters in China are mainly located in the fluvial/alluvial-lacustrine plains and basins located in arid/semi-arid regions and alluvial plains/basins and river deltas in humid/semi-humid regions [87]. Most areas of the world are at risk from excessive arsenic levels, for example, 59 endemic arsenic villages with approximately 215,600 people at high risk have been identified among the inhabitants of the Datong Basin [24]; approximately 1.56 to 19.8 million people in Bangladesh are thought to be at risk from drinking arsenic-contaminated water [90]. Arsenic contamination is more severe in the Ganges Basin, where several people have developed skin and nerve cancer and most children have been diagnosed with arsenic poisoning [91].

In arid and semi-arid areas, groundwater is an important source of drinking water for domestic use. In the study region, almost every household relies on private wells for drinking water because of the low cost. This means that there is an urgent need to improve public awareness and to understand the relationship between oral ingestion and adverse health effects. The timely disclosure of this information is an effective management approach. Maintaining the safety of drinking water should be the highest priority for the decision makers. Due to the potential adverse health effects, groundwater abstracted from private wells is not recommended for drinking purposes in the study region. Compared with phreatic water, confined water is of a relatively good quality for the multi-layered aquifer system in Yinchuan Plain [92]. However, due to intensive pumping activities, the aquifer leakage between the phreatic aquifer and the confined aquifer caused the downward migration of pollutants. As reported by Chen, high NH_4-N concentrations were detected in confined water for pumping wells, while the corresponding contaminant concentration was below the detection limit for water in monitoring wells [40]. Therefore, a well-developed groundwater monitoring network is of great importance to identify the variation of chemical constituents in groundwater and to elucidate the migration of contaminants. It will be helpful to optimize groundwater abstraction and provide more insight for further management.

5. Conclusions

This study assessed the groundwater quality and hydrochemistry and the associated risks to human health in the Shizuishan area. The results showed that the groundwater was slightly alkaline. TDS values ranged from 232 to 18448 mg/L, with half of the samples exceeding the permissible limit for drinking water. The ranking of groundwater cations according to mean content was $Na^+ > Ca^{2+} > Mg^{2+} > K^+$, while that of anions was $SO_4{}^{2-} > Cl^- > HCO_3{}^-$. Approximately 20% and 5% of the samples had NO_3-N concentrations above the WHO and Chinese drinking water standards, respectively. Nearly 25% of samples had high NH_4-N concentrations above the drinking water threshold and showed a wide spatial distribution. Among the groundwater samples, the fluoride concentrations of 2 samples and 16 samples were higher than the WHO and Chinese standards for drinking water, respectively. While the measured water quality parameters of most groundwater samples were within the permissible limits for drinking water, 10 samples from the central region had levels of As concentrations above the permissible limit. The groundwater hydrochemical facies were mainly SO_4·Cl-Ca·Mg and SO_4·Cl-Na. Based on the principal component analysis, four principal components in groundwater were the salinity factor, carbonate factor, alkaline factor, and pollution factor in controlling groundwater chemistry in the study region.

The value of the WQI was in the range of 28–968. The WQI of about 60% of the groundwater samples indicated the suitability of their use as drinking water. The water samples of excellent quality are distributed along the Helan Mountains. This may be related to the good quality of lateral recharge in the groundwater system. There were three samples of poor quality. Six samples (Q1, Q5, Q8, Q11, Q19, and Q41) were of extremely poor quality. These samples were not suitable for direct drinking. The HI values in the study were in the range 0.05–10.62 for adults and 0.09–20.65 for children, respectively. Children showed a high risk of exposure to groundwater contaminants upon the long-term consumption of contaminated drinking water. The impact of the pollutants in the groundwater for residents decreased in the order As > F^- > NO_3-N > NH_4-N. Meanwhile, the carcinogenic risk value of the oral intake of As was 0–7.37×10^{-4} for children and 0–1.89×10^{-4} for adults. There were 6 and 12 samples located in the middle east area with levels beyond the permissible acceptable levels. This study can further support groundwater management in the study area. More effective approaches should be considered in the future.

Author Contributions: Conceptualization, J.C.; data curation, S.X., Y.G. and P.X.; funding acquisition, J.C.; investigation, Y.F., Z.Z., X.J. and M.R.; methodology, Y.G. and P.X.; resources, Y.F.; software, S.X.; validation, S.X. and J.C.; writing—original draft, S.X.; writing—review and editing, J.C. All authors have read and agreed to the published version of the manuscript.

Funding: This research is supported by grants from the National Natural Science Foundation of China (42007184, 42272289, 41790441, and 41931285), China Postdoctoral Science Foundation (300204000181), the Natural Science Foundation of Shaanxi Province (2022JQ-243), and Youth Talents Promotion Plan of Shaanxi Association for Science and Technology (20220711).

Data Availability Statement: All processed data generated or used during the study appear in the submitted article. Raw data may be provided on request from the corresponding author.

Acknowledgments: Anonymous reviewers and the Editor are sincerely acknowledged for their useful comments.

Conflicts of Interest: The authors declare that they have no conflict of interest.

References

1. Shrestha, S.; Kafle, R.; Pandey, V.P. Evaluation of index-overlay methods for groundwater vulnerability and risk assessment in Kathmandu Valley, Nepal. *Sci. Total Environ.* **2017**, *575*, 779–790. [CrossRef] [PubMed]
2. Zhong, C.; Wang, H.; Yang, Q. Hydrochemical interpretation of groundwater in Yinchuan basin using self-organizing maps and hierarchical clustering. *Chemosphere* **2022**, *309*, 136787. [CrossRef] [PubMed]
3. Han, S.; Zhang, F.; Zhang, H. Spatial and temporal patterns of groundwater arsenic in shallow and deep groundwater of Yinchuan Plain, China. *J. Geochem. Explor.* **2013**, *135*, 71–78. [CrossRef]
4. Moyé, J.; Picard-Lesteven, T.; Zouhri, L.; El Amari, K.; Hibti, M.; Benkaddour, A. Groundwater assessment and environmental impact in the abandoned mine of Kettara (Morocco). *Environ. Pollut.* **2017**, *231*, 899–907. [CrossRef] [PubMed]
5. Ahmed, N.; Bodrud-Doza, M.; Towfiqul Islam, A.R.M.; Hossain, S.; Moniruzzaman, M.; Deb, N.; Bhuiyan, M.A.Q. Appraising spatial variations of As, Fe, Mn and NO3 contaminations associated health risks of drinking water from Surma basin, Bangladesh. *Chemosphere* **2019**, *218*, 726–740. [CrossRef]
6. Chen, R.; Teng, Y.; Chen, H.; Hu, B.; Yue, W. Groundwater pollution and risk assessment based on source apportionment in a typical cold agricultural region in Northeastern China. *Sci. Total Environ.* **2019**, *696*, 133972. [CrossRef]
7. Abdalla, O.; Al-Abri, R.b.Y. Factors affecting groundwater chemistry in regional arid basins of variable lithology: Example of Wadi Umairy, Oman. *Arab. J. Geosci.* **2014**, *7*, 2861–2870. [CrossRef]
8. Busico, G.; Cuoco, E.; Kazakis, N.; Colombani, N.; Mastrocicco, M.; Tedesco, D.; Voudouris, K. Multivariate statistical analysis to characterize/discriminate between anthropogenic and geogenic trace elements occurrence in the Campania Plain, Southern Italy. *Environ. Pollut.* **2018**, *234*, 260–269. [CrossRef]
9. Reddy, A.G.S.; Kumar, K.N. Identification of the hydrogeochemical processes in groundwater using major ion chemistry: A case study of Penna–Chitravathi river basins in Southern India. *Environ. Monit. Assess.* **2010**, *170*, 365–382. [CrossRef]
10. Xueli, L. *Hydrogeochemistry*; Atomic Energy Press: Beijing, China, 1988; p. 446. (In Chinese)
11. Wali, S.U.; Alias, N.; Bin Harun, S. Hydrogeochemical evaluation and mechanisms controlling groundwater in different geologic environments, Western Sokoto Basin, Northwestern Nigeria. *SN Appl. Sci.* **2020**, *2*, 1808. [CrossRef]
12. Okkonen, J.; Kløve, B. Assessment of temporal and spatial variation in chemical composition of groundwater in an unconfined esker aquifer in the cold temperate climate of Northern Finland. *Cold Reg. Sci. Technol.* **2012**, *71*, 118–128. [CrossRef]
13. Li, Z.-J.; Song, L.-L.; Gui, J.; Li, Z.-X. Hydrochemical patterns indicating hydrological processes with the background of changing climatic and environmental conditions in China: A review. *Environ. Sci. Pollut. Res.* **2022**, *29*, 15364–15379. [CrossRef]
14. Schot, P.P.; van der Wal, J. Human impact on regional groundwater composition through intervention in natural flow patterns and changes in land use. *J. Hydrol.* **1992**, *134*, 297–313. [CrossRef]
15. Adams, S.; Titus, R.; Pietersen, K.; Tredoux, G.; Harris, C. Hydrochemical characteristics of aquifers near Sutherland in the Western Karoo, South Africa. *J. Hydrol.* **2001**, *241*, 91–103. [CrossRef]
16. Zhang, B.; Song, X.; Zhang, Y.; Han, D.; Tang, C.; Yu, Y.; Ma, Y. Hydrochemical characteristics and water quality assessment of surface water and groundwater in Songnen plain, Northeast China. *Water Res.* **2012**, *46*, 2737–2748. [CrossRef] [PubMed]
17. Güler, C.; Kurt, M.A.; Alpaslan, M.; Akbulut, C. Assessment of the impact of anthropogenic activities on the groundwater hydrology and chemistry in Tarsus coastal plain (Mersin, SE Turkey) using fuzzy clustering, multivariate statistics and GIS techniques. *J. Hydrol.* **2012**, *414–415*, 435–451. [CrossRef]
18. Maheswari, M.; Murthy, A.N.G.; Shanker, A.K. Nitrogen Nutrition in Crops and Its Importance in Crop Quality. In *The Indian Nitrogen Assessment*; Abrol, Y.P., Adhya, T.K., Aneja, V.P., Raghuram, N., Pathak, H., Kulshrestha, U., Sharma, C., Singh, B., Eds.; Elsevier: Amsterdam, The Netherlands, 2017; pp. 175–186. [CrossRef]
19. Bodirsky, B.L.; Popp, A.; Lotze-Campen, H.; Dietrich, J.P.; Rolinski, S.; Weindl, I.; Schmitz, C.; Müller, C.; Bonsch, M.; Humpenöder, F.; et al. Reactive nitrogen requirements to feed the world in 2050 and potential to mitigate nitrogen pollution. *Nat. Commun.* **2014**, *5*, 3858. [CrossRef]
20. Zahoor; Ahmad, W.; Hira, K.; Ullah, B.; Khan, A.-u.; Shah, Z.; Khan, F.; Naz, R. Role of nitrogen fertilizer in crop productivity and environmental pollution. *Int. J. Agric. For.* **2014**, *4*, 201–206.
21. Ahmed, M.; Rauf, M.; Mukhtar, Z.; Saeed, N.A. Excessive use of nitrogenous fertilizers: An unawareness causing serious threats to environment and human health. *Environ. Sci. Pollut. Res.* **2017**, *24*, 26983–26987. [CrossRef]
22. Kumar, M.; Goswami, R.; Patel, A.K.; Srivastava, M.; Das, N. Scenario, perspectives and mechanism of arsenic and fluoride Co-occurrence in the groundwater: A review. *Chemosphere* **2020**, *249*, 126126. [CrossRef]
23. Li, P.; He, X.; Guo, W. Spatial groundwater quality and potential health risks due to nitrate ingestion through drinking water: A case study in Yan'an City on the Loess Plateau of northwest China. *Hum. Ecol. Risk Assess. Int. J.* **2019**, *25*, 11–31. [CrossRef]
24. He, X.; Li, P.; Wu, J.; Wei, M.; Ren, X.; Wang, D. Poor groundwater quality and high potential health risks in the Datong Basin, northern China: Research from published data. *Environ. Geochem. Health* **2021**, *43*, 791–812. [CrossRef] [PubMed]
25. Cheng, P.S.; Weng, S.F.; Chiang, C.H.; Lai, F.J. Relationship between arsenic-containing drinking water and skin cancers in the arseniasis endemic areas in Taiwan. *J. Dermatol.* **2016**, *43*, 181–186. [CrossRef]
26. Astolfi, E.; Maccagno, A.; García Fernández, J.C.; Vaccaro, R.; Stímola, R. Relation between arsenic in drinking water and skin cancer. *Biol. Trace Elem. Res.* **1981**, *3*, 133–143. [CrossRef] [PubMed]

27. Yang, L.; Peterson, P.J.; Williams, W.P.; Wang, W.; Hou, S.; Tan, J. The Relationship Between Exposure to Arsenic Concentrations in Drinking Water and the Development of Skin Lesions in Farmers from Inner Mongolia, China. *Environ. Geochem. Health* **2002**, *24*, 293–303. [CrossRef]
28. Şener, Ş.; Şener, E.; Davraz, A. Evaluation of water quality using water quality index (WQI) method and GIS in Aksu River (SW-Turkey). *Sci. Total Environ.* **2017**, *584–585*, 131–144. [CrossRef] [PubMed]
29. Lumb, A.; Sharma, T.C.; Bibeault, J.-F. A Review of Genesis and Evolution of Water Quality Index (WQI) and Some Future Directions. *Water Qual. Expo. Health* **2011**, *3*, 11–24. [CrossRef]
30. Vasanthavigar, M.; Srinivasamoorthy, K.; Vijayaragavan, K.; Rajiv Ganthi, R.; Chidambaram, S.; Anandhan, P.; Manivannan, R.; Vasudevan, S. Application of water quality index for groundwater quality assessment: Thirumanimuttar sub-basin, Tamilnadu, India. *Environ. Monit. Assess.* **2010**, *171*, 595–609. [CrossRef] [PubMed]
31. Xiao, J.; Wang, L.; Deng, L.; Jin, Z. Characteristics, sources, water quality and health risk assessment of trace elements in river water and well water in the Chinese Loess Plateau. *Sci. Total Environ.* **2019**, *650*, 2004–2012. [CrossRef] [PubMed]
32. Hasan, M.S.U.; Rai, A.K. Groundwater quality assessment in the Lower Ganga Basin using entropy information theory and GIS. *J. Clean. Prod.* **2020**, *274*, 123077. [CrossRef]
33. Radfard, M.; Yunesian, M.; Nabizadeh, R.; Biglari, H.; Nazmara, S.; Hadi, M.; Yousefi, N.; Yousefi, M.; Abbasnia, A.; Mahvi, A.H. Drinking water quality and arsenic health risk assessment in Sistan and Baluchestan, Southeastern Province, Iran. *Hum. Ecol. Risk Assess.* **2019**, *25*, 949–965. [CrossRef]
34. Adimalla, N.; Li, P. Occurrence, health risks, and geochemical mechanisms of fluoride and nitrate in groundwater of the rock-dominant semi-arid region, Telangana State, India. *Hum. Ecol. Risk Assess. Int. J.* **2018**, *25*, 81–103. [CrossRef]
35. Rasool, A.; Farooqi, A.; Masood, S.; Hussain, K. Arsenic in groundwater and its health risk assessment in drinking water of Mailsi, Punjab, Pakistan. *Hum. Ecol. Risk Assess.* **2016**, *22*, 187–202. [CrossRef]
36. Li, P.; Wu, J.; Qian, H. Hydrogeochemistry and Quality Assessment of Shallow Groundwater in the Southern Part of the Yellow River Alluvial Plain (Zhongwei Section), Northwest China. *Earth Sci. Res. J.* **2014**, *18*, 27–38. [CrossRef]
37. Chen, J.; Gao, Y.; Qian, H.; Ren, W.; Qu, W. Hydrogeochemical evidence for fluoride behavior in groundwater and the associated risk to human health for a large irrigation plain in the Yellow River Basin. *Sci. Total Environ.* **2021**, *800*, 149428. [CrossRef] [PubMed]
38. Ningxia Statistical Bureau. *Ningxia Statistical Yearbook*; China Statistics Press: Beijing, China, 2022.
39. Wu, J.; Li, P.; Qian, H. Hydrochemical characterization of drinking groundwater with special reference to fluoride in an arid area of China and the control of aquifer leakage on its concentrations. *Environ. Earth Sci.* **2015**, *73*, 8575–8588. [CrossRef]
40. Chen, J.; Qian, H.; Wu, H. Nitrogen contamination in groundwater in an agricultural region along the New Silk Road, northwest China: Distribution and factors controlling its fate. *Environ. Sci. Pollut. Res.* **2017**, *24*, 13154–13167. [CrossRef]
41. Ministry of Water Resources of PRC. *Technical Regulation of Water Quality Sampling (SL 187–96)*; Standards Press of China: Beijing, China, 1997. (In Chinese)
42. State Environmental Protection Administration. *The Technical Specification for Environmental Monitoring of Groundwater*; China Environmental Science Press: Beijing, China, 2004. (In Chinese)
43. Adimalla, N.; Qian, H. Groundwater quality evaluation using water quality index (WQI) for drinking purposes and human health risk (HHR) assessment in an agricultural region of Nanganur, south India. *Ecotoxicol. Environ. Saf.* **2019**, *176*, 153–161. [CrossRef]
44. Verma, D.K.; Bhunia, G.S.; Shit, P.K.; Tiwari, A.K. Assessment of Groundwater Quality of the Central Gangetic Plain Area of India Using Geospatial and WQI Techniques. *J. Geol. Soc. India* **2018**, *92*, 743–752. [CrossRef]
45. USEPA. *Risk Assessment Guidance for Superfund*; Human Health Evaluation Manual (Part A); Office of Emergency and Remedial Response: Washington, DC, USA, 1989; Volume 1.
46. Feng, B.; Ma, Y.; Qi, Y.; Zhong, Y.; Sha, X. Health risk assessment of groundwater nitrogen pollution in Yinchuan plain. *J. Contam. Hydrol.* **2022**, *249*, 104031. [CrossRef]
47. Qaiser, F.U.R.; Zhang, F.; Pant, R.R.; Zeng, C.; Khan, N.G.; Wang, G. Characterization and health risk assessment of arsenic in natural waters of the Indus River Basin, Pakistan. *Sci. Total Environ.* **2023**, *857*, 159408. [CrossRef] [PubMed]
48. Li, P.; Li, X.; Meng, X.; Li, M.; Zhang, Y. Appraising Groundwater Quality and Health Risks from Contamination in a Semiarid Region of Northwest China. *Expo. Health* **2016**, *8*, 361–379. [CrossRef]
49. Hunter, P.R.; Fewtrell, L. *Acceptable Risk 10.1 Introduction*; IWA Publishing: London, UK, 2001.
50. Bureau of Quality and Technical Supervision of China. National Standard of the People's Republic of China:Quality Standard for Groundwater. In *GB/T 14848-2017*; Bureau of Quality and Technical Supervision of China: Beijing, China, 2017. (In Chinese)
51. World Health Organization (WHO). *Guidelines for Drinking Water Quality*; World Health Organization: Geneva, Switzerland, 2011.
52. Su, Y.H.; Zhu, G.F.; Feng, Q.; Li, Z.Z.; Zhang, F.P. Environmental isotopic and hydrochemical study of groundwater in the Ejina Basin, northwest China. *Environ. Geol.* **2009**, *58*, 601–614. [CrossRef]
53. Li, P.; Qian, H.; Wu, J.; Zhang, Y.; Zhang, H. Major Ion Chemistry of Shallow Groundwater in the Dongsheng Coalfield, Ordos Basin, China. *Mine Water Environ.* **2013**, *32*, 195–206. [CrossRef]
54. Marghade, D.; Malpe, D.B.; Zade, A.B. Major ion chemistry of shallow groundwater of a fast growing city of central India. *Environ. Monit. Assess.* **2012**, *184*, 2405–2418. [CrossRef] [PubMed]

55. Zhang, X.; Qian, H.; Wu, H.; Chen, J.; Qiao, L. Multivariate Analysis of Confined Groundwater Hydrochemistry of a Long-Exploited Sedimentary Basin in Northwest China. *J. Chem.* **2016**, *2016*, 3812125. [CrossRef]
56. Rahman, A.; Mondal, N.C.; Tiwari, K.K. Anthropogenic nitrate in groundwater and its health risks in the view of background concentration in a semi arid area of Rajasthan, India. *Sci. Rep.* **2021**, *11*, 9279. [CrossRef]
57. Galloway, J.N.; Leach, A.M.; Bleeker, A.; Erisman, J.W. A chronology of human understanding of the nitrogen cycle. *Philos. Trans. R. Soc. Lond. Ser. B Biol. Sci.* **2013**, *368*, 20130120. [CrossRef]
58. Morrissy, J.G.; Currell, M.J.; Reichman, S.M.; Surapaneni, A.; Megharaj, M.; Crosbie, N.D.; Hirth, D.; Aquilina, S.; Rajendram, W.; Ball, A.S. Nitrogen contamination and bioremediation in groundwater and the environment: A review. *Earth-Sci. Rev.* **2021**, *222*, 103816. [CrossRef]
59. Jha, S.K.; Singh, R.K.; Damodaran, T.; Mishra, V.K.; Sharma, D.K.; Rai, D. Fluoride in groundwater: Toxicological exposure and remedies. *J. Toxicol. Environ. Health Part B Crit. Rev.* **2013**, *16*, 52–66. [CrossRef]
60. Vithanage, M.; Bhattacharya, P. Fluoride in the environment: Sources, distribution and defluoridation. *Environ. Chem. Lett.* **2015**, *13*, 131–147. [CrossRef]
61. Kechiched, R.; Nezli, I.E.; Foufou, A.; Belksier, M.S.; Benhamida, S.A.; Djeghoubbi, R.; Slamene, N.; Ameur-zaimeche, O. Fluoride-bearing groundwater in the complex terminal aquifer (a case study in Hassi Messaoud area, southern Algeria): Hydrochemical characterization and spatial distribution assessed by indicator kriging. *Sustain. Water Resour. Manag.* **2020**, *6*, 54. [CrossRef]
62. Wang, S.X.; Wang, Z.H.; Cheng, X.T.; Li, J.; Sang, Z.P.; Zhang, X.D.; Han, L.L.; Qiao, X.Y.; Wu, Z.M.; Wang, Z.Q. Arsenic and fluoride exposure in drinking water: Children's IQ and growth in Shanyin county, Shanxi province, China. *Environ. Health Perspect.* **2007**, *115*, 643–647. [CrossRef] [PubMed]
63. Villalba, E.; Tanjal, C.; Borzi, G.; Páez, G.; Carol, E. Geogenic arsenic contamination of wet-meadows associated with a geothermal system in an arid region and its relevance for drinking water. *Sci. Total Environ.* **2020**, *720*, 137571. [CrossRef] [PubMed]
64. Rahman, A.; Persson, L.A.; Nermell, B.; El Arifeen, S.; Ekstrom, E.C.; Smith, A.H.; Vahter, M. Arsenic Exposure and Risk of Spontaneous Abortion, Stillbirth, and Infant Mortality. *Epidemiology* **2010**, *21*, 797–804. [CrossRef]
65. Piper, A.M. A graphic procedure in the geochemical interpretation of water-analyses. *Eos Trans. Am. Geophys. Union* **1944**, *25*, 914–928. [CrossRef]
66. Liu, F.X.; Qian, H.; Shi, Z.W.; Wang, H.K. Long-term monitoring of hydrochemical characteristics and nitrogen pollution in the groundwater of Yinchuan area, Yinchuan basin of northwest China. *Environ. Earth Sci.* **2019**, *78*, 885–897. [CrossRef]
67. Wang, Y.X.; Pi, K.F.; Fendorf, S.; Deng, Y.M.; Xie, X.J. Sedimentogenesis and hydrobiogeochemistry of high arsenic Late Pleistocene-Holocene aquifer systems. *Earth-Sci. Rev.* **2019**, *189*, 79–98. [CrossRef]
68. Huq, M.E.; Fahad, S.; Shao, Z.; Sarven, M.S.; Khan, I.A.; Alam, M.; Saeed, M.; Ullah, H.; Adnan, M.; Saud, S.; et al. Arsenic in a groundwater environment in Bangladesh: Occurrence and mobilization. *J. Environ. Manag.* **2020**, *262*, 110318. [CrossRef]
69. Böhlke, J.K.; Smith, R.L.; Miller, D.N. Ammonium transport and reaction in contaminated groundwater: Application of isotope tracers and isotope fractionation studies. *Water Resour. Res.* **2006**, *42*. [CrossRef]
70. Sun, Y.; Sun, J.; Nghiem, A.A.; Bostick, B.C.; Ellis, T.; Han, L.; Li, Z.; Liu, S.; Han, S.; Zhang, M.; et al. Reduction of iron (hydr)oxide-bound arsenate: Evidence from high depth resolution sampling of a reducing aquifer in Yinchuan Plain, China. *J. Hazard. Mater.* **2021**, *406*, 124615. [CrossRef] [PubMed]
71. Gibbs, R.J. Mechanisms Controlling World Water Chemistry. *Science* **1970**, *170*, 1088–1090. [CrossRef] [PubMed]
72. Maurya, S.; Saxena, A. Spatiotemporal assessment of groundwater quality in the Central Ganga Plain, India, using multivariate statistical tools. *Environ. Monit. Assess.* **2022**, *194*, 865. [CrossRef] [PubMed]
73. Naseem, S.; Rafique, T.; Bashir, E.; Bhanger, M.I.; Laghari, A.; Usmani, T.H. Lithological influences on occurrence of high-fluoride groundwater in Nagar Parkar area, Thar Desert, Pakistan. *Chemosphere* **2010**, *78*, 1313–1321. [CrossRef] [PubMed]
74. Li, P.; Wu, J.; Qian, H. Hydrochemical appraisal of groundwater quality for drinking and irrigation purposes and the major influencing factors: A case study in and around Hua County, China. *Arab. J. Geosci.* **2015**, *9*, 15. [CrossRef]
75. Guo, X.; Wang, X.; Shi, X.; Yu, H.; Huirong, Z.; Fang, Z. Hydrochemical characteristics and sources of chemical constituents in groundwater in Hunchun River Basin, Northeast China. *Arab. J. Geosci.* **2022**, *15*, 694. [CrossRef]
76. Chen, J.; Huang, Q.; Lin, Y.; Fang, Y.; Qian, H.; Liu, R.; Ma, H. Hydrogeochemical Characteristics and Quality Assessment of Groundwater in an Irrigated Region, Northwest China. *Water* **2019**, *11*, 96. [CrossRef]
77. Chen, J.; Qian, H.; Gao, Y.; Wang, H.; Zhang, M. Insights into hydrological and hydrochemical processes in response to water replenishment for lakes in arid regions. *J. Hydrol.* **2020**, *581*, 124386. [CrossRef]
78. Rahman, M.A.T.M.T.; Majumder, R.K.; Rahman, S.H.; Halim, M.A. Sources of deep groundwater salinity in the southwestern zone of Bangladesh. *Environ. Earth Sci.* **2011**, *63*, 363–373. [CrossRef]
79. Schoeller, H. *Qualitative Evaluation of Groundwa-Ter Resources*; UNESCO: Paris, France, 1965; Volume 1965, pp. 54–83.
80. Ehrlich, H.L.; Newman, D.K. *Geomicrobiology*, 5th ed.; CRC Press: Boca Raton, FL, USA, 2009.
81. Chen, J.; Qian, H.; Wu, H.; Gao, Y.; Li, X. Assessment of arsenic and fluoride pollution in groundwater in Dawukou area, Northwest China, and the associated health risk for inhabitants. *Environ. Earth Sci.* **2017**, *76*, 314. [CrossRef]
82. Feng, S.; Guo, H.; Sun, X.; Han, S. Limited roles of anthropogenic activities on arsenic mobilization in groundwater from the Yinchuan Basin, China. *J. Hydrol.* **2022**, *610*, 127910. [CrossRef]
83. Adimalla, N.; Marsetty, S.K.; Xu, P.P. Assessing groundwater quality and health risks of fluoride pollution in the Shasler Vagu (SV) watershed of Nalgonda, India. *Hum. Ecol. Risk Assess.* **2020**, *26*, 1569–1588. [CrossRef]

84. Kanagaraj, G.; Elango, L. Chromium and fluoride contamination in groundwater around leather tanning industries in southern India: Implications from stable isotopic ratio delta Cr-53/delta Cr-52, geochemical and geostatistical modelling. *Chemosphere* **2019**, *220*, 943–953. [CrossRef] [PubMed]
85. Chung, J.Y.; Yu, S.D.; Hong, Y.S. Environmental source of arsenic exposure. *J. Prev. Med. Public Health Yebang Uihakhoe Chi* **2014**, *47*, 253–257. [CrossRef] [PubMed]
86. Morales, K.H.; Ryan, L.; Kuo, T.L.; Wu, M.M.; Chen, C.J. Risk of internal cancers from arsenic in drinking water. *Environ. Health Perspect.* **2000**, *108*, 655–661. [CrossRef]
87. He, X.; Li, P.; Ji, Y.; Wang, Y.; Su, Z.; Elumalai, V. Groundwater Arsenic and Fluoride and Associated Arsenicosis and Fluorosis in China: Occurrence, Distribution and Management. *Expo. Health* **2020**, *12*, 355–368. [CrossRef]
88. Zhang, Y.; Wu, J.; Xu, B. Human health risk assessment of groundwater nitrogen pollution in Jinghui canal irrigation area of the loess region, northwest China. *Environ. Earth Sci.* **2018**, *77*, 273. [CrossRef]
89. Bengtsson, G.r.; Annadotter, H. Nitrate Reduction in a Groundwater Microcosm Determined by 15N Gas Chromatography-Mass Spectrometry. *Appl. Environ. Microbiol.* **1989**, *55*, 2861–2870. [CrossRef]
90. Edmunds, W.M.; Ahmed, K.M.; Whitehead, P.G. A review of arsenic and its impacts in groundwater of the Ganges-Brahmaputra-Meghna delta, Bangladesh. *Environ. Science. Process. Impacts* **2015**, *17*, 1032–1046. [CrossRef]
91. Chakraborti, D.; Singh, S.K.; Rahman, M.M.; Dutta, R.N.; Mukherjee, S.C.; Pati, S.; Kar, P.B. Groundwater Arsenic Contamination in the Ganga River Basin: A Future Health Danger. *Int. J. Environ. Res. Public Health* **2018**, *15*, 180. [CrossRef]
92. Zhou, Y.; Wei, A.; Li, J.; Yan, L.; Li, J. Groundwater Quality Evaluation and Health Risk Assessment in the Yinchuan Region, Northwest China. *Expo. Health* **2016**, *8*, 443–456. [CrossRef]

water MDPI

Article

Groundwater Hydrochemical Characteristics and Water Quality in Egypt's Central Eastern Desert

Ahmed Saleh [1,*], Ahmed Gad [2,3], Alaa Ahmed [2], Hasan Arman [2] and Hassan I. Farhat [4]

1 National Research Institute of Astronomy and Geophysics (NRIAG), Cairo 11421, Egypt
2 Geosciences Department, College of Science, United Arab Emirates University, Al Ain 15551, United Arab Emirates
3 Geology Department, Faculty of Science, Ain Shams University, Cairo 11566, Egypt
4 Geology Department, Faculty of Science, Suez University, El Salam City 43518, Egypt
* Correspondence: ahmed.saleh@nriag.sci.eg

Abstract: The rapid expansion of economic activities in Egypt's Central Eastern Desert has resulted in poorly coordinated groundwater development, having a negative impact on the resource. This study was conducted to assess the hydrochemical characteristics of the different aquifers in the Central Eastern Desert, with an emphasis on the impact of seawater intrusion and groundwater quality for different purposes. A total of 21 groundwater samples were collected representing the three main aquifers (Eocene Carbonate, Nubian Sandstone, and Fractured Basement) in the Central Eastern Desert, and analyzed for major ions and trace elements. The majority of the samples had electrical conductivity values that exceeded the salinity limit for natural water. Groundwater saline load is primarily influenced by sodium, calcium, chloride, and sulfate concentrations. The groundwater in the Central Eastern Desert mainly consists of Na-Cl, Ca-Cl, and Na-SO_4 water-types. Saltwater intrusion and water-rock interactions via cation exchange and minerals weathering are the primary controlling factors of groundwater hydrochemistry. The high salinity of this groundwater renders it unsuitable for irrigation or consumption. Additionally, it is unfit for domestic use based on total hardness values. Furthermore, the Al, Cd, Fe, Mn, and Ni concentrations in the investigated groundwater exceed the allowable limits for human consumption. Proper mitigation measures and adaptation strategies are required for groundwater sustainability in the Central Eastern Desert.

Keywords: groundwater; hydrogeochemistry; hydrochemical facies; seawater intrusion; water quality; Eastern Desert

Citation: Saleh, A.; Gad, A.; Ahmed, A.; Arman, H.; Farhat, H.I. Groundwater Hydrochemical Characteristics and Water Quality in Egypt's Central Eastern Desert. *Water* **2023**, *15*, 971. https://doi.org/10.3390/w15050971

Academic Editors: Imokhai Theophilus Tenebe and Domenico Cicchella

Received: 21 February 2023
Revised: 26 February 2023
Accepted: 27 February 2023
Published: 3 March 2023

1. Introduction

In semiarid, arid climate, and coastal regions, communities rely mostly on groundwater as their source of drinking water [1,2]. Generally, groundwater is a finite natural resource of fresh water on Earth. The daily drinking and domestic water need of about 33% of the world's population are satisfied by groundwater [1,3,4]. Almost 2.1 billion people worldwide don't have access to safe drinking water [5]. Groundwater quality has received more attention in recent decades as a result of growing urbanization, intense agricultural activities, reclamation of new agricultural lands, and industrialization, which, in addition to an increasing population, pose an increased risk of groundwater and soil pollution [1,6–8]. Groundwater quality is affected by rainfall rates, the nature of recharge water, and surface-water resources, in addition to hydrogeochemical processes occurring in an aquifer [2,9]. Groundwater hydrogeochemistry is affected by geochemical natures for oxidation-reduction, ionic exchange, mineral weathering, authigenic mineral dissolution, and precipitation, in addition to groundwater abodes of time [1,9,10].

Intrusion processes are considered a major factor in influencing groundwater chemistry [11–14]. Salinization presents an outstanding problem threatening groundwater resources in coastal basins of arid and semi−arid climates [15]. Salinization process is directly related to the effects of many factors (e.g., marine intrusion, climatic conditions, aquifers characteristics, and anthropogenic activities) that can exacerbate or decrease these problems [14,16–18]. The origin of the salinization of groundwater in coastal areas has been investigated by implementing many approaches [19,20]. The balance of groundwater discharge and recharge clarify how far saline water infiltrates, the aquifer−piezometric level, and the distance between the groundwater aquifer and the saline water sources, in addition to its geological structure [21]. Accordingly, it is indispensable to grasp the chemical processes leading to the salinity of coastal groundwater aquifers to create a valuable plan for the sustainable management of all vulnerable groundwater resources [20]. In addition, continuous groundwater extraction leads to a lowering in the groundwater table leading to increases in domestic water seepage and seawater intrusion, especially in the coastal areas [22,23].

A porous medium and permeable matrix are two terms that are frequently used to describe sandstone aquifers. Such aquifers, however, may be heavily fractured if the characteristics of the rock and the time of the deformation encourage brittle failure and crack opening [24–27]. Fractured carbonate aquifers are a common source of water supply [28,29]. In most sedimentary rocks, rapid solute transport of contaminants occurs within bedding plane fractures and joints rather than through intergranular porosity [24,28]. Groundwater availability in igneous and metamorphic rock is extremely rare and is directly controlled by geological processes (weathering and fracturing) [1]. The current consciousness is focused on preserving existing constrained drinking water resources in the context of rising demand [13,30]. So, continuous monitoring and assessment of both quantity and quality of existing limited water sources are extremely vital [30,31]. Effective continuous monitoring surveys are needed for discovering the mechanisms which control groundwater quality in coastal provinces. Considering the factors that contribute to groundwater quality deterioration will be critical for future management plans [32]. Many health and agriculture organizations establish standards used extensively comparing various components (major ions concentrations) and pollutants (e.g., potentially toxic elements and radionuclides) for irrigation and drinking water [5,33,34]. In addition, a lot of water quality indices and statistical analyses equally be utilized to reduce reliable data for assessing water-quality appropriateness within single and multiple−digit tools [32,35].

Nowadays, Egypt faces serious water shortage challenges coinciding with high population growth and climate change. Many focuses have recently been placed on agricultural developments and sustainable growth in the Egyptian deserts, and how they must meet most crucial objectives of the Governmental Strategy for Sustainable Agricultural Development 2030 [36,37]. Therefore, the resources of groundwater in Egypt play critical roles in satisfying part of water requirements for different uses, especially in coastal and arid provinces [8,37,38]. This study's main objectives are (i) to recognize the mechanisms governing groundwater chemistry in Egypt's Central Eastern Desert (CED), (ii) to evaluate the adequacy of groundwater quality for various uses.

2. Materials and Methods

2.1. The Study Area

The study area lies between 25°50′–27°00′ N and 32°30′–34°25′ E in the CED, Egypt (Figure 1). The geologic setting of the CED is represented mainly by the Precambrian basement rocks (igneous and metamorphic), which form the Red Sea Hills, and it is bordered on the eastern side along the coast by a narrow strip of sedimentary succession, and on the western side, it is also bounded by sedimentary rocks that extend up to Nile Valley. Basement rocks includes gneisses, serpentinites, metagabbros, and metabasalts that have been intruded by volcanic and younger granitic rocks and are overlain by molasse sediments (metasediments) [39,40]. After a long period of tectonic activity, the sedimentary rocks were deposited inside the subsidence blocks (basins) of the basement rocks near the coastal plain and on the western side until the Nile Valley [41]. The Nubian Sandstone Formation of Lower Cretaceous age, the oldest rock unit in the sedimentary succession, is distinguished by fine- to coarse-grained sandstone with some shale and clay beds [42]. It is overlain by Late Cretaceous to Eocene shale and carbonate marine deposits [43]. The post-rift sediments are comprised of alluvial and alluvial fans deposits. The groundwater in the Central Eastern Desert is available at different depths and lithology. Based on the lithology, three main regional aquifer systems are distinguished; they are the Eocene Carbonate, the Nubian Sandstone (Upper Cretaceous sandstone, shale, and marl layers), and the underlying Fractured Basement aquifer (Precambrian igneous and metamorphic rocks).

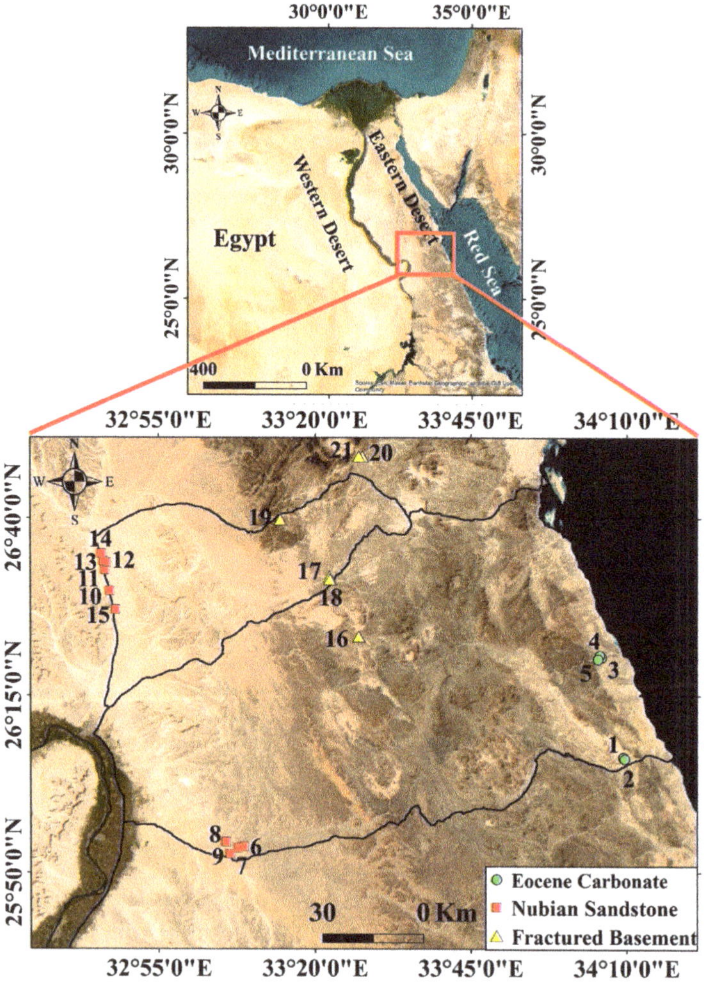

Figure 1. The study area and sampling site's locations.

2.2. Sampling and Analyses

Twenty-one groundwater samples in all were collected in 2019–2020 from the wells located throughout the Golden Triangle region of the CED (Figure 1). Samples were taken from the three mean aquifers in the studied area Eocene Carbonate (EOC), Nubian Sandstone (NS), and Fractured Basement (FB) aquifers. To avoid the effects of static water, water was flushed for about 10 min before collecting samples. 1 Litre of water was collected in new and rinsed two polyethylene bottles for each sample. One bottle was filled with only fresh well water, while the other was acidified with HNO_3 to a pH < 2 in order to reduce trace elements adsorption to bottle walls and biological activity. Titration with H_2SO_4 (0.01 N) was used to determine the concentrations of carbonate and bicarbonate. An ion chromatography system (Dionex, ICS-1100, Thermo Fisher Scientific Inc., Waltham, MA, USA) was used to measure Ca^{2+}, Mg^{2+}, Na^{+}, K^{+}, SO_4^{2-}, and Cl^{-}. The accuracy and reliability of the chemical analysis were examined by the Electrical Balance (EB%) between the cations and anions [44]. Trace elements (Al, Cd, Cr, Cu, Co, Fe, Mn, Mo, Ni, Pb, V, and Zn) were measured using inductively coupled plasma mass spectrometry (ICP, POEMSIII, thermo Jarrell elemental company, Waltham, MA, USA) with standard solutions containing 1000 mg/L (Merck). The major ions and trace elements were analyzed in the Desert Research Centre's hydro-geochemistry laboratories in Cairo, Egypt.

2.3. *Groundwater Quality*

The geochemical characteristic of water is an essential component in the process of evaluation of water quality. The current study attempted to determine the suitability of existing groundwater for various purposes. Chloride content, Electrical Conductivity (EC) [45], Soluble Sodium Percentage (SSP) [46], Sodium Adsorption Ratio (SAR) [47], Residual Sodium Carbonate (RSC) [47], Magnesium Adsorption Ratio (MAR) [47], TDS [48], Total Hardness (TH) [49], and Corrosivity Ratio (CR) [50], and toxic metal content are used to assess its suitability for irrigation, domestic uses, and drinking. Table S1 summarizes the formulae used to calculate the various indicators (in Supplementary Materials).

2.4. *Data Treatment*

The map of sampling locations in the study area was created using Google Earth and Arc-Map software (Arc-Map 10.3). Descriptive statistics and box and whisker plot charts (boxplots) have been presented by OriginLab (version OriginPro 2021). Bivariate X–Y plots (bivariate diagrams) between major ion have been presented by Excel (Microsoft Office 365) to precisely determine the origin of each element and separate different mechanisms that contribute to groundwater evolution. OriginLab, also, was used to calculate and present multivariate statistical analyses in the form of Pearson Correlation Coefficient (PCC) and R mode Hierarchical Cluster Analysis (HCA), to elucidate the interrelation between major ions and their effect in groundwater chemistry. Piper diagram is presented using OriginLab and HFE-Diagram is created using the spreadsheet software package, Microsoft Excel (Excel Macro) provided by Giménez-Forcada and Sánchez San Román [51].

3. Results and Discussion

3.1. *General Hydrogeochemistry*

The physicochemical parameters (pH, EC, and TDS) and the major ion concentrations of investigated groundwater samples within the CED representing different aquifer types (EOC, NS, and FB) are presented in Table 1 and Figure 2. The EB% in this study was within 5%, indicating that the chemical analysis was accurate [4,6,13]. From Figure 2 it is evident that the groundwater in the investigated wells samples is characterized by a narrow range of pH values from slightly acidic to neutral (varied between 6 to 7.2). EC ranged from 4270.0–7980.0, 2730.0–12,550.0, and 690.0–5460.0 in the EOC, NS, and FB aquifers, respectively. These values exceeded the recommended EC limit for natural water (1300 uS/cm at 25 °C; WHO [5]), except for FB aquifer (samples 16, 17, and 20). TDS, like EC, varied greatly, with the higher the value recorded in the NS aquifer and the least values recorded in the FB aquifer. The marked differences in the EC and TDS values reflect significant variation in the hydrochemical features in the EOC, NS, and FB aquifers. The concentration of the major ions in the groundwater samples varies significantly; Ca^{2+} (20.0–888.0), Mg^{2+} 9.11–214.8), Na^{+} (105.0–1400.0), K^{+} (2.0–17.0), HCO_3^{-} (108.6–791.8), SO_4^{2-} (50.0–1900.0), and Cl^{-} (55.0–2900.0 mg/L). Data suggested that the chemistry of groundwater was heterogeneous and governed by a variety of mechanisms, such as evaporation, water-rock interaction mixing processes, and saltwater intrusions [19,38,52]. In a recent study by Sherif et al. [53], groundwater mixing between different aquifers in Egypt's Eastern Desert was proven utilizing ^{36}Cl abundances and $^{37}Cl/^{35}Cl$, $^{2}H/^{1}H$, and $^{18}O/^{16}O$ isotope ratios. In most cases, the three aquifers are dominated by Na^{+}, followed by Ca^{2+}, Mg^{2+}, K^{+}, and Cl^{-}, followed by SO_4^{2-} and HCO_3^{-}. Some fluctuations are observed between Ca^{2+} and Na^{+} in NS and FB aquifers. The anionic distribution in the FB aquifer shows complex fluctuation between HCO_3^{-}, SO_4^{2-}, and Cl^{-}.

Table 1. Physicochemical parameters and major ions concentrations (mg/L) in EOC, NS, and FB aquifers.

NO.	Aquifer	pH	EC uS/cm	TDS mg/L	Ca^{2+} mg/L	Mg^{2+} mg/L	Na^{+} mg/L	K^{+} mg/L	HCO_3^{-} mg/L	SO_4^{2-} mg/L	Cl^{-} mg/L	EB %
1		6.4	7980.0	4734.0	557.4	141.5	860.0	14.0	207.2	1251.9	1800.0	−1.88
2		6.4	7330.0	4317.0	530.0	214.8	640.0	15.0	134.2	900.0	1950.0	−2.40
3	EOC	6.1	5340.0	3188.0	299.5	121.3	600.0	9.0	109.8	1028.1	1075.0	−2.16
4		6.1	4320.0	2523.0	310.8	116.3	380.0	7.0	195.2	736.8	875.0	−1.68
5		6.3	4270.0	2496.0	267.1	176.5	300.0	17.0	268.4	926.3	675.0	−1.68
Min.	EOC	6.1	4270.0	2496.0	267.1	116.3	300.0	7.0	109.8	736.8	675.0	−2.4
Max.	(N = 5)	6.4	7980.0	4734.0	557.4	214.8	860.0	17.0	268.4	1251.9	1950.0	−1.7
6		6.8	3770.0	2219.0	208.0	65.7	480.0	8.0	316.4	475.5	800.0	−1.03
7		6.9	12,550.0	7424.0	888.0	202.2	1400.0	13.0	194.4	1900.0	2900.0	−0.93
8		7.2	3810.0	2229.0	91.5	35.4	680.0	7.0	316.4	407.7	825.0	0.41
9		6.8	4730.0	2893.0	75.5	17.7	920.0	7.0	292.8	826.8	900.0	−2.13
10	NS	6.0	9640.0	5699.0	609.9	165.8	1100.0	6.0	122.0	1556.6	2200.0	−2.30
11		6.2	4270.0	2521.0	358.1	85.9	360.0	5.0	134.2	794.7	850.0	−2.36
12		6.4	3010.0	1768.0	114.9	100.5	340.0	3.0	146.4	586.1	550.0	−2.14
13		6.2	3710.0	2209.0	279.0	75.9	360.0	9.0	134.2	688.1	730.0	−1.43
14		6.6	2730.0	1609.0	95.9	35.2	420.0	4.0	183.0	382.7	580.0	−2.38
15		6.8	3340.0	2078.0	45.6	30.3	640.0	5.0	292.4	698.8	500.0	−1.08
Min.	NS	6.0	2730.0	1609.0	45.6	17.7	340.0	3.0	122.0	382.7	500.0	−2.4
Max.	(N = 10)	7.2	12,550.0	7424.0	888.0	202.2	1400.0	13.0	316.4	1900.0	2900.0	0.4
16		7.1	1057.0	642.0	95.7	17.7	105.0	2.0	268.4	175.3	112.5	−1.67
17		7.0	690.0	405.0	20.0	9.1	120.0	2.0	243.6	65.0	55.0	0.84
18		6.8	3650.0	2255.0	235.0	20.0	500.0	5.0	280.7	804.2	550.0	−2.21
19	FB	6.3	5460.0	3041.0	700.0	96.0	240.0	5.0	108.6	259.8	1650.0	−0.22
20		6.9	1128.0	638.0	28.0	13.0	200.0	2.0	390.0	50.0	137.5	−0.43
21		7.0	2595.0	1600.0	125.4	25.3	390.0	13.0	791.8	90.6	380.0	0.12
Min.	FB	6.3	690.0	405.0	20.0	9.1	105.0	2.0	108.6	50.0	55.0	−2.2
Max.	(N = 6)	7.1	5460.0	3041.0	700.0	96.0	500.0	13.0	791.8	804.2	1650.0	0.8
Min.	All Samples	6.0	690.0	405.0	20.0	9.1	105.0	2.0	108.6	50.0	55.0	−2.4
Max.	(N = 21)	7.2	12,550.0	7424.0	888.0	214.8	1400.0	17.0	791.8	1900.0	2900.0	0.8
Mean		6.6	4541.9	2689.9	282.6	84.1	525.5	7.5	244.3	695.5	956.9	−1.4

Table 2 displays R values of the PCC between physicochemical parameters and major ions of the investigated groundwater samples. The correlation matrix shows a significant positive correlation of TDS with Cl^{-} (R = 0.937), SO_4^{2-} (R = 0.915), Na^{+} (R = 0.880), Ca^{2+} (R = 0.879), Mg^{2+} (R = 0.794), and K^{+} (R = 0.556). Moreover, it was observed strong positive correlation of Cl^{-} with Ca^{2+} (R = 0.930), SO_4^{2-} (R = 0.817), Mg^{2+} (R = 0.804), Na^{+} (R = 0.797), and K^{+} (R = 0.507). likewise, SO_4^{2-} with Na^{+} (R = 0.848), Mg^{2+} (R = 0.755), Ca^{2+} (R = 0.716), and K^{+} (R = 0.504). Ca^{2+} exhibits a positive correlation with Mg^{2+} (R = 0.798), which is most likely owing to the dissolution of dolomite and calcite. HCO_3^{-} exhibits a negative and weak correlation with the other ions as well as TDS. This means that the majority of cations are strongly associated with Cl^{-} and SO_4^{2-}. It may indicate the impact of seawater intrusion on groundwater because the major anions in seawater are Cl^{-} and SO_4^{2-}. The R-Mode (variables) dendrogram created using HCA (Figure 3), supports the inferred relationship between TDS and the major ions.

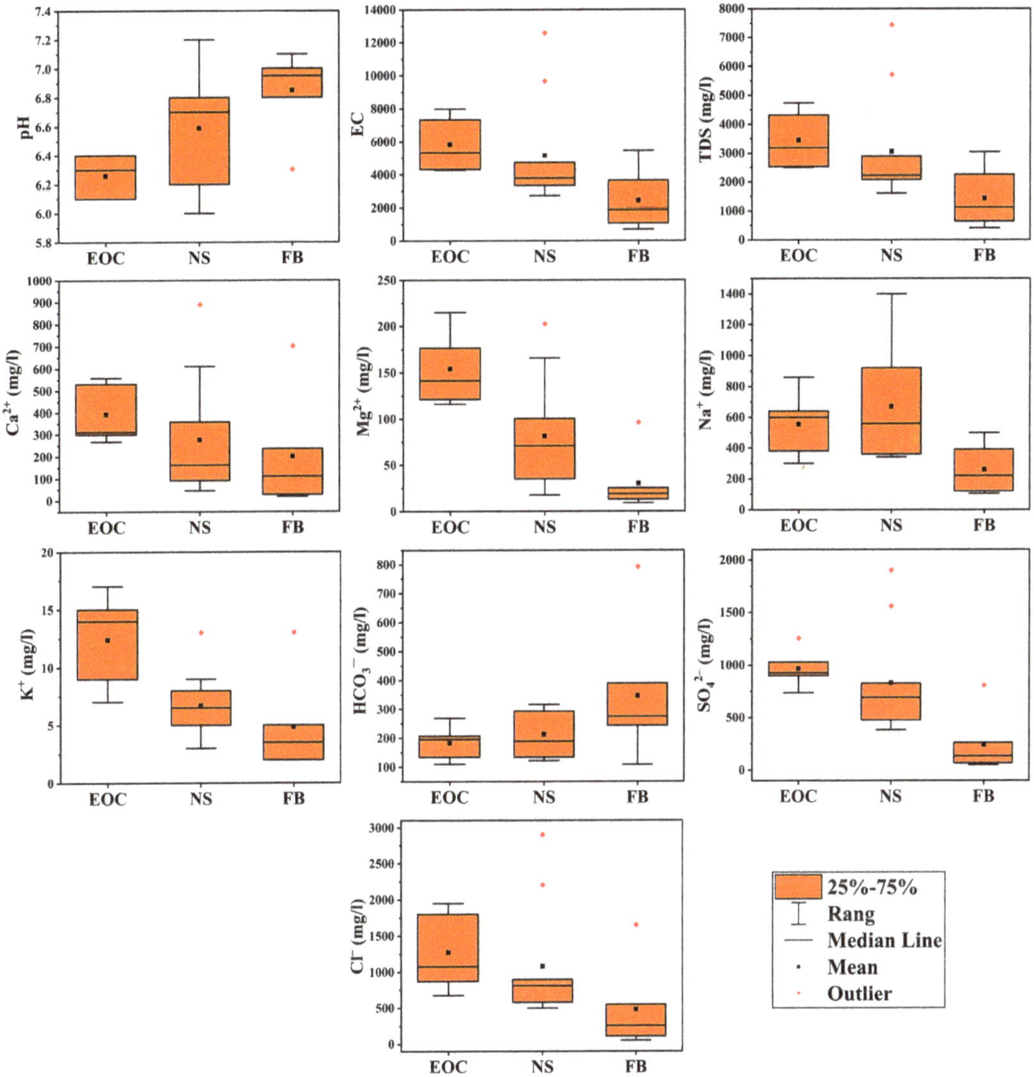

Figure 2. Boxplots of physicochemical parameters and major ions in EOC, NS, and FB aquifers.

Table 2. PCC for pH, TDS (mg/L), and major ions (mg/L) in EOC, NS, FB aquifers (n = 21).

	TDS	Ca^{2+}	Mg^{2+}	Na^+	K^+	HCO_3^-	SO_4^{2-}	Cl^-
pH	−0.354	−0.455	−0.603	−0.065	−0.217	0.614	−0.439	−0.367
TDS		0.879	0.794	0.880	0.556	−0.355	0.915	0.973
Ca^{2+}			0.798	0.572	0.473	−0.437	0.716	0.930
Mg^{2+}				0.493	0.678	−0.473	0.755	0.804
Na^+					0.413	−0.151	0.848	0.797
K^+						0.137	0.504	0.507
HCO_3^-							−0.441	−0.416
SO_4^{2-}								0.817

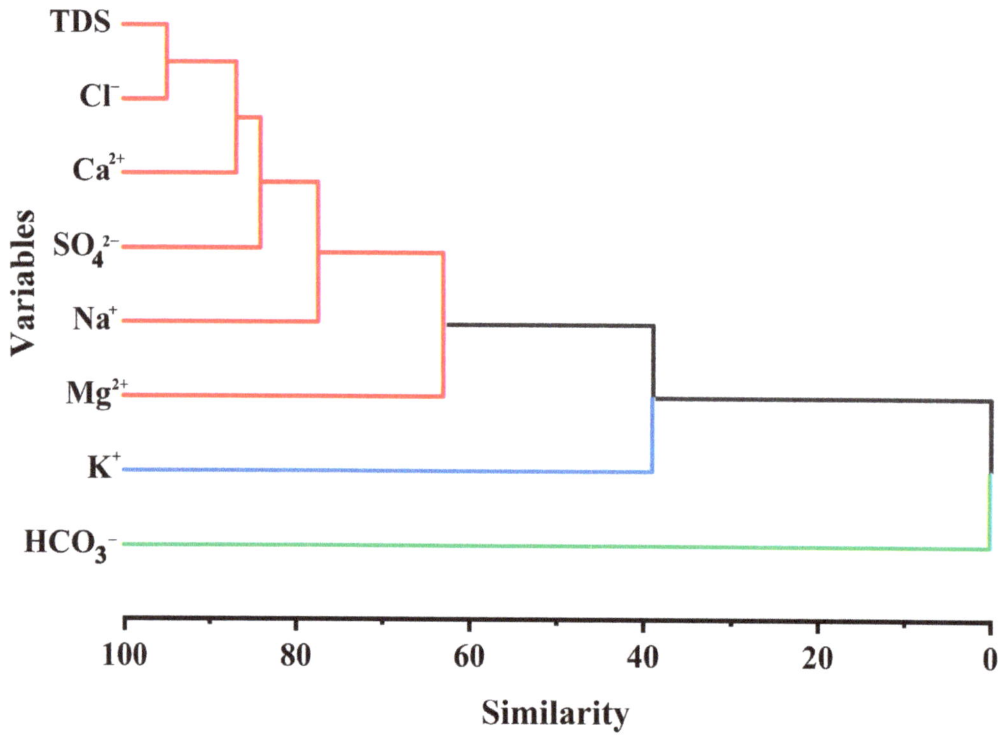

Figure 3. CA dendrogram in R-mode (Variables) for TDS, and major ions in EOC, NS, FB aquifers (*n* = 21).

3.2. *Hydrochemical Facies*

Many hydrogeochemical arithmetical diagrams may be used for distinguishing between differences and similarities in the composition of groundwater as well as for classifying them into specific chemical categories [54,55]. The hydrochemical components of studied aquifers groundwater were applied on a Piper Diagram (Figure 4) to highlight and clarify the main hydrochemical facies, and characteristics and for illustrating the different processes that control groundwater [52]. The excess of Cl^- type is clear in the anionic triangle for both EOC and NS aquifers while in FB aquifer bicarbonate type was dominant in most samples. Some samples exhibit sodium dominance in the cation's triangle, while others exhibit pole mixed calcium, except in FB (Sample 19) aquifers the calcium was dominant (Figure 4). In the anion's triangle, samples show a chloride predominance with a slight propensity sulfate pole (FB aquifer). The dominant water types in EOC and NS aquifers are Na-Cl type. The FB aquifer recorded many water types; Na-Cl (sample 18), Na-Ca-HCO_3 (samples 17 and 20), Ca-Mg-Cl (sample 16), and Ca-Cl (sample 19). The dominance of Na-Cl water type is confirmed by the HFE-Diagram [56] (Figure 5), indicating that this groundwater may be affected by seawater intrusion and leaching out of residual evaporative salts (i.e., gypsum/anhydrite and halite) during movement [53]. Generally, high values of Cl^-, Na^+, and $SO_4{}^{2-}$ corresponding with seawater intrusion, and Ca^{2+}, Mg^{2+}, and K^+ are not useful when distinguishing between different types of saline water.

The hydrogeochemical processes and hydrogeochemistry of groundwater vary depending on the mineralogy and geochemistry of the aquifer. The chemistry of groundwater is heavily influenced by the mineralogical composition of the aquifer through which it flows [38,57]. Groundwater major ion chemistry and interrelations are effective in determining solute sources and describing groundwater evolution [58,59]. The Na^+ vs. Cl^- relationship has frequently been serves to identify the processes that cause water salinity in arid and semi-arid regions [38,60]. The relationship between these two ions (Figure 6a) shows that most of the samples are near the 1:1 Line (halite dissolution line). Equal amounts of Na^+ and Cl^- are released into the solution when halite dissolves in water [61]. The recorded high Cl^- relative to Na^+ ion indicates the effect of saltwater intrusion [62], especially in EOC aquifer samples (Figures 5 and 6a). The decreasing trend in the concentration of Cl^- relative to the concentration of Na^+ in some NS aquifer samples likely reflects the cation exchange processes and weathering of the silicate minerals [62]. The additional Na^+ originates from the cation exchange in the clay minerals that present as shale intercalation of the NS aquifer leading to the adsorption of Ca^{2+} and the simultaneous releasing of Na^+ ions [38]; these expected cation exchange confirmed by plotting Na^+ vs. Ca^{2+} (Figure 6b), since the sampling points lie far below the uniline of Na^+-Ca^{2+}. Regarding FB aquifer samples, the excess sodium may be ascribed to the dissolution of Na-Ca-Al-Silicates (albite-plagioclase).

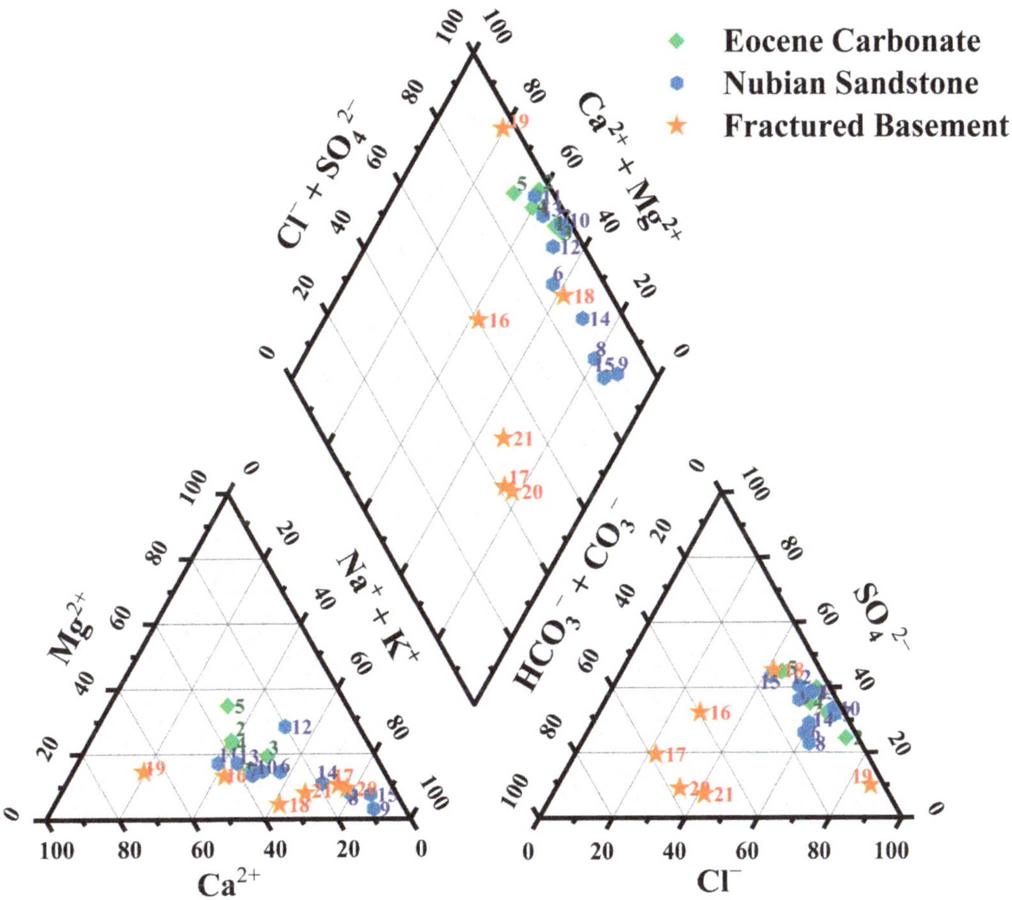

Figure 4. Piper plot of groundwater samples in EOC, NS, and FB aquifers.

The dissolution of carbonate minerals like calcite ($CaCO_3$) and dolomite ($CaMg(CO_3)_2$) which are responsible for enriching of Ca^{2+} and Mg^{2+} was explored by (Ca^{2+} + Mg^{2+}) vs. (HCO_3^- + SO_4^{2-}) scatter diagram. Most of the NS and FB sample points lie below the equiline with few along it and above (Figure 6c), which indicates the predominance of silicate weathering. On the other hand, EOC sampling points lie above that line that indicates the predominance of carbonate weathering. Given the abundance of carbonate and dolomite rocks in the EOC aquifer, it is likely that the groundwater will become much more Ca^{2+} and Mg^{2+} enriched as a result of the dissolution of these minerals. The effects of silicate weathering and carbonate mineral dissolution on groundwater hydrochemistry are well documented [27,37,38]. The relation between (Ca^{2+} + Mg^{2+})-(HCO_3^- + SO_4^{2-}) vs. Na^+ + K^+-Cl^- [61] referred to the effect of reverse ion exchange. Enrichment of Na^+ and K^+ when compared to the Ca^{2+} and Mg^{2+} in the FB aquifer and some samples of the NS aquifer is noted in Figure 6d (second quadrant negative ordinate and positive abscissa) [37]. As a result, the cation exchange sites preferentially absorb Ca^{2+} and Mg^{2+} while releasing Na^+ and K^+ (direct cationic exchange). In EOC aquifer samples, the relative depletion of Na^+ and K^+ regarding Ca^{2+} and Mg^{2+} is noted, which suggests reverse ion exchange (fourth quadrant positive ordinate and negative abscissa).

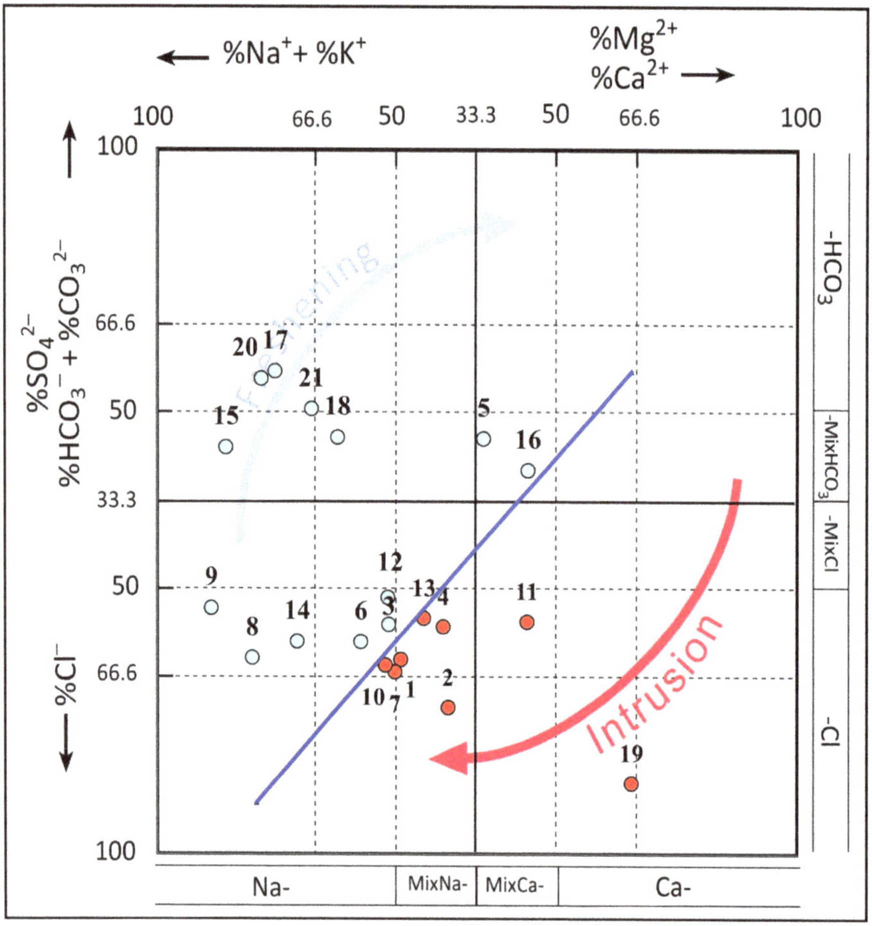

Figure 5. HFE-Diagram of the studied groundwater samples.

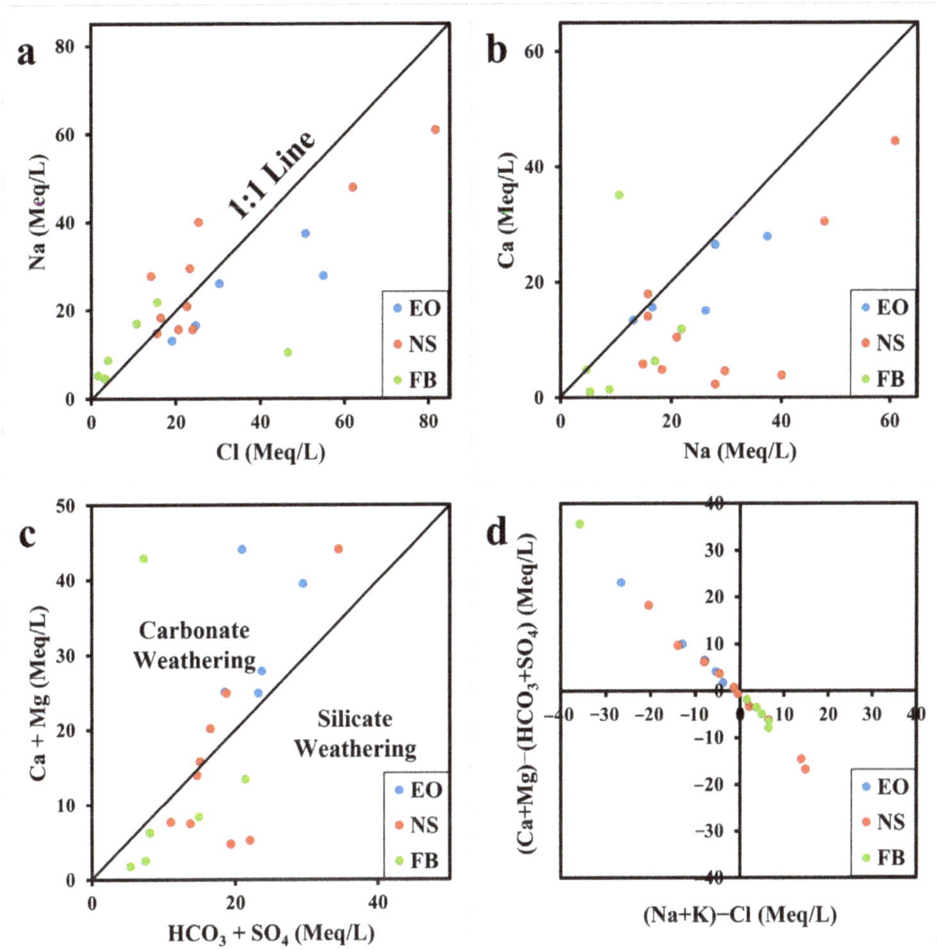

Figure 6. Plots of the EOC, NS, and FB groundwater samples on scatter diagrams: (**a**) Na^{+} vs. Cl^{-}; (**b**) Na^{+} vs. Ca^{2+}; (**c**) (Ca^{2+} + Mg^{2+}) vs. (HCO_3^{-} + SO_4^{2-}); (**d**) (Ca^{2+} + Mg^{2+})-(HCO_3^{-} + SO_4^{2-}) vs. Na^{+} + K^{+}-Cl^{-}.

3.3. Trace Elements

Trace element concentrations clarify highly significant differences between the studied aquifers (Table 3). Overall, variations in the distribution of trace and heavy metals in groundwater are caused by a variety of factors such as cation exchange, evaporation, the leaching and disintegration of marine water seepage, climate, bedrock type, pH, redox potential, and mixing capacity [37,44]. Mostly, the monitoring results of the studied aquifers indicate that the rock type has a considerable impact on trace element distributions. EOC aquifer samples show the recorded maximum concentrations of Cd, Fe, Mn, Ni, V, and Zn. Particularly these may be owing to the anthropogenic impact of industrial and tourism activities near the coastline, reflecting the effects of water flows on trace elements allocation in these aquifers [63]. Both Fe and Mn are the most distributed metallic elements in Earth crust. Fe concentrations in groundwater is mostly regulated by many factors including flow rate, redox agents, dissolved organic matter, pH, leaching and disintegration of marine water seepage, and bedrock type [16,52]. The dissolution of Fe-bearing minerals commonly found in aquifer sediments is one of its main sources in groundwater. As a result of industrialization progresses, the Ni presence in water systems increased, and many Ni compounds were introduced for industrial products [64].

Table 3. Trace element concentrations (mg/L) in EOC, NS, and FB aquifers.

NO.	Aquifer	Al	Cd	Co	Cr	Cu	Fe	Mn	Mo	Ni	Pb	V	Zn
1	EOC	0.546	<0.0006	0.018	<0.01	<0.006	7.074	0.793	<0.001	0.070	<0.008	0.074	0.063
2		0.052	<0.0006	0.019	<0.01	<0.006	0.608	0.140	0.009	0.039	<0.008	<0.01	0.015
3		0.032	0.004	<0.001	<0.01	<0.006	<0.02	0.010	0.014	0.082	<0.008	<0.01	0.094
4		0.181	0.026	0.009	<0.01	<0.006	<0.02	<0.004	<0.001	<0.002	<0.008	<0.01	0.114
5		<0.01	<0.0006	<0.001	<0.01	<0.006	1.929	0.884	0.113	0.057	<0.008	<0.01	0.100
6	NS	<0.01	<0.0006	0.025	<0.01	<0.006	1.781	0.110	<0.001	0.073	<0.008	<0.01	0.017
8		<0.01	<0.0006	0.024	0.010	<0.006	<0.02	0.136	0.078	0.026	<0.008	<0.01	0.049
9		0.136	<0.0006	<0.001	<0.01	<0.006	0.696	0.168	0.028	0.011	<0.008	<0.01	<0.0005
10		0.354	0.013	0.073	<0.01	<0.006	0.065	0.007	<0.001	0.002	<0.008	<0.01	0.037
11		1.376	<0.0006	<0.001	<0.01	<0.006	4.384	0.091	<0.001	0.018	<0.008	<0.01	0.089
14		0.087	<0.0006	<0.001	<0.01	<0.006	<0.02	0.121	<0.001	0.022	<0.008	<0.01	<0.0005
15		1.078	<0.0006	0.032	<0.01	<0.006	0.454	0.093	<0.001	<0.002	<0.008	<0.01	0.099
17	FB	1.267	<0.006	<0.001	<0.01	<0.006	1.162	0.090	0.089	<0.002	<0.008	0.032	0.029
18		0.155	<0.006	<0.001	<0.01	0.006	<0.02	<0.004	0.126	0.060	<0.008	0.024	0.011
19		0.487	<0.006	<0.001	<0.01	<0.006	0.857	0.091	0.186	0.020	<0.008	<0.01	0.050
20		<0.01	0.009	0.010	<0.01	<0.006	0.281	0.046	0.203	0.028	<0.008	0.020	0.023
21		0.628	0.001	<0.001	<0.01	<0.006	0.637	0.197	0.015	<0.002	<0.008	<0.01	0.012
Min.	All Samples (N = 17)	0.032	0.001	0.009	0.010	0.006	0.065	0.007	0.009	0.002	-	0.020	0.011
Max.		1.376	0.026	0.073	0.010	0.006	7.074	0.884	0.203	0.082	-	0.074	0.114
Mean		0.491	0.010	0.026	-	-	1.661	0.198	0.086	0.039	-	0.038	0.053
WHO [5]		0.9	0.003	-	0.05	2	-	0.4	-	0.07	0.01	-	3
ESDW [34]		-	0.003	-	0.05	2	0.3	0.4	-	0.02	0.01	-	3
USEPA [65]		-	0.003	-	0.05	-	0.3	-	-	-	-	0.07	3
FAO [33]		5	0.01	-	0.1	0.2	5	0.2	-	0.2	-	0.1	2
CCME [66]		5	0.01	-	-	-	5	0.2	-	0.2	0.2	0.1	-

3.4. Assessmnet of Groundwater Quality for Different Purposes

3.4.1. Assessment for Irrigation Use

Irrigation water quality refers to its suitability for use in the irrigation of different crops. Under good soil and water management practices, good quality water has the potential to result in maximum yield [67]. The concentration and composition of dissolved constituents in water determine their quality and viability for irrigation. The results of the calculated indicators used for assessing the quality of groundwater samples for irrigation use are presented in Table 4. The studied groundwater samples have pH values within the permissible limit (6.5–8.8; FAO [33]). The measured values of EC indicate that the groundwater of EOC and NS and 50% of FB aquifers samples are unsuitable for irrigation. The main effect of high EC water on crop productivity is an increase in the osmotic pressure of the nutrient solution in the soil, which can result in reduced water uptake and nutrient deficiencies [33,68]. The combined SSP, SAR, RSC, and MAR computed results showed that there is no hazard with sodium and magnesium for irrigation use, except for a few samples which record high values for these indicators. The sodium hazards of irrigation water are significantly related to the degree to which the sodium is adsorbed by the soil. If the irrigation water contains an excessive concentration of Na^+ but little Ca^{2+}, the cation-exchange intricate could become saturated with sodium and destroy the soil structure [33,62]. Water infiltration is also influenced by SAR. Although Cl^- is crucial to crops at low levels, it can cause toxicity (leaf burns or leaf tissue deaths) in sensitive crops at elevated concentrations [69,70]. Only 50% of FB aquifer groundwater samples fall below the recommended chloride limit (140 mg/L; FAO [33]; Bouselsal and Saibi [70]) (Table 1), and the rest of the samples fall above this limit and can cause Cl^- toxicity to corps if used in irrigation.

Table 4. Groundwater quality indicators for irrigation and domestic uses.

NO.	Aquifer	SSP		SAR		RSC		MAR		TH		CR	
1	EOC	48.9	Safe	8.4	Excellent	−36.07	Excellent	29.4	Suitable	1973.7	V. Hard	15.4	Unsafe
2		39.0	Safe	5.9	Excellent	−41.91	Excellent	39.9	Suitable	2205.7	V. Hard	24.0	Unsafe
3		51.4	Safe	7.4	Excellent	−23.12	Excellent	39.9	Suitable	1246.2	V. Hard	18.7	Unsafe
4		40.0	Safe	4.7	Excellent	−21.87	Excellent	38.0	Suitable	1253.6	V. Hard	8.3	Unsafe
5		32.6	Safe	3.5	Excellent	−23.42	Excellent	52.0	Unsuit.	1391.5	V. Hard	5.3	Unsafe
6	NS	57.2	Safe	7.4	Excellent	−10.60	Excellent	34.1	Suitable	789.4	V. Hard	4.3	Unsafe
7		50.1	Safe	11.0	Good	−57.79	Excellent	27.2	Suitable	3049.0	V. Hard	26.1	Unsafe
8		79.9	Unsafe	15.3	Good	−2.29	Excellent	38.8	Suitable	373.9	V. Hard	4.3	Unsafe
9		88.5	Unsafe	24.7	Doubtful	−0.43	Excellent	27.8	Suitable	261.3	Hard	5.8	Unsafe
10		52.1	Safe	10.2	Good	−42.08	Excellent	30.8	Suitable	2204.4	V. Hard	32.0	Unsafe
11		38.7	Safe	4.4	Excellent	−22.75	Excellent	28.2	Suitable	1247.5	V. Hard	12.0	Unsafe
12		51.5	Safe	5.6	Excellent	−11.58	Excellent	58.9	Unsuit.	699.4	V. Hard	7.4	Unsafe
13		44.1	Safe	4.9	Excellent	−17.97	Excellent	30.8	Suitable	1008.5	V. Hard	10.3	Unsafe
14		70.5	Unsafe	9.3	Excellent	−4.68	Excellent	37.5	Suitable	383.9	V. Hard	5.6	Unsafe
15		85.4	Unsafe	18.0	Good	0.03	Excellent	52.2	Unsuit.	238.4	Hard	3.7	Unsafe
16	FB	42.5	Safe	2.6	Excellent	−1.84	Excellent	23.3	Suitable	311.8	V. Hard	0.9	Safe
17		75.1	Unsafe	5.6	Excellent	2.25	Doubtful	42.7	Suitable	87.3	Moderately	0.5	Safe
18		62.0	Unsafe	8.4	Excellent	−8.79	Excellent	12.2	Suitable	669.5	V. Hard	4.3	Unsafe
19		19.8	Safe	2.3	Excellent	−41.09	Excellent	18.4	Suitable	2143.6	V. Hard	22.6	Unsafe
20		78.0	Unsafe	7.8	Excellent	3.93	Unsuitable	43.2	Suitable	123.3	Moderately	0.6	Safe
21		67.4	Unsafe	8.3	Excellent	4.64	Unsuitable	24.9	Suitable	417.2	V. Hard	0.7	Safe

When evaluating an irrigation water supply, the hazardous metal concentrations of the irrigation waters should be carefully examined [69]. A comparison with several international standards and guidelines for irrigation water [33,66] has been done to evaluate trace elements content in the investigated groundwater (Table 3). The results show that there is no problem with trace elements concentration for using this groundwater for irrigation except for some samples which contain elevated concentrations of Cd, Fe, and Mn.

3.4.2. Assessment for Domestic Use

High TDS values may be associated with excessive corrosion and scaling in pipes, fittings, and household appliances. Comparison between the TDS values of the investigated groundwater with the classification proposed by Bruvold and Daniels [48], disclosed that the majority of the samples can be classified as unacceptable for domestic use (Tables 1 and S1). According to Total Hardness (TH) values, this water is very hard and hard (Table 4). The presence of Ca^{2+} and Mg^{2+} in the aquifer rocks contributes significantly to the TH of groundwater. Hard water is not a health risk, but it can be infuriating in the home. Hard water is unfit for domestic use, and its industrial applications have been limited due to the degree of hardness of the water, which causes the scaling of pots, boilers, and irrigation pipelines [5]. Groundwater in the CED is unfit for domestic use on the basis of TH values.

3.4.3. Assessment for Drinking Use

Although the pH values of this groundwater are neutral to slightly acidic, and it is well within the acceptable WHO [5] range, the high TDS values render it invalid for drinking. Some of the major ions, including Na^+, Ca^{2+}, Mg^{2+}, SO_4^{2-}, and Cl^-, have concentrations that are higher than that that are recommended for taste and aesthetics [5]. Furthermore, some samples recorded Al, Cd, Fe, Mn, and Ni concentrations exceeding the permissible limits in drinking water [5,34,65] (Table 3).

4. Conclusions

Findings of this study indicated that there were significant differences in the chemical composition between the studied three aquifers. The highest salinity values were recorded in the NS aquifer and the lowest values were recorded in the FB aquifer. Sodium, calcium, sulfate, and chloride concentrations are the main contributors to the elevated salinity of the groundwater in these aquifers. The cation contents are strongly associated with Cl^- and SO_4^{2-} through Na-Cl, Ca-Cl, and Na-SO_4 water types. The salinity of groundwater from the EOC, NS, and FB aquifers of the CED of Egypt is being regulated by two main factors. Seawater intrusion is the most important factor. The second factor is the water-rock interactions through direct and reverse cationic exchange and carbonate and silicate weathering. The groundwater of these aquifers is not safe and unsuitable for all purposes. It's very high salinity values prevent its use for irrigation, domestic, and drinking, with an exception for some FB samples. Groundwater resources in the CED are predicted to become more salinized as a consequence of increasing climate change effects, hence it is critical to prevent saltwater intrusion. These necessitate mitigation measures and adaptation strategies, which include actions such as regulating groundwater exploitation, monitoring groundwater quality, and better capacity to buffer heavy rainfall to diminish the effects of climate change and protect ecosystems from potential negative consequences.

Supplementary Materials: The following supporting information can be downloaded at: https://www.mdpi.com/article/10.3390/w15050971/s1, Table S1: Summary of the formulae used to calculate the water quality indicators.

Author Contributions: Conceptualization, A.S., A.G., A.A. and H.I.F.; methodology, A.S., A.G. and H.I.F.; software, A.S., A.G. and H.I.F.; validation, A.S., A.G., A.A., H.A. and H.I.F.; formal analysis, A.S. and A.G.; investigation, A.S., A.G. and H.I.F.; resources, A.S.; data curation, A.S., A.G., A.A., H.A and H.I.F.; writing—original draft preparation, A.S., A.G. and H.I.F.; writing—review and editing A.S., A.G., A.A. and H.A.; supervision, A.S., A.G. and H.A.; project administration, A.S.; funding acquisition, A.S. All authors have read and agreed to the published version of the manuscript.

Funding: This paper is part of the work supported by the Science, Technology & Innovation Funding Authority (STDF), Egypt, Under Grant Number: 30116.

Institutional Review Board Statement: Not applicable.

Informed Consent Statement: Not applicable.

Data Availability Statement: Not applicable.

Acknowledgments: This paper is part of the work supported by the Science, Technology & Innovation Funding Authority (STDF), Egypt, Under Grant Number: 30116. The authors thank El-Shlemy E. (Ain Shams University, Cairo, Egypt) for her valuable assistant.

Conflicts of Interest: The authors declare no conflict of interest. The funders had no role in the design of the study; in the collection, analyses, or interpretation of data; in the writing of the manu-script; or in the decision to publish the results.

References

1. Etikala, B.; Golla, V.; Li, P.; Renati, S. Deciphering groundwater potential zones using MIF technique and GIS: A study from Tirupati area, Chittoor District, Andhra Pradesh, India. *HydroResearch* **2019**, *1*, 1–7. [CrossRef]
2. Ma, Q.; Ge, W.; Tian, F. Geochemical Characteristics and Controlling Factors of Chemical Composition of Groundwater in a Part of the Nanchang Section of Ganfu Plain. *Sustainability* **2022**, *14*, 7976. [CrossRef]
3. Saana, S.; Martin, B.; Fosu, S.A.; Sebiawu, G.E.; Jackson, N.; Karikari, T. Assessment of the quality of groundwater for drinking purposes in the Upper West and Northern regions of Ghana. *SpringerPlus* **2016**, *5*, 2001. [CrossRef]
4. Iqbal, J.; Su, C.; Rashid, A.; Yang, N.; Baloch, M.Y.J.; Talpur, S.A.; Ullah, Z.; Rahman, G.; Rahman, N.U.; Earjh; et al. Hydrogeochemical Assessment of Groundwater and Suitability Analysis for Domestic and Agricultural Utility in Southern Punjab, Pakistan. *Water* **2021**, *13*, 3589. [CrossRef]
5. WHO World Health Organization. *Guidelines for Drinking-Water Quality: First Addendum to the Fourth Edition 2017*; WHO: Geneva, Switzerland, 2017. [CrossRef]

6. Mansouri, Z.; Leghrieb, Y.; Kouadri, S.; Al-Ansari, N.; Najm, H.M.; Mashaan, N.S.; Eldirderi, M.M.A.; Khedher, K.M. Hydro-Geochemistry and Groundwater Quality Assessment of Ouargla Basin, South of Algeria. *Water* **2022**, *14*, 2441. [CrossRef]
7. Chen, W.; Zhang, Y.; Shi, W.; Cui, Y.; Zhang, Q.; Shi, Y.; Liang, Z. Analysis of Hydrogeochemical Characteristics and Origins of Chromium Contamination in Groundwater at a Site in Xinxiang City, Henan Province. *Appl. Sci.* **2021**, *11*, 11683. [CrossRef]
8. Ramadan, R.S.; Dawood, Y.H.; Yehia, M.M.; Gad, A. Environmental and health impact of current uranium mining activities in southwestern Sinai, Egypt. *Environ. Earth Sci.* **2022**, *81*, 213. [CrossRef]
9. Awad, E.S.; Imran, N.S.; Albayati, M.M.; Snegirev, V.; Sabirova, T.M.; Tretyakova, N.A.; Alsalhy, Q.F.; Al-Furaiji, M.H.; Salih, I.K.; Majdi, H.S. Groundwater Hydrogeochemical and Quality Appraisal for Agriculture Irrigation in Greenbelt Area, Iraq. *Environments* **2022**, *9*, 43. [CrossRef]
10. Kumar, L.D.; Dhakate, R.; Guguloth, S.; Srinivas, B. Hydrochemical appraisal of groundwater quality for drinking and agricultural utility in a granitic terrain of Maheshwaram area of Ranga Reddy district, Telnagana State, India. *HydroResearch* **2021**, *4*, 11–23. [CrossRef]
11. Eissa, M.A.; Shawky, H.; Samy, A.; Khalil, M.M.H.; El Malky, M. Geochemical and Isotopic Evidence of Groundwater Salinization Processes in El Dabaa Area, Northwestern Coast, Egypt. *Geosciences* **2018**, *8*, 392. [CrossRef]
12. Bhagat, C.; Puri, M.; Mohapatra, P.K.; Kumar, M. Imprints of seawater intrusion on groundwater quality and evolution in the coastal districts of South Gujarat, India. Case Stud. Chem. *Environ Eng.* **2021**, *3*, 100101. [CrossRef]
13. Dieu, L.P.; Cong-Thi, D.; Segers, T.; Ho, H.H.; Nguyen, F.; Hermans, T. Groundwater Salinization and Freshening Processes in the Luy River Coastal Aquifer, Vietnam. *Water* **2022**, *14*, 2358. [CrossRef]
14. Sarker, M.M.R.; Hermans, T.; Van Camp, M.; Hossain, D.; Islam, M.; Ahmed, N.; Bhuiyan, M.A.Q.; Karim, M.M.; Walraevens, K. Identifying the Major Hydrogeochemical Factors Governing Groundwater Chemistry in the Coastal Aquifers of Southwest Bangladesh Using Statistical Analysis. *Hydrology* **2022**, *9*, 20. [CrossRef]
15. Abu Salem, H.S.; Gemail, K.S.; Junakova, N.; Ibrahim, A.; Nosair, A.M. An Integrated Approach for Deciphering Hydrogeochemical Processes during Seawater Intrusion in Coastal Aquifers. *Water* **2022**, *14*, 1165. [CrossRef]
16. Bahir, M.; Ouazar, D.; Ouhamdouch, S.; Zouari, K. Assessment of groundwater mineralization of alluvial coastal aquifer of essaouira basin (Morocco) using the hydrochemical facies evolution diagram (HFE-Diagram). *Groundwater Sustain. Dev.* **2020**, *11*, 100487. [CrossRef]
17. Ouhamdouch, S.; Bahir, M.; Ouazar, D. Climate change impact assessment on a carbonate aquifer under semi-arid climate; example of the Cenomanian-Turonian aquifer within Meskala-Ouazzi region (Essaouira Basin, Morocco). *Arab. J. Geosci.* **2020**, *13*, 173. [CrossRef]
18. Sefelnasr, A.; Ebraheem, A.A.; Faiz, M.A.; Shi, X.; Alghafli, K.; Baig, F.; Al-Rashed, M.; Alshamsi, D.; Ahamed, M.B.; Sherif, M. Enhancement of Groundwater Recharge from Wadi Al Bih Dam, UAE. *Water* **2022**, *14*, 3448. [CrossRef]
19. Charfi, S.; Zouari, K.; Feki, S.; Mami, E. Study of variation in groundwater quality in a coastal aquifer in north-eastern Tunisia using multivariate factor analysis. *Quat. Int.* **2013**, *302*, 199–209. [CrossRef]
20. El Yousfi, Y.; Himi, M.; El Ouarghi, H.; Elgettafi, M.; Benyoussef, S.; Gueddari, H.; Aqnouy, M.; Salhi, A.; Alitane, A. Hydrogeochemical and statistical approach to characterize groundwater salinity in the Ghiss-Nekkor coastal aquifers in the Al Hoceima province, Morocco. *Groundwater Sustain. Dev.* **2022**, *19*, 100818. [CrossRef]
21. Kumar, B. Integrated hydrogeochemical, isotopic and geomorphological depiction of the groundwater salinization in the aquifer system of Delhi, India. *J. Asian Earth Sci.* **2015**, *111*, 936–947. [CrossRef]
22. Adepelumi, A.A.; Ako, B.D.; Ajayi, T.R.; Afolabi, O.; Omotoso, E.J. Delineation of saltwater intrusion into the freshwater aquifer of Lekki Peninsula, Lagos, Nigeria. *Environ. Geol.* **2009**, *56*, 927–933. [CrossRef]
23. Kura, N.U.; Ramli, M.F.; Sulaiman, W.N.A.; Ibrahim, S.; Aris, A.Z.; Mustapha, A. Evaluation of factors influencing the groundwater chemistry in a small tropical island of Malaysia. *Int. J. Environ. Res. Public Health* **2013**, *10*, 1861–1881. [CrossRef] [PubMed]
24. Medici, G.; West, L.J.; Mountney, N.P. Characterizing flow pathways in a sandstone aquifer: Tectonic vs sedimentary heterogeneities. *J. Contam. Hydrol.* **2016**, *194*, 36–58. [CrossRef] [PubMed]
25. Medici, G.; West, L.J. Review of groundwater flow and contaminant transport modelling approaches for the Sherwood Sandstone aquifer, UK; insights from analogous successions worldwide. *Q. J. Eng. Geol. Hydrogeol.* **2022**, *55*, 176. [CrossRef]
26. Tellam, J.H.; Barker, R.D. Towards prediction of saturated-zone pollutant movement in groundwaters in fractured permeable-matrix aquifers: The case of the UK Permo-Triassic sandstones. In *Fluid Flow and Solute Movement in Sandstones: The Onshore UK Permo-Triassic Red Bed Sequence*; Barker, R.D., Tellam, J.H., Eds.; Geological Society of London: London, UK, 2006; pp. 1–48. [CrossRef]
27. Worthington, S.R.; Davies, G.J.; Alexander, E.C., Jr. Enhancement of bedrock permeability by weathering. *Earth-Sci. Rev.* **2016**, *160*, 188–202. [CrossRef]
28. Medici, G.; West, L.J.; Banwart, S.A. Groundwater flow velocities in a fractured carbonate aquifer-type: Implications for contaminant transport. *J. Contam. Hydrol.* **2019**, *222*, 1–16. [CrossRef]
29. Terzić, J.; Frangen, T.; Borović, S.; Reberski, J.L.; Patekar, M. Hydrogeological Assessment and Modified Conceptual Model of a Dinaric Karst Island Aquifer. *Water* **2022**, *14*, 404. [CrossRef]
30. Kaur, T.; Bhardwaj, R.; Arora, S. Assessment of groundwater quality for drinking and irrigation purposes using hydrochemical studies in Malwa region, southwestern part of Punjab, India. *Appl. Water Sci.* **2017**, *7*, 3301–3316. [CrossRef]

31. Suneetha, N.; Gautam, G.; Rambabu, S.; Khan, T.; Vinit, C.E. Hydrogeochemical processes regulating the groundwater quality and its suitability for drinking and irrigation purpose in parts of coastal Sindhudurg district, Maharashtra. *J. Geol. Soc. India* **2021**, *97*, 173–185. [CrossRef]
32. Rao, N.S.; Das, R.; Gugulothu, S. Understanding the factors contributing to groundwater salinity in the coastal region of Andhra Pradesh, India. *J. Contam. Hydrol.* **2022**, *250*, 104053. [CrossRef]
33. FAO (Food and Agriculture Organization). *Water Quality for Agriculture. FAO Irrigation and Drainage Paper 29*; FAO: Rome, Italy, 1985.
34. ESDW (Egypt Standard for Drinking Water). *Standards and Specifications of Water for Drinking and Domestic Use*; Decision of the Minister of Health and Population No. 458 (2007); Ministry of Health and Population: Cairo, Egypt, 2007. (In Arabic)
35. Perera, T.A.N.T.; Herath, H.; Piyadasa, R.U.K.; Jianhui, L.; Bing, L. Spatial and physicochemical assessment of groundwater quality in the urban coastal region of Sri Lanka. *Environ. Sci. Pollut. Res.* **2021**, *29*, 16250–16264. [CrossRef] [PubMed]
36. Saber, A.A.; Ullah Bhat, S.; Hamid, A.; Gabrieli, J.; Garamoon, H.; Gargini, A.; Cantonati, M. Chemical Quality and Hydrogeological Settings of the El-Farafra Oasis (Western Desert of Egypt) Groundwater Resources in Relation to Human Uses. *Appl. Sci.* **2022**, *12*, 5606. [CrossRef]
37. Abd El-Wahed, M.; El-Horiny, M.M.; Ashmawy, M.; El Kereem, S.A. Multivariate Statistical Analysis and Structural Sovereignty for Geochemical Assessment and Groundwater Prevalence in Bahariya Oasis, Western Desert, Egypt. *Sustainability* **2022**, *14*, 6962. [CrossRef]
38. Yehia, M.; Baghdady, A.; Howari, F.M.; Awad, S.; Gad, A. Natural radioactivity and groundwater quality assessment in the northern area of the Western Desert of Egypt. *J. Hydrol. Reg. Stud.* **2017**, *12*, 331–344. [CrossRef]
39. Stern, R.J.; Johnson, P.; Kröner, A.; Yibas, B. Neoproterozoic Ophiolites of the ArabianNubian Shield. In *Precambrian Ophiolites and Related Rocks*; Kusky, T.M., Ed.; Elsevier: Amsterdam, The Netherlands, 2004.
40. Fowler, A.; Baghdady, A.; Abdelmalik, K.; Gad, A. Remote sensing-guided stratigraphic dissection of an Ediacaran terrestrial molasse basin (Kareim basin, Egypt), with implications for sedimentary evolution. *Precambrian Res.* **2020**, *338*, 105589. [CrossRef]
41. Sultan, M.; Yousef, A.; Metwally, S.; Becker, R.; Milewski, A.; Sauck, W.; Sturchio, N.; Mohamed, A.; Wagdy, A.; El Alfy, Z.; et al. Red Sea rifting controls on aquifer distribution: Constraints from geochemical, geophysical, and remote sensing data. *GSA Bull.* **2011**, *123*, 911–924. [CrossRef]
42. Said, R. *The Geology of Egypt*; Balkema Publishers: Rotterdam, The Netherlands, 1990.
43. Montenat, C.; D'estevou, P.; Purser, B.; Burollet, P.; Jarrige, J.; Sperber, F.; Philobbos, F.; Plaziat, J.; Part, P.; Richert, J.; et al. Tectonic and sedimentary evolution of the Gulf of Suez and the Northwestern Red Sea. *Tectophysics* **1988**, *153*, 161–177. [CrossRef]
44. Appelo, C.A.J.; Postma, D. *Geochemistry, Groundwater and Pollution*, 2nd ed.; CRC Press: London, UK, 2005.
45. Ragunath, H.M. *Groundwater*; Wiley Eastern Ltd.: New Delhi, India, 1987.
46. Eaton, F.M. Significance of carbonate in irrigation water. *Soil Sci.* **1950**, *69*, 123–134. [CrossRef]
47. Todd, D.K. *Groundwater Hydrology*, 2nd ed.; Jhon Wiley and Sons: New York, NY, USA, 1980.
48. Bruvold, W.H.; Daniels, J.I. Standards for mineral content in drinking water. *J. Am. Water Works Assoc.* **1990**, *82*, 59–65. [CrossRef]
49. Sawyer, G.N.; McCarthy, D.L. *Chemistry of Sanitary Engineers*, 2nd ed.; McGraw Hill: New York, NY, USA, 1967.
50. Ryznre, J.W. A new index for determining amount of calcium carbonate scale formed by water. *J. Am. Water Works Assoc.* **1944**, *36*, 472–486. [CrossRef]
51. Giménez-Forcada, E.; Sánchez San Román, F.J. An Excel Macro to Plot the HFE-Diagram to Identify Sea Water Intrusion Phases. *Groundwater* **2015**, *53*, 819–824. [CrossRef] [PubMed]
52. Askri, B. Hydrochemical processes regulating groundwater quality in the coastal plain of Al Musanaah, Sultanate of Oman. *J. Afr. Earth Sci.* **2015**, *106*, 87–98. [CrossRef]
53. Sherif, M.I.; Sultan, M.; Sturchio, N.C. Chlorine Isotopes as Tracers of Solute Origin and Age of Groundwaters from the Eastern Desert of Egypt. *Earth Planet. Sci. Lett.* **2019**, *510*, 37–44. [CrossRef]
54. Selvam, S.; Jesuraja, K.; Roy, P.D.; Venkatramanan, S.; Chung, S.Y.; Elzain, H.E.; Muthukumar, P.; Nath, A.V.; Karthik, R. Assessment of groundwater from an industrial coastal area of South India for human health risk from consumption and irrigation suitability. *Environ. Res.* **2021**, *200*, 111461. [CrossRef]
55. Zolekar, R.B.; Todmal, R.S.; Bhagat, V.S.; Bhailume, S.A.; Korade, M.S.; Das, S. Hydrochemical characterization and geospatial analysis of groundwater for drinking and agricultural usage in Nashik district in Maharashtra, India. *Environ. Dev. Sustain.* **2021**, *23*, 4433–4452. [CrossRef]
56. Giménez-Forcada, E. Use of the Hydrochemical Facies Diagram (HFE-D) for the evaluation of salinization by seawater intrusion in the coastal Oropesa Plain: Comparative analysis with the coastal Vinaroz Plain, Spain. *HydroResearch* **2019**, *2*, 76–84. [CrossRef]
57. Lakshmanan, E.; Kannan, R.; Kumar, M. Major ion chemistry and identification of hydrogeochemical processes of ground water in a part of Kancheepuram district, Tamil Nadu, India. *Environ. Geosci.* **2003**, *10*, 157–166. [CrossRef]
58. Petrides, B.; Cartwright, I. The hydrogeology and hydrogeochemistry of the Barwon Downs Graben aquifer, southwestern Victoria, Australia. *Hydrogeol. J.* **2006**, *14*, 809–826. [CrossRef]
59. Petalas, C.P. A preliminary assessment of hydrogeological features and selected anthropogenic impacts on an alluvial fan aquifer system in Greece. *Environ. Earth Sci.* **2013**, *70*, 439–452. [CrossRef]
60. Abid, K.; Zouari, K.; Dulinski, M.; Chkir, N.; Abidi, B. Hydrologic and geologic factors controlling groundwater geochemistry in the Turonian aquifer (southern Tunisia). *Hydrogeol. J.* **2011**, *19*, 415–427. [CrossRef]

61. García, M.G.; Hidalgo, M.V.; Blesa, M.A. Geochemistry of groundwater in the alluvial plain of Tucumán province, Argentina. *Hydrogeol. J.* **2001**, *9*, 597–610. [CrossRef]
62. Yidana, S.M.; Yidana, A. An assessment of the origin and variation of groundwater salinity in southeastern Ghana. *Environ. Earth Sci.* **2010**, *61*, 1259–1273. [CrossRef]
63. Farhat, H.I.; Gad, A.; Saleh, A.; Abd El Bakey, S.M. Risks Assessment of Potentially Toxic Elements' Contamination in the Egyptian Red Sea Surficial Sediments. *Land* **2022**, *11*, 1560. [CrossRef]
64. Abdel-Satar, A.M.; Ali, M.H.; Goher, M.E. Indices of water quality and metal pollution of Nile River, Egypt. *Egypt. J. Aquat. Res.* **2017**, *43*, 21–29. [CrossRef]
65. USEPA (United States Environmental Protection Agency). *Drinking Water Standards and Health Advisories Table*; USEPA: San Francisco, CA, USA, 2007.
66. CCME (Canadian Council of Ministers of the Environment). *Canadian Environmental Quality Guidelines*; Publication No. 1299; Canadian Council of Ministers of the Environment: Ottawa, ON, Canada, 1999.
67. Jasmin, I.; Mallikarjuna, P. Evaluation of Groundwater Suitability for Irrigation in the Araniar River Basin, South India—A Case Study Using Gis Approach. *Irrig. Drain.* **2015**, *64*, 600–608. [CrossRef]
68. Subramani, T.; Elango, L.; Damodarasamy, S. Groundwater quality and its suitability for drinking and agricultural use in Chithar River Basin, Tamil Nadu, India. *Environ. Geol.* **2005**, *47*, 1099–1110. [CrossRef]
69. Simsek, C.; Gunduz, O. IWQ index: A GIS-integrated technique to assess irrigation water quality. *Environ. Monit. Assess.* **2007**, *128*, 277–300. [CrossRef]
70. Bouselsal, B.; Saibi, H. Evaluation of groundwater quality and hydrochemical characteristics in the shallow aquifer of El-Oued region (Algerian Sahara). *Groundwater Sustain. Dev.* **2022**, *17*, 100747. [CrossRef]

MDPI
St. Alban-Anlage 66
4052 Basel
Switzerland
www.mdpi.com

Water Editorial Office
E-mail: water@mdpi.com
www.mdpi.com/journal/water

www.ingramcontent.com/pod-product-compliance
Lightning Source LLC
Chambersburg PA
CBHW061519210926
51850CB00008B/1977
9783725804481